Signals & Systems

FOR

DUMMIES®

A Wiley Brand

by Mark Wickert, PhD

Signals & Systems For Dummies®

Published by
John Wiley & Sons, Inc.
111 River St.
Hoboken, NJ 07030-5774
www.wiley.com

Library of Congress Control Number: 2013934418

ISBN 978-1-118-47581-2 (pbk); ISBN 978-1-118-47566-9 (ebk); ISBN 978-1-118-47582-9 (ebk); ISBN 978-1-118-47583-6 (ebk)

Manufactured in the United States of America

10 9 8 7 6 5 4 3 2 1

About the Author

Mark Wickert is a professor of electrical and computer engineering at the University of Colorado, Colorado Springs, Colorado. His teaching focus is signals and systems with an emphasis in communications and signal processing. Mark was previously a board-level designer at Motorola Government Electronics, now a division of General Dynamics.

Mark also works as an industry consultant in digital communications and signal processing for Amergint Technologies LLC. He's worked with Real Time Logic and developed algorithms for a ZIGBEE radio chip at Atmel Corporation as well.

Mark earned BS and MS degrees in electrical engineering from Michigan Technological University and PhD from Missouri University of Science and Technology (then UMR). He is a member of the Institute of Electrical and Electronics Engineers.

Dedication

To Becki, David, and Paul — my family.

To God be the glory!

Author's Acknowledgments

This project started with an invitation from my agent, Matt, and a strong dose of encouragement from my wife, Becki, who endured a lot of changes that came with my focus on developing material for this book. My sons, David and Paul, had to accept seeing less of me, too, especially during our summer vacation. Thank you all for your support and encouragement throughout this process.

Thanks, too, to my faculty peers — Greg, Kalkur, and Charlie — for all your encouragement. And I appreciate that Amergint Technologies allowed me do some of the writing at its offices and provided an environment that let me explore the capabilities of Python. Thanks Jeff, Sean, Mark, and Mark.

The staff at Wiley was also very encouraging along the way. I especially want to thank my project editor, Jenny Brown, and my acquisitions editor, Erin Mooney. I know I was stubborn at times, and I am thankful you kept me going. Thanks, too, to copy editor Jennette ElNaggar for making sure my *t*'s are crossed and *i*'s are dotted.

Great appreciation goes to Christopher L. Felton, electrical engineer at the Mayo Clinic, for providing the technical review to ensure that the information provided in this book is valuable to the people for whom it's written. I am also grateful for the help provided by PhD student McKenna Lovejoy, who was willing to jump onboard and provide a fresh eye on this material during the final phases of editing.

To veteran book author, dissertation advisor, fellow faculty member prior to retirement, and still close friend, Rodger Ziemer, thank you for your invaluable support.

Finally, to my dad. Thanks for encouraging my middle-school interest in building electronic gadgets. You spurred my lifelong love for building hardware, software, and countless other things — I'll always be grateful.

Publisher's Acknowledgments

We're proud of this book; please send us your comments at http://dummies.custhelp.com. For other comments, please contact our Customer Care Department within the U.S. at 877-762-2974, outside the U.S. at 317-572-3993, or fax 317-572-4002.

Some of the people who helped bring this book to market include the following:

Acquisitions, Editorial, and Vertical Websites

Project Editor: Jenny Larner Brown

Acquisitions Editor: Erin Calligan Mooney

Copy Editor: Jennette ElNaggar

Assistant Editor: David Lutton

Editorial Program Coordinator: Joe Niesen

Technical Editors: Christopher L. Felton

Editorial Manager: Christine Meloy Beck

Editorial Assistant: Alexa Koschier

Cover Photos: © agsandrew/iStockphoto.com

Composition Services

Project Coordinator: Patrick Redmond

Layout and Graphics: Carrie A. Cesavice, Brent Savage, Christin Swinford, Erin Zeltner

Proofreaders: John Greenough, Lauren Mandelbaum, Wordsmith Editorial

Indexer: Steve Rath

Illustrations courtesy of Mark Wickert, PhD

Publishing and Editorial for Consumer Dummies

 Kathleen Nebenhaus, Vice President and Executive Publisher

 David Palmer, Associate Publisher

 Kristin Ferguson-Wagstaffe, Product Development Director

Publishing for Technology Dummies

 Andy Cummings, Vice President and Publisher

Composition Services

 Debbie Stailey, Director of Composition Services

Contents at a Glance

Table of Contents

Introduction

S ignals and systems is one of the toughest classes you'll take as an engineering student. But struggling to figure out this material doesn't necessarily mean you need to sprout early-onset gray hairs and resign yourself to frown lines in your college years. And you definitely don't want to give up on engineering over this stuff because becoming an engineer is, in my opinion, one of the best career choices you can make. See, you're no dummy!

This book can help you make sense of the fundamental concepts of signals and systems that may be giving you some static — or even frying your brain. Even better, you can apply the tips and tricks I provide in this book to the courses you'll take down the line — and right into the real world of computer and electrical engineering!

About This Book

Like all other *For Dummies* books, *Signals & Systems For Dummies* isn't a tutorial. It's a reference book that you can use as you need it. You don't need to read each chapter cover to cover (but you may find all the material utterly mesmerizing). You can jump right to the topics or concepts that are giving you trouble, get the help you need, and be on your way with helpful insight to real-world examples of electrical concepts that may be tough to imagine in your textbook of equations.

Conventions Used in This Book

I use the following conventions throughout the text to make things consistent and easy to follow:

- New terms appear in *italic* and are closely followed by an easy-to-understand definition. Variables also appear in italic.

- **Bold** highlights keywords in bulleted lists and the action parts of numbered steps.

- Lowercase variables indicate signals that change with time, and uppercase variables indicate signals that are constant. For example, $v(t)$ and $i(t)$ denote voltage and current signals that change with time. If, however, V and I are capitalized, these signals don't vary in time.

What You're Not to Read

Although I'm sure you want to read every word of this book, I realize you have other reading material to get through. When you're short on time and need to just get through the basics, you can skip the sidebars (the shaded boxes sprinkled throughout the book) and paragraphs flagged with a Technical Stuff icon.

Foolish Assumptions

I know you're a unique kind of brilliant and have one-of-a-kind skills and attributes, but as I wrote this book, I had to make some assumptions about my readers. Here's what I assume about you:

- ✔ You're currently taking an introductory signals and systems course as part of your computer or electrical engineering major, and you need help with certain concepts and techniques. Or you're planning to take a signals and systems course next semester, and you want to prepare by checking out some supplementary material.
- ✔ You have a solid handle on algebra and calculus.
- ✔ You've taken an introductory physics class, which exposed you to the concepts of voltage, current, and power in circuits.
- ✔ You're familiar with linear differential equations with constant coefficients.

How This Book Is Organized

The study of signals and systems integrates a handful of specific topics from your math and physics courses, and it introduces new techniques to design and manage electrical systems. To help you grasp the core concepts of this electrifying field (sorry, I couldn't resist) in manageable bites, I've split the book into several parts, each consisting of chapters on related topics. Chapters are laid out in an alternation of continuous- and discrete-time topics, starting with the time domain, moving to the frequency domain, and then covering the s- and z-domains.

Additional content, including case studies, is available online at www.dummies.com/extras/signalsandsystems.

Part 1: Getting Started with Signals and Systems

This part gives you the signals and systems lingo and an overview of the basic concepts and techniques necessary for tackling your signals and systems course. If you're already familiar with the fundamentals of how signals and systems operate in the continuous- and discrete-time domains, you can use this part as a refresher.

Part 11: Exploring the Time Domain

The focus of these chapters narrows to more closely examine the time domain of signals and systems. In Chapter 7, I introduce differential and difference equation system models, which are used to represent electronic circuits, the audio equalizer on your MP3 music player, filters that separate signals from one another, hybrid systems composed of electrical and mechanical components, and more. I also describe signal and system classifications and properties in these chapters.

Part 111: Picking Up the Frequency Domain

The chapters in this part drill down on the frequency domain and the world of system design, particularly wireless systems. Bridging the gap between the continuous- and discrete-time worlds is sampling theory, which is covered in Chapter 10.

Part 1V: Entering the s- and z-Domains

This part gets tougher because you're dealing with the *s*- and *z*-domains — a third domain system that engineers use to view the world. *Poles* and *zeros* rule here. Signal processing and control systems designers are fond of the *s*- and *z*-domains because, for starters, they reduce the mathematics of passing a signal through a system to rather simple algebraic manipulation. From the poles and zeros, you can easily discern system stability and the impact they have on the frequency domain. Great stuff.

Part V: The Part of Tens

Here, get hip to more than ten common mistakes people make when solving problems for signals and systems. Also find a list of ten properties you never want to forget. You may want to print these lists and keep 'em within view.

Icons Used in This Book

To make this book easier to read and simpler to use, I include some icons to help you find key information.

Anytime you see this icon, you know the information that follows is so important that it's worth recalling after you close this book — even if you don't remember anything else you read.

This icon appears next to information that's interesting but not essential. Don't be afraid to skip these paragraphs.

This bull's-eye points out advice that can save you time when managing signals and systems.

This icon tries to prevent you from making fatal mistakes in your analysis.

This icon flags worked-through examples in the content so you can find the most practical stuff fast if you're especially pressed for time.

Where to Go from Here

This book isn't a novel — although it just may be as intriguing as one. You can start at the beginning and read through to the end, or you can jump in at any chapter to get the information you need on a specific topic. If you need help with calculus and other math basics before dishing out the heartier fare of signals and systems, then pick through Chapter 2 for a quick review. If you just can't wait another second to find out how the Fourier transform works with different types of signals, then by all means flip to Chapters 9 and 11 right away.

If you're not sure where to start, or you don't know enough about signals and systems yet to even wonder about specific topics, no problem — that's exactly what this book is for. I recommend starting with the chapters in Part I and moving forward from there if you really are a newbie. Then, keep on reading; you'll be charged up with nitty-gritty details of signals and systems in no time.

Part I

Getting Started with Signals and Systems

In this part . . .

✔ Find out why computer and electrical engineers need to understand signals and systems analysis.

✔ See how signals and systems function in the worlds of continuous- and discrete-time.

✔ Discover alternative domains used for modeling signals and systems.

✔ Refresh your mathematical know-how and see how algebra, calculus, and trig apply to signals and systems work.

✔ Explore the basic means for assessing the performance of technology-based solutions.

Chapter 1

Introducing Signals and Systems

*W*hich came first: the signal or the system? Before you answer, you may want to know that by *system,* I mean a structure or design that operates on signals. You live and breathe in a sea of signals, and systems harness signals and put them to work. So which came first, you think? It may not really matter, but I'm guessing — as I smooth out a long imaginary philosopher-type beard — that signals came first and then began passing through systems.

But I digress. The study of signals and systems as portrayed in this book centers on the *mathematical modeling* of both signals and systems. Mathematical modeling allows an engineer to explore a variety of product design approaches without committing to costly prototype hardware and software development. After you tune your model to produce satisfactory results, you can implement your design as a prototype. And at some point, real signals (and sometimes math-based simulations) test the system design before full implementation.

When studying signals and systems, it's easy to get mired in mathematical details and lose sight of the big picture — the functional systems of your end result. So try to remember that, at its best, signals and systems is all about designing and working with products through applied math. Math is the means, not the star of the show.

Two broad classes of signals are those that are *continuous* functions of time t and those that are *discrete* functions of time index n. Throughout this book, I separate information on continuous- and discrete-time signals and systems. In this chapter, I introduce simple continuous and discrete signals and the corresponding systems. I also point out some of the distinguishing characteristics of signal types.

Before getting started, I want to mention that signals as functions of time are how most people experience the real world of computer and electronic engineering, yet transforming signals and systems to other domains — specifically, the frequency, s-, and z-domains — and back again is quite beneficial in some situations. I touch on the transformation of signals and systems in this chapter and dig into the details in Parts III and IV.

In this chapter, I also cover the important role of computer tools in signals and systems problem solving and tell you how to use a few specific open-source programs. If you want to set up these freely available tools on your computer, you can follow along when I describe specific functions that enable you to check your work or work more efficiently — after you get a handle on core concepts and techniques.

Applying Mathematics

Anyone aspiring to a working knowledge of signals and systems needs a solid background in math, including these specific concepts:

- Calculus of one variable
- Integration and differentiation
- Differential equations

To actually implement designs that center on signals and systems, you also need a background in these subjects:

- Electrical/electronic circuits
- Computer programming fundamentals, such as C/C++ and Java
- Analysis, design, and development software tools
- Programmable devices

Many signals and systems designers rely on modeling tools that use a matrix/vector language or class library for numerics and a graphics visualization capability to allow for rapid prototyping. I use numerical Python for examples in this book; other languages with similar syntax include MATLAB and NI LabVIEW MathScript.

Finding perspective on analog processing

Once upon a time, the implementation path for signals and systems was purely analog circuit design. As technology has advanced, solutions based on digital signal processing (discrete-time signals and systems) through powerful low-cost and low-power digital hardware has become the mainstay. Digital hardware solutions are programmable and can be reconfigured through software updates after products ship.

The signals you're likely to work with in the real world are analog in nature, but you'll almost always process them digitally. Knowing programming languages is important in this environment. Yet analog signal processing is alive and well — it's vital to your working knowledge of signals and systems — but the overall role of analog processing in current design is less formidable than it's been in the past.

With so many electrical engineering solutions being software-based today — versus a matter of analog circuitry (see nearby sidebar "Finding perspective on analog processing") — a system designer can also be the implementer. This leap requires only simulation code to be transformed into the implementation language, such as Verilog or C/C++.

Working pencil-and-paper solutions for signals and systems coursework requires a good scientific calculator. I recommend a calculator that supports complex arithmetic operations, using the minimum number of keystrokes. At minimum, your calculator needs to have trig, log, and exponential functions for signals and systems work.

Getting Mixed Signals . . . and Systems

Signals come in two flavors: continuous and discrete. It's the same story with systems. In other words, some signals — and some systems — are active all the time; others aren't. In this section, I describe continuous and discrete signals along with the corresponding systems. I also tell you how to classify certain signals and systems based on their most basic properties.

Going on and on and on

Continuous-time signals and systems never take a break. When a circuit is wired up, a signal is there for the taking, and the system begins working — and doesn't stop. Keep in mind that I use the term *signal* here loosely; any one specific signal may come and go, but a signal is always present at each and every time instant imaginable in a continuous-time system.

Continuous-time signals

Continuous signals function according to time t. A sinusoidal function of time is one of the most basic signals. The mathematical model for a sinusoid signal is $x(t) = A\cos(2\pi f_0 t - \phi)$, $-\infty < t < \infty$, where A is the signal amplitude, f_0 is the signal frequency, and ϕ is the signal phase shift. The independent variable is time t. If you're curious about the first peak of $x(t)$ occuring at 3/16, notice that this occurs when the argument of the cosine is 0 — the is, $2\pi \cdot 2 \cdot t - 3\pi/4 = 0$ or $t = 3\pi/4 \cdot 1/(4\pi) = 3/16$.

I cover this signal in detail in Chapter 3, but to help you get acquainted, check out the plot of a sinusoid signal in Figure 1-1.

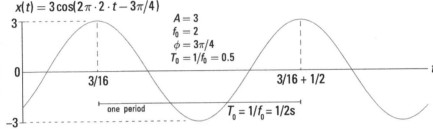

$$x(t) = 3\cos(2\pi \cdot 2 \cdot t - 3\pi/4)$$

$A = 3$
$f_0 = 2$
$\phi = 3\pi/4$
$T_0 = 1/f_0 = 0.5$

3/16

one period

3/16 + 1/2

$T_0 = 1/f_0 = 1/2s$

Figure 1-1: The plot of a sinusoidal signal.

The amplitude of this signal is 3, the frequency is 2 Hz, and the phase shift is $3\pi/4$ rad.

Continuous-time systems

Systems operate on signals. In mathematical terms, a *system* is a function or operator, $T\{\ \}$, that maps the input signal $x(t)$ to output signal $y(t) = T\{x(t)\}$.

An example of a continuous-time system is the electronic circuits in an amplifier, which has gain 5 and level shift 2: $y(t) = T\{x[n]\} = 5x(t) + 2$.

See a block diagram representation of this simple system in Figure 1-2.

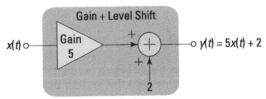

Figure 1-2: A simple continuous-time system model.

Gain + Level Shift

$x(t)$

Gain 5

$y(t) = 5x(t) + 2$

2

Building an amplifier that corresponds to this mathematical model is another matter entirely. You can create a simple electronic circuit, but it will have limitations that the math model doesn't have. It's up to you, as an electronic engineer, to refine the model to accurately reflect the level of detail needed to assess overall performance of a design candidate.

Working in spurts: Discrete-time signals and systems

Discrete-time signals and systems march along to the tick of a clock. Mathematical modeling of discrete-time signals and systems shows that activity occurs with whole number (integer) spacing, but signals in the real world operate according to periods of time, or the update rate also known as the *sampling rate*. Discrete-time signals, which can also be viewed as sequences, only exist at the ticks, and the systems that process these signals are, mathematically speaking, resting in the periods between signal activity.

Systems take inputs and produce outputs with the same clock tick, generally speaking. Depending on the nature of the digital hardware and the complexity of the system, calculations performed by the system continue — between clock ticks — to ensure that the next system output is available at the next tick when a new signal sample arrives at the input.

Discrete-time signals

Discrete-time signals are a function of time index n. Discrete-time signal $x[n]$, unlike continuous-time signal $x(t)$, takes on values only at integer number values of the independent variable n. This means that the signal is active only at specific periods of time. Discrete-time signals can be stored in computer memory because the number of signal values that need to be stored to represent a finite time interval is finite.

The following simple signal, a pulse sequence, is shown in Figure 1-3 as a *stem plot* — a plot where you place vertical lines, starting at 0 to the sample value, along with a marker such as a filled circle. The stem plot is also known as a *lollipop plot* — seriously.

$$x[n] = \begin{cases} 5, & 0 \le n < 10 \\ 0, & \text{otherwise} \end{cases}$$

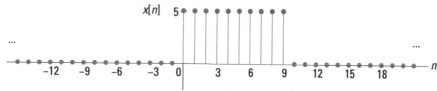

Figure 1-3: A simple discrete-time signal.

The stem plot shows only the discrete values of the sequence. Find out more about discrete-time signals in Chapter 4.

Discrete-time systems

A discrete-time system, like its continuous-time counterpart, is a function, $T\{\}$, that maps the input $x[n]$ to the output $y[n] = Y\{x[n]\}$. An example of a discrete-time system is the *two-tap* filter:

$$y[n] = T\{x[n]\} = \frac{3}{4}x[n] + \frac{1}{4}x[n-1]$$

The term *tap* denotes that output at time instant n is formed from two time instants of the input, n and $n-1$. Check out a block diagram of a two-tap filter system in Figure 1-4.

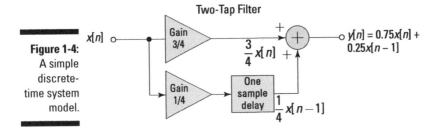

Figure 1-4: A simple discrete-time system model.

In words, this system scales the present input by 3/4 and adds it to the past value of the input scaled by 1/4. The notion of the past input comes about because $x[n-1]$ is lagging one sample value behind $x[n]$. The term *filter* describes the output as an *averaging* of the present input and the previous input. *Averaging* is a form of filtering.

Classifying Signals

Signals, both continuous and discrete, have attributes that allow them to be classified into different types. Three broad categories of signal classification are periodic, aperiodic, and random. In this section, I briefly describe these classifications (find details in Chapters 3 and 4).

Periodic

Signals that repeat over and over are said to be *periodic.* In mathematical terms, a signal is periodic if

$$x(t+T) = x(t) \text{ (continuous-time)}$$
$$x[n+N] = x[n] \text{ (discrete-time)}$$

The smallest T or N for which the equality holds is the signal period. The sinusoidal signal of Figure 1-1 is periodic because of the $\mod 2\pi$ property of cosine. The signal of Figure 1-1 has period 0.5 seconds (s), which turns out to be the reciprocal of the frequency $f_0 = 2$ Hz. The *square wave* signal of Figure 1-5a is another example of a periodic signal.

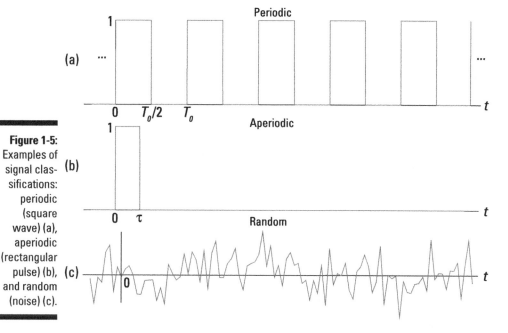

Figure 1-5: Examples of signal classifications: periodic (square wave) (a), aperiodic (rectangular pulse) (b), and random (noise) (c).

Aperiodic

Signals that are *deterministic* (completely determined functions of time) but not periodic are known as *aperiodic*. Point of view matters. If a signal occurs infrequently, you may view it as aperiodic. The rectangular pulse of duration τ shown in Figure 1-5b is an aperiodic signal.

Random

A signal is *random* if one or more signal attributes takes on unpredictable values in a probability sense (you love statistics, right?).

The full mathematical description of random signals is outside the scope of this book, but here are two good examples of a random signal:

- ✔ The *noise* you hear when you're between stations on an FM radio. See a waveform representation of this noise in Figure 1-5c.

- ✔ Speech: If you try to capture audio samples on a computer of someone speaking the word *hello* over and over, you'll find that each capture looks a little different.

Engineers working with communication receivers are concerned with random signals, especially noise.

Signals and Systems in Other Domains

Most of the signals you encounter on a daily basis — in computers, in wireless devices, or through a face-to-face conversation — reside in the time domain. They're functions of independent variable *t* or *n*. But sometimes when you're working with continuous-time signals, you may need to transform away from the time domain (*t*) to either the frequency domain (*f* or *ω*) or the *s*-domain (*s*). Similarly, for discrete-time signals, you may need to transform from the discrete-time domain (*n*) to the frequency domain ($\hat{\omega}$) or the *z*-domain (*z*).

Systems, continuous and discrete, can also be transformed to the frequency and *s*- and *z*-domains, respectively. Signals can, in fact, be passed through systems in these alternative domains. When a signal is passed through a system in the frequency domain, for example, the frequency domain output signal can later be returned to the time domain and appear just as if the time-domain version of the system operated on the signal in the time domain.

This section briefly explores the world of signals and systems in the frequency, *s*-, and *z*-domains. Find more on these alternative domains in Chapters 13 and 14.

Viewing signals in the frequency domain

The time domain is where signals naturally live and where human interaction with signals occurs, but the full information for a signal isn't always visible in that space. Consider the sum of a two-sinusoids signal (as depicted in Figure 1-6):

$$x(t) = \underbrace{A_1 \cos(2\pi f_1 t)}_{s_1} + \underbrace{A_2 \cos(2\pi f_2 t)}_{s_2}$$

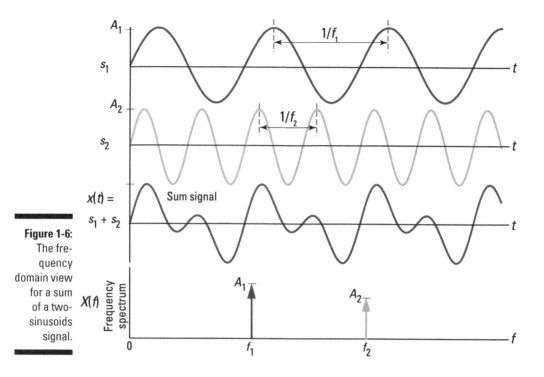

The top waveform plot, denoted s_1, is a single sinusoid at frequency f_1 and peak amplitude A_1. The waveform repeats every period $T_1 = 1/f_1$. The second waveform plot, denoted s_2, is a single sinusoid at frequency $f_2 > f_1$ and peak amplitude $A_2 < A_1$. The sum signal, $s_1 + s_2$, in the time domain is a squiggly line (third waveform plot), but the amplitudes and frequencies (periods) of the sinusoids aren't clear here as they are in the first two plots. The frequency spectrum (bottom plot) reveals that $x(t)$ is composed of just two sinusoids, with both the frequencies and amplitudes discernible.

Think about tuning in a radio station. Stations are located at different center frequencies. The stations don't interfere with one another because they're separated from each other in the frequency domain. In the frequency spec-trum plot at the bottom of Figure 1-6, imagine that f_1 and f_2 are the signals from two radio stations, viewed in the frequency domain. You can design a receiving system to *filter* s_1 from $s_1 + s_2$. The filter is designed to pass s_1 and block s_2. (I cover filters in Chapter 9.)

Use the *Fourier transform* to move away from the time domain and into the frequency domain. To get back to the time domain, use the *inverse Fourier transform*. (Find out more about these transforms in Chapter 9.)

Traveling to the s- or z-domain and back

From the time domain to the frequency domain, only one independent variable, $t \to f$, exists. When a signal is transformed to the s-domain, it becomes a function of a complex variable $s = \sigma + j\omega$. The two variables (real and imaginary parts) describe a location in the s-plane.

In addition to visualization properties, the s-domain reduces differential equation solving to algebraic manipulation. For discrete-time signals, the z-transform accomplishes the same thing, except differential equations are replaced by difference equations. Did you think going to the z-domain meant taking a nap? Details on difference equations begin in Chapter 7.

Testing Product Concepts with Behavioral Level Modeling

Computer and electrical engineers provide society with a vast array of products — ranging from cellphones and high-definition televisions to powerful computers with high resolution displays that are small and lightweight. The mystery of how brilliant people come up with world-changing ideas may never be solved, but after an idea is out there, engineers work through a process that allows them to test, or *model,* potential solutions to find out whether the idea is likely to work in the real world. For products that rely on signal processing, engineers use signals and system modeling and analysis to reveal what's possible.

When you're trying to quickly prove a solution approach, you'll often turn to *behavioral level* modeling of certain elements of the overall system to avoid low-level implementation details. For example, a subsystem design may require knowledge of a signal parameter (such as amplitude or frequency) to function. At first, you may assume that the parameter is well known. Later, you add low-level details to estimate (not perfectly) the parameter. As your confidence and understanding grows, you represent the low-level details in the model and actual implementation becomes possible.

Behavioral level modeling also applies when you need to model physical environments that lie outside a design but are needed to evaluate performance under realistic scenarios.

In this section, I describe the role of abstraction as a means to generate preliminary concepts and then work those concepts into a top-level design. The top-level design becomes a detailed plan as you work down to implementation specifics. Mathematical modeling is a thread running through the entire process, so you come to rely on it.

Staying abstract to generate ideas

Behavioral level modeling isn't void of hardware constraints and realities, but it requires a certain level of abstraction to allow preliminary concept solutions to materialize quickly. Behavioral level models depend on applied mathematics.

In other words, computer and electronic engineers don't frequently handle actual hardware and devices used for an implementation. The model of the hardware is what's important at this point. The engineer's job is to conceptualize systems and subsystems through a framework of mathematical concepts, and abstraction provides great creative freedom to explore the possibilities.

Suppose you seek a new design for an existing system to improve performance. You hope to make such improvements with new device technology. You don't want to get bogged down in all the details of how to interface this device into the current design, so you move up in abstraction with a model to quickly find out how much you can improve performance with a new design. If adequate improvement potential doesn't exist, then you settle down and investigate other options. Rinse, lather, and repeat.

Keep in mind that improved performance isn't always the primary objective of signals and systems modeling. Sometimes, a design is driven by cost, availability of materials, manufacturing processes, and time to market, or some other consideration.

Working from the top down

A design that relies on signals and systems starts from a top-level view and works down to the nitty-gritty details of final implementation. Analysis and simulation performed at the top level depends on behavioral level modeling. The model is ultimately broken into subsystems for testing and refinement, and then the system comes together again before implementation.

Typically, your task as an electronic engineer is to create some new or enhanced functionality for a computer- or electrical-based product. For example, you may need to support a new radio interface due to recent standard updates. At first, the changes may seem simple and straightforward, but as you dig into the work, you may begin to see that the changes require significant adjustments in signal processing algorithms. This means that the new radio interface will require a few totally new designs, so you need to model and simulate various implementation approaches to find out what's likely to work best.

Relying on mathematics

Many people write off signals and systems as a pile of confusing math, and they run for the hills. True, the math can be intimidating at first, but the rewards of seeing your finely crafted mathematical model lead the way to a shipping product is worth the extra effort — at least I think so. In the end, the math is on your side. It's the only way to model concepts that function properly in the real world.

My go-to approach when a problem seems unsolvable: Take it slow and steady. If a solution isn't clear after you think about the problem for a while, walk away and come back to it later. Practice and experience with various problem-solving techniques and options help, so try to work as many types of problems as you can — especially in the areas you feel the most discomfort. Eventually, a solution reveals itself.

When possible, verify your solutions by using computer analysis and simulation tools. In this book, I use Python with the numerical support and visualization capabilities of PyLab (NumPy, SciPy, matplotlib) and the IPython environment to perform number-crunching analysis and simulations. For problems involving more symbolic mathematics, I use the computer algebra system (CAS) provided by Maxima.

Exploring Familiar Signals and Systems

I'm guessing you have some level of familiarity with consumer electronics, such as MP3 music players, smartphones, and tablet devices, and realize that these products rely on signals and systems. But you may take for granted the cruise control in your car. In this section, I point out the signals and systems framework in familiar devices at the *block diagram* level — a system diagram that identifies the significant components inside rectangular boxes, interconnected with arrows that show the direction of signal flow. The block diagram expresses the overall concept of a system without intimate implementation details.

MP3 music player

Signals and systems are operating in all the major peripherals of the music player — even in the processor. In reality, signals are in every part of the system, but I exclude pure digital signals in this example, so I don't address memory. The processor runs an operating system (OS); under that OS, *tasks* perform *digital signal processing* (DSP) algorithms for streaming audio and image data. Note that this book is focused on one-dimensional signals only.

Find a top-level block diagram of an MP3 device in Figure 1-7. All the peripheral blocks (the blocks that sit outside the processor block) contain a combination of continuous- and discrete-time systems. You stream digital music in real time from memory in a compressed format. The processor has to decompress the audio stream into signal sample values (a discrete-time signal) to send to the audio codec. The audio codec contains a digital-to-analog converter (DAC) that converts the discrete-time signal to a continuous-time signal.

The Wi-Fi and Bluetooth radios (blocks with antennas) interface to the processor with digital data but interface to the antenna by using a continuous-time signal at a frequency of 2.4 GHz. The sensors' block acquires analog signals from the environment, temperature, light level, and acceleration in three dimensions.

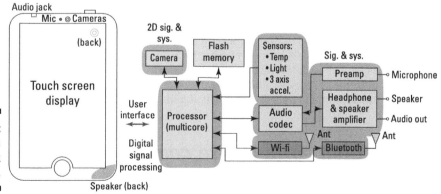

Figure 1-7:
MP3 music player block diagram.

Smartphone

The structure of a smartphone is similar to an MP3 music player, but a smartphone also has a global positioning receiver (GPS) and multiband radio blocks that send and receive continuous-time signals from base stations (antenna sites) of a cellular network. The GPS receiver acquires signals from multiple satellites to get your latitude and longitude. The primary purpose of the GPS in most smartphones is to provide location information when placing an emergency call (E911).

Check out a block diagram of a smartphone in Figure 1-8. Four antennas are shown, but only a single multiband antenna is employed in most models, so only a single antenna structure is really needed.

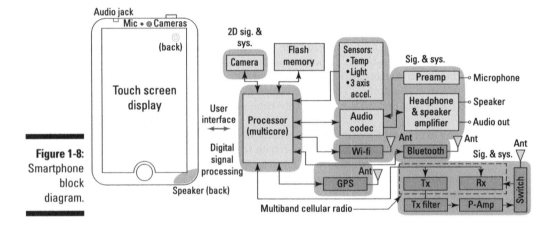

Figure 1-8: Smartphone block diagram.

The multiband cellular radio subsystem is thick with signals and systems. The multiband digital communications transmitter (tx) and receiver (rx) allows the smartphone to be backward compatible with older technologies as well as with the newest high-speed wireless data technologies. This transmitter and receiver enable the product to operate throughout the world. A smartphone is overflowing with signals and systems examples!

Automobile cruise control

I think all new automobiles come equipped with a cruise control system now. This is good news because this feature may keep you from getting a speeding ticket when you're driving long distances on the interstate. It's also great for getting better gas mileage. But I'm no sales guy for cruise control. I just think this product is interesting from a signals and systems standpoint.

Figure 1-9 shows a block diagram of a cruise control system. Cruise control involves both electrical and mechanical signals and systems. The controller is electrical and the *plant,* the system being controlled, is the car. Wind and hills are *disturbance* signals, which thwart the normal operation of the control system. The controller puts out a compensating signal to the throttle to overcome wind resistance (an opposing force) and the force of gravity when going up and down hills. The error signal that follows the summing block is driven to a very small value by the action of the feedback loop. This means that the output velocity *tracks* the reference velocity. This is exactly what you want. For a more detailed look at cruise control, check out the case studies at www.dummies.com/extras/signalsandsystems.

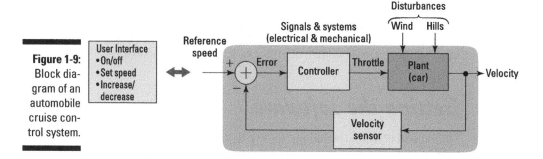

Figure 1-9: Block diagram of an automobile cruise control system.

Using Computer Tools for Modeling and Simulation

Today's technology-based solutions are rarely built without the use of some form of computer tool. Signals and systems research and product development is no exception. Throughout this book, I show you how to solve problems by hand calculation and how to check your work with computer tools. Hand calculation is vital for building concepts. Computer tools help ensure that you don't make mistakes. And why wouldn't you use the best tools available for your work?

A variety of commercial and open-source tools are available for signals and systems problem solving. Two broad categories are computer algebra system (CAS) programs, such as Mathematica, Maple, and Maxima, and those that excel at vector/matrix problem solving, such as MATLAB, NI LabVIEW MathScript, Octave, and Python. Both types of computer programs offer function libraries that are tailored to the needs of the signals and systems analysis and simulation.

The examples in this book feature two open-source tools:

- ✔ Scientific Python via *PyLab* and the shell *IPython*

  Python becomes scientific Python with the inclusion of NumPy and SciPy for vector/matrix number crunching and matplotlib for graphics.

- ✔ CAS *Maxima* via *wxMaxima*

I've chosen open-source tools because I want to provide an easy on-ramp for users everywhere. Both Mac and Windows OS computers can run these software products via free downloads. Specifically for this book, I wrote the code module ssd.py, which provides additional signals and systems functions. After you import this module into your IPython session, you can run all the examples in this book. I prefer to use the QT console version of IPython (see www.ipython.org). Similarly for wxMaxima, the notebook ssd.wxm contains all the example code from this book, organized by chapter.

Getting the software

Python and IPython (including NumPy, SciPy, and matplotlib) from Enthought Python Distribution (EPD) is a free download for the 32-bit version (www.enthought.com/products/epd_free.php). Python(x,y) is also very good, especially under Windows (http://code.google.com/p/pythonxy). If you're running Linux, in particular Ubuntu Linux, the Ubuntu Software Center is a good starting place. If you're an experienced open-source user, you can do a custom install as opposed to the monolithic distributions. If you're looking for a full integrated development environment (IDE) for debugging Python, I suggest the open-source IDE Eclipse (www.eclipse.org) with the plug-in PyDev (http://pydev.org). Eclipse is supported on Mac, Windows, and Linux. I developed the module ssd.py by using this setup.

Find wxMaxima for Windows and Mac at http://andrejv.github.com/wxmaxima. Under Ubuntu Linux, you can find wxMaxima in the Ubuntu Software Center.

To get files specific to this book go to www.dummies.com/extras/signalsandsystems for the Python code module ssd.py and the Maxima notebook ssd.wxm along with some tutorial screencasts and documents.

Exploring the interfaces

Take a quick tour of the interfaces of these computer programs when you get them installed. I provide a peek of how the program looks on the Mac in Figures 1-10, 1-11, and 1-12. The appearance and functionality for Windows is virtually the same.

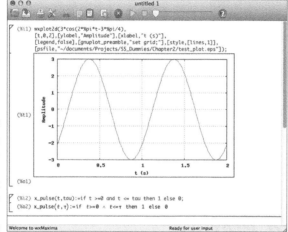

Figure 1-10: The wxMaxima notebook interface to Maxima.

You can send Maxima plots to a file in a variety of formats or display them directly in the notebook, as shown in Figure 1-11.

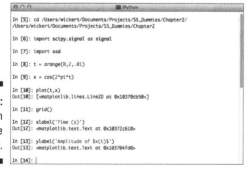

Figure 1-11: The IPython QT console window.

You can write and debug functions right from the console window, as shown in Figure 1-12.

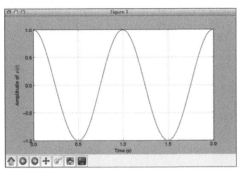

Figure 1-12:
matplotlib
plot window
resulting
from a call
to plot
(x,y) in
IPython.

You can manipulate plots by using the controls you see at the bottom of the figure window. Plot cursors are also available. You can save plots from the command line or from the figure window. Many of the plots found in this book were created with matplotlib.

Seeing the Big Picture

Figure 1-13 illustrates the content organization of this book as an unfolding of core topics, starting from the time domain and moving to the frequency domain before exploring the s- and z-domains. Continuous (left side) and discrete (right side) signals and systems topics parallel each other every step of the way — with some continuous- and discrete-time topics shared (center) within a few chapters. The last four chapters, which follow the z-domain chapter, emphasize applications, including signal processing, wireless communications, and control systems.

I start with the time domain because this is where signals originate and where systems operate on signals (with the exception of transform domain processing, which is covered in Chapter 12). The frequency domain augments a base knowledge of both signals and systems and is important to grasping sampling theory, which leads to the processing of continuous-time signals in the discrete-time domain. The s- and z-domain are the last of the core topics, but by no means are they any less important than the topics that come before them. The s- and z-domains are particularly powerful when working with linear time-invariant systems described by differential and difference equations.

After covering the core topics, you can appreciate the chapter that focuses on how to work across domains (Chapter 15). Get a taste of how signals and systems fit into the real world of electrical engineering by reading the case studies at www.dummies.com/extras/signalsandsystems. Take a look at the application examples to get inspired when you're struggling to see the forest for the trees of the dense study of signals and systems.

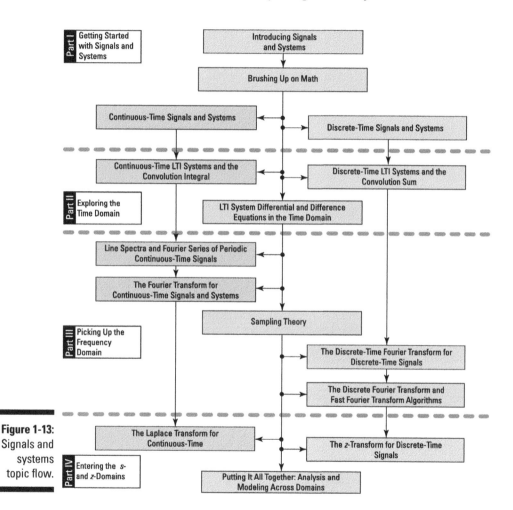

Figure 1-13: Signals and systems topic flow.

Chapter 2

Brushing Up on Math

In This Chapter

▶ Reviewing basic algebra
▶ Finding the love for trig
▶ Calculating complex arithmetic
▶ Recalling calculus
▶ Rooting for polynomials

Can you believe it? A mathematics review chapter in a signals and systems book! Well, yes, you saw it coming. Mathematics plays a starring role in the study of signals and systems, which is basically a discipline of applied math.

In this chapter, I cover the mathematical concepts that you need for studying signals and systems, including specific aspects of algebra, trigonometry, complex arithmetic, calculus, and polynomial roots. So grab what you need from this chapter, and come back as necessary when you're trying to decipher the core concepts of signals and systems. If you need a more rigorous review of mathematical tools than what I provide here, check out *Calculus For Dummies* (Wiley) and other *For Dummies* resources available at www.dummies.com.

Revealing Unknowns with Algebra

Algebraic manipulation is at the core of many engineering calculations, especially in the study of signals and systems. After all, when three variables are at play, you can solve for any one by fixing the other two, using algebra. This is helpful because in design problems, you often need to meet some requirement but have one or more design values to choose. For the case of two design values, just pick a value for one and satisfy the requirement by solving for the remaining design value.

In the following sections, I review fundamental algebra concepts for two equations and two unknowns. I also cover *partial fraction expansion* (PFE), which is an algebraic technique for splitting up a ratio of polynomials into a sum of fractions, each having a denominator containing a single or repeated root.

Solving for two variables

To restart your algebraic manipulation engine, consider finding the values for two variables given two equations.

Example 2-1: A linear chirp signal involves changing the *pitch*, or frequency, of a single tone signal linearly with time. The instantaneous signal frequency is $f_i(t) = 2\mu t + f_0$ (Hz), where μ and f_0 are constants. The units of $f_i(t)$ are hertz (Hz), or cycles per second. A design task is to produce a linear chirp signal from 200 Hz to 3,500 Hz over the time interval $0 \le t \le 10$ s.

Notice that this problem requires you to solve for two unknowns. Here's how to find the solutions:

1. **Create two equations by using the start and stop time for the chirp as boundary conditions:**

$$f_i(0) = 200 = \underbrace{2\mu \cdot 0}_{0} + f_0$$

$$f_i(10) = 3,500 = 2\mu \cdot 10 + f_0$$

2. **In the first equation, note that f_0 = 200 Hz. In the second equation, $3,500 = 20\mu + f_0 = 20\mu + 200$.**

3. **Solve for μ:**

$$\mu = \frac{3,500 - 200}{20} = \frac{3,300}{20} = 165 \text{ Hz/s}$$

Checking solutions with computer tools

A computer algebra system (CAS) can help you solve problems and check your work. One CAS option is the open-source software tool Maxima (see Chapter 1 for details).

Note: I use the GUI environment *wxMaxima* throughout this book for *computer algebra-oriented calculations* — problems that involve symbols and numbers.

Create this command line solution with wxMaxima:

```
(%i1) solve([200 = 2*mu*0+f0,3500 = 2*mu*10+f0],[mu,f0]);
(%o1) [[μ= 165, f0= 200]]
```

In wxMaxima, each *cell* is an individual calculation. The first line of a cell is the input you type into Maxima; the second line is the output you get after using the key stroke Shift + Enter.

In solving for μ and f_0, the input consists of two equations entered as a list followed by the variables to be solved for, which are also a list. A *list* is simply quantities placed between brackets and separated by commas, such as [mu, f0].

Exploring partial fraction expansion

Partial fraction expansion (PFE) techniques are important to the inverse Laplace transform (covered in Chapter 13) and the inverse z-transform (described in Chapter 14). These techniques make it possible to use a set of transform tables instead of contour integration from complex variable theory. With PFE, you split up a ratio of polynomials into a sum of fractions, each having a denominator containing a single or repeated root.

Consider the *N*th-order polynomial:

$$D(s)=a_N s^N +a_{N-1}s^{N-1}+\cdots+a_1 s+a_0$$
$$=(s-p_1)(s-p_2)\cdots(s-p_{N-1})(s-p_N)$$

And the *M*th-order polynomial:

$$N(s)=b_M s^M +b_{M-1}s^{M-1}+\cdots+b_0$$

In the case of $M < N$, $N(s)/D(s)$ is a *proper rational* function, meaning the degree of the numerator is less than the degree of the denominator. If you assume that p_i, $i = 1, \ldots, N$ are distinct roots of $D(s)$ — that is, no roots are repeated — then the PFE of $N(s)/D(s)$ is

$$\frac{N(s)}{D(s)}=\frac{A_1}{s-p_1}+\frac{A_2}{s-p_2}+\cdots+\frac{A_N}{s-p_N}$$

In the general case, again $M < N$, each of *r* roots of $D(s)$ may have multiplicity N_i, such that $N_1+N_2+\cdots+N_r = N$.

The PFE now takes the more complex form

$$\frac{N(s)}{D(s)} = \sum_{k=1}^{N_1} \frac{A_{1k}}{(s-p_1)^k} + \sum_{k=1}^{N_2} \frac{A_{2k}}{(s-p_2)^k} + \cdots + \sum_{k=1}^{N_r} \frac{A_{rk}}{(s-p_r)^k}$$

WARNING!

When $M \geq N$, you need to use long division to reduce the order of $N(s)$ to be less than $D(s)$ and then continue with the expansion. However, not taking this step to reduce the order of $N(s)$ doesn't prevent you from getting A_i values. But the values are incorrect, and you have no way of knowing this unless you recombine your expansion terms over a common denominator to see whether you return to the starting point.

With polynomial long division, your goal is

$$\frac{N(s)}{D(s)} = K(s) + \frac{R(s)}{D(s)}$$

where $R(s)/D(s)$ is proper rational. Here's how polynomial long division works:

$$\frac{N(s)}{D(s)} = \frac{s^2 + 3s + 1}{s(s-2)}, M = N = 2$$

Carrying out long division requires two steps. One is a setup, and the other is the long division itself, which repeats $M - N + 1$ times. Note that $M - N$ long division brings numerator/denominator order parity, so you need to do one additional long division to make the numerator order one less than the denominator.

1. **For the setup, multiply out the denominator into one big polynomial;** for example, $s(s - 2) = s^2 - 2s$.

2. **Perform long division one time because $M - N + 1 = 2 - 2 + 1 = 1$:**

$$
\begin{array}{r}
1 \\
s^2 - 2s \enclose{longdiv}{s^2 + 3s + 1} \\
\underline{s^2 - 2s} \\
5s + 1
\end{array}
$$

3. **Notice that the numerator order is reduced to $M = 1$ because the remainder is $R(s) = 5s + 1$ and $K(s) = 1$, making**

$$\frac{N(s)}{D(s)} = \frac{s^2 + 3s + 1}{s(s-2)} \overset{also}{=} 1 + \frac{5s+1}{s(s-2)}$$

4. **The roots of $D(s)$ are 0 and 2, so the PFE of $R(s)/D(s)$ is**

$$\frac{5s+1}{s(s-2)} = \frac{A_1}{s} + \frac{A_2}{s-2}$$

With the conversion from an improper rational function to a proper rational function now complete, you can move on to finding the expansion coefficients.

For the case of distinct roots, the formula for the coefficients is

$$A_i = \frac{N(s)}{D(s)} \cdot (s - p_i) \bigg|_{s=p_i} , i = 1, 2, ..., N$$

When working with repeated roots, the formula is

$$A_{ik} = \frac{1}{(N_i - k)!} \cdot \frac{d^{(N_i-k)}}{ds^{(N_i-k)}} \left[(s - p_i)^{N_i} \frac{N(s)}{D(s)} \right] \bigg|_{s=p_i}$$

Example 2-2: To find the expansion coefficients for $H_1(s) = \dfrac{s+1}{s(s+2)(s+3)}$, follow this process:

1. **Verify that the function is proper rational.**

   Because 2 < 3, it's proper rational.

2. **Expand the formula:**

   $$H_1(s) = \frac{s+1}{s(s+2)(s+3)} = \frac{A_1}{s} + \frac{A_2}{s+2} + \frac{A_3}{s+3}$$

3. **One by one, solve for the coefficients:**

   $$A_1 = \frac{s+1}{s(s+2)(s+3)} \cdot s \bigg|_{s=0} = \frac{(0+1)}{(0+2)(0+3)} = \frac{1}{2 \cdot 3} = \frac{1}{6}$$

   $$A_2 = \frac{s+1}{s(s+2)(s+3)} \cdot (s+2) \bigg|_{s=-2} = \frac{(-2+1)}{-2(-2+3)} = \frac{-1}{-2} = \frac{1}{2}$$

   $$A_3 = \frac{s+1}{s(s+2)(s+3)} \cdot (s+3) \bigg|_{s=-3} = \frac{(-3+1)}{-3(-3+2)} = \frac{-2}{-3 \cdot -1} = -\frac{2}{3}$$

4. **Put the numerical coefficients in place to get the PFE representation:**

   $$H_1(s) = \frac{1/6}{s} + \frac{1/2}{s+2} - \frac{2/3}{s+3}$$

5. **Check your calculations with the Maxima function `partfrac()`:**

   ```
   (%i7) partfrac((s+1)/(s*(s+2)*(s+3)), s);

   (%o7)  - ────── + ────── + ───
           3 (s+3)   2 (s+2)   6 s
              2         1        1
   ```

Note: Maxima chooses the expansion coefficient order, so you need to verify that the same terms are present in both solutions, independent of the order you chose for the hand calculation.

Example 2-3: Consider a partial fraction expansion for $D(s)$ having one distinct root and a single root repeated, for a total of three roots. The function is proper rational ($M = 1, N = 3$), so use the repeated roots formulation of the PFE:

$$H_2(s) = \frac{s}{(s+1)(s+2)^2} = \frac{A_1}{s+1} + \frac{A_{21}}{s+2} + \frac{A_{22}}{(s+2)^2}$$

Because the first root is distinct, I made the connection that $A_{11} \equiv A_1$ in the coefficient formulas.

Here's how to determine the expansion coefficients:

1. **Find the coefficients A_1 and A_{22} by using the PFE coefficient formulas:**

$$A_1 = \frac{s}{(s+1)(s+2)^2} \cdot (s+1) \Big|_{s=-1} = \frac{-1}{(-1+2)^2} = -1$$

$$A_{22} = \frac{s}{(s+1)(s+2)^2} \cdot (s+2)^2 \Big|_{s=-2} = \frac{-2}{(-2+1)} = 2$$

2. **Find the coefficient A_{21}.**

The coefficient formula in this case involves taking a derivative of a ratio of polynomials. You can do this all right, but it's tedious and error prone. For this type of problem, I prefer to find the coefficient by writing one equation to solve for the one unknown, A_{21} in this case. Choose a value for s where denominator terms are never 0. By choosing $s = 1$, you can work the algebra to solve for A_{21}:

$$H_2(-1) = \frac{s}{(s+1)(s+2)^2} \Big|_{s=1} = \frac{-1}{s+1} \Big|_{s=1} + \frac{A_{21}}{s+2} \Big|_{s=1} + \frac{2}{(s+2)^2} \Big|_{s=1}$$

$$= \frac{1}{(2)(3)^2} = \frac{-1}{2} + \frac{A_{21}}{3} + \frac{2}{(3)^2}$$

$$A_{21} = 3 \cdot \left[\frac{1}{18} + \frac{1}{2} - \frac{2}{9}\right] = \left[\frac{1}{6} + \frac{3}{2} - \frac{2}{3}\right] = \frac{1+9-4}{6} = 1$$

3. **Put the numerical coefficients in place to establish the PFE representation:**

$$H_2(s) = \frac{-1}{s+1} + \frac{1}{s+2} + \frac{2}{(s+2)^2}$$

4. **Check your calculations with the Maxima function `partfrac()`:**

```
(%i8) partfrac(s/((s+1)*(s+2)^2), s);

(%o8)  1      2       1
      ---- + -------- - ----
      s+2    (s+2)²    s+1
```

I explore the SciPy package `signal` and its functions `residue` and `residuez` in Chapters 15 and 14, respectively, as tools to numerically find the expansion coefficients.

Making Nice Signal Models with Trig Functions

When someone says the word *trigonometry,* what comes to mind? That ulcer you had freshman year, maybe? Okay, perhaps trig isn't an exciting subject for you. Nevertheless, signal modeling and signal analysis using trig functions is a fact of life in electrical engineering.

That's right; trigonometry is part of signal modeling (see Chapters 3 and 4), and trig functions are the basis of line spectra and Fourier series (Chapter 8), the Fourier transform (Chapter 9), and the discrete Fourier transform (Chapter 11). I know, there's never a dull moment here; trig is the gig.

I explain how trig functions in complex arithmetic in the next section; but before we get into all of that, take a look at some of the most beloved trig identities in Table 2-1. Come back here when you need an identity or two as you work through trig functions.

Table 2-1	Useful Trig Identities
$\sin u = \cos(u - \pi/2)$	$\cos u = \sin(u + \pi/2)$
$\cos(u \pm 2\pi k) = \cos u, \ k = \pm 1, \pm 2, \ldots$	$\cos(-u) = \cos u$
$\sin(-u) = -\sin(u)$	$\sin^2 u + \cos^2 u = 1$
$\cos^2 u - \sin^2 u = \cos 2u$	$2\sin u \cos u = \sin 2u$
$\cos^2 u = \frac{1}{2}(1 + \cos 2u)$	$\sin^2 u = \frac{1}{2}(1 - \cos 2u)$
$\sin(u \pm v) = \sin u \cos v \pm \cos u \sin v$	$\cos(u \pm v) = \cos u \cos v \mp \sin u \sin v$
$\cos u \cos v = \frac{1}{2}\left[\cos(u-v) + \cos(u+v)\right]$	$\sin u \sin v = \frac{1}{2}\left[\cos(u-v) + \cos(u+v)\right]$
$\sin u \cos v = \frac{1}{2}\left[\sin(u-v) + \sin(u+v)\right]$	

These signal model examples can get you started with some common trig functions.

Example 2-4: A single continuous-time sinusoid takes the form $x(t) = A\cos(2\pi f_0 t + \theta)$, $-\infty < t < \infty$, where A is the amplitude, f_0 is the frequency in hertz, and θ is the phase in radians. If, say, $\theta = -\pi/2$, then using the identity $\sin u = \cos(u - \pi/2)$, you can write $x(t) = A\sin(2\pi f_0 t)$.

Example 2-5: Pressing a key on a telephone touchpad generates two tones. If you press 5, the row and column sinusoid frequencies are 770 Hz and 1,336 Hz, respectively. The signal you generate and hear is in this form:

$$x(t) = A\{\cos[2\pi(770)t] + \cos[2\pi(1,336)t]\}, \ T_{start} \le t \le T_{stop}$$

Go ahead and press a key on a telephone right now and see whether you can discern two tones being played. This reveals a practical application of trig; it's at work when you make a simple phone call! Trig and signal generation go hand in hand.

Manipulating Numbers: Essential Complex Arithmetic

Complex arithmetic is at the core of both analysis and design tasks performed by signals and systems engineers. Being able to manipulate complex numbers by using paper, pencil, and a pocket calculator on the fly is imperative. So I strongly suggest that you work the examples in this section on your pocket calculator.

Knowing how to use the tools available to you is also important. Using computer programs to manipulate complex numbers and check your answers is a part of everyday signals and systems work, so I give numerical examples in Python to back up hand-calculator calculations.

You need a scientific calculator to work complex arithmetic problems efficiently. Don't expect Apple's Siri to be there, spouting out answers for you. You need to get to know your scientific calculator — before you actually need it. So spend time digging into the details of how the calculator works, and practice! The payoff is quick calculations and peace of mind.

Believing in imaginary numbers

Imaginary numbers are indeed real, but they have a special multiplicative scale factor that makes them imaginary. The need for imaginary numbers comes about when you try to solve an equation involving the square root of a negative number. The scale factor represents the square root of -1 but is denoted with the letters i and j.

A complex number is composed of a real part x and $\sqrt{-1}$ or i times the imaginary part, y: $z = x + jy$, where $j = \sqrt{-1} \overset{also}{=} i$.

Electrical engineers typically use j as the imaginary part scale factor because i usually denotes current. When viewed in the complex plane, the real part lies along the horizontal axis, and the imaginary part lies along the vertical or j axis. Therefore, a complex number is *like* a point (x, y) in a 2D Cartesian coordinate system.

As you may have heard in a physics and/or an analytic geometry course, the location of an object or point in two dimensions can be viewed as an order pair (x, y) and also as $x\hat{i} + y\hat{j}$, where $\hat{i}$ and $\hat{j}$ are unit vectors pointing along the positive x- and y-axis, respectively. The same idea holds in the complex plane, but the operator rules are different. The imaginary part, y, is a real number just like the real part, x, is a real number. Also keep in mind that $\sqrt{-1} \cdot \sqrt{-1} = 1$, so $j \cdot j = -1$.

The *rectangular* (real/imaginary part) form of a complex number is $z = (x, y) = x + jy$. The corresponding *polar* form (magnitude and angle form) is $z = (r, \theta) = r \cdot e^{j\theta} = |z|e^{j\arg(z)} = r\angle\theta$, where, just like in 2D vector analysis, $r = \sqrt{x^2 + y^2}$ is the magnitude, and $\theta = \arctan(y/x)$ is the angle in radians ($\arctan = \tan^{-1}$). Voilà! I just described the rectangular-to-polar conversion operation. To get the angle in degrees from radians, you just need to convert, using the factor 180 degrees/π radians.

Finding the angle of a complex number by using arctan can be tough. The arctan function produces the proper angle only if the complex number lies in Quadrants I or IV. But $\arctan(y/x)$ can't discern where the negative values occur in the ratio y/x, so you find complex numbers that lie in Quadrants II and III by using $\pi + \arctan(y/x)$.

If you're using a calculator that supports complex number operations, use the built-in complex number capability. If you're using a tool such as Python with PyLab, use abs(z) to find the magnitude and angle(z) or arctan2(y, x) to directly get the angle. See Examples 2-6 and 2-7 for details.

A polar-to-rectangular conversion takes you back to rectangular form: $(r,\theta) \Rightarrow (x,y)$ via $x = r\cos\theta$ and $y = r\sin\theta$.

Example 2-6: Consider $z_a = 3 + j4$. Find z_a in polar form, following these steps:

1. **Calculate the magnitude of z_a.**

   The magnitude is $r_a = |z_a| = \sqrt{3^2 + 4^2} = \sqrt{25} = 5$.

2. **Find the angle of z_a.**

   Because 3 and 4 are both positive, you know that the point lies in Quadrant I. The angle of z_a is given by the arctan alone: $\theta_a = \angle z_a = \arctan(4/3) = 0.9273$ radians.

   Putting these two steps together gives you $z_a = 5e^{j0.9273}$, which is the polar form of z_a.

Check out a graphical depiction of resolving the proper quadrant for $\angle z_a$ when using arctan in Figure 2-1.

Figure 2-1:
Finding the proper quadrant for the angles of complex numbers z_a and z_b when using arctan.

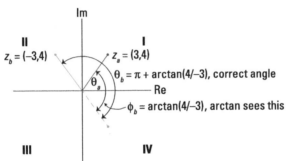

Using Python with PyLab, you can convert the complex number to magnitude and angle:

```
In [28]: z_a = 3 + 4j
In [29]: abs(z_a)
Out[29]: 5.0
In [30]: angle(z_a)   # angle in radians by default
Out[30]: 0.9272
In [31]: angle(z_a, deg=True) # gives angle in degrees
Out[31]: 53.130
```

Example 2-7: Test your ability to resolve the proper quadrant when using arctan by considering $z_b = -3 + j4$. Find z_b in polar form in two steps:

1. **Calculate the magnitude of z_b.**

   The magnitude of z_b is $r_b = |z_b| = \sqrt{(-3)^2 + 4^2} = \sqrt{25} = 5$.

2. **Find the angle of z_b.**

   Because 3 is negative and 4 is positive, you know that the point lies in Quadrant II. The angle of z_b is given by
   $\theta_b = \angle z_b = \pi + \arctan\left[4/(-3)\right] = \pi - 0.9273 = 2.2143$ radians.

   Therefore, the polar form of is $z_b = 5e^{j2.2143}$.

The details of the angle calculation are depicted in Figure 2-1. Using Python with PyLab, you can convert the complex number to magnitude and angle:

```
In [32]: z_b = -3 + 4j
In [33]: abs(z_b)
Out[33]: 5.0
In [34]: angle(z_b)
Out[34]: 2.2143
```

Example 2-8: Consider $z_c = 2e^{j\pi/3}$. Find z_c in rectangular coordinates in one step:

Looking at z_c reveals that $r_c = 2$ and $\theta_c = \pi/3$, so
$$z_c = \underbrace{2\cos(\pi/3)}_{x} + \underbrace{j2\sin(\pi/3)}_{y} = 1 + j\sqrt{3} = 1.0 + j1.7321.$$

Using Python, the calculation is direct:

```
In [41]: z_c = 2*exp(1j*pi/3.) #  enter in polar form
In [42]: z_c
Out[42]: (1.0000+1.7321j)
```

Operating with the basics

The basic arithmetic operations with complex numbers are addition, subtraction, multiplication, and division. Addition and subtraction are easiest in rectangular form, and multiplication and division are easiest in polar form. Table 2-2 shows the formulas for the four operations, starting from $z_1 = x_1 + jy_1$ and $z_2 = x_2 + jy_2$.

Table 2-2	Basic Complex Math Operations
Operation	*Formula*
Add: $z_3 = z_1 + z_2$	$(x_1 + x_2) + j(y_1 + y_2)$
Subtract: $z_3 = z_1 - z_2$	$(x_1 - x_2) + j(y_1 - y_2)$
Multiply: $z_3 = z_1 z_2$	$(x_1 x_2 - y_1 y_2) + j(x_1 y_2 + y_1 x_2)$
(polar form)	$r_1 r_2 e^{j(\theta_1 + \theta_2)}$
Divide: $z_3 = z_1 / z_2$	$\dfrac{(x_1 x_2 + y_1 y_2) - j(x_1 y_2 - y_1 x_2)}{x_2^2 + y_2^2}$
(polar form)	$\dfrac{r_1}{r_2} e^{j(\theta_1 - \theta_2)}$

Another basic operation is complex conjugation: $\text{conjugate}(z) \equiv z^* = x - jy$ when $z = x + jy$.

Note that $z \cdot z^* = (x + jy)(x - jy) = x^2 + y^2 = |z|^2$.

At times, you may want to deal with only the real or imaginary part of an expression. The notation for the *real part* (real part operator) is $\text{Re}\{z\} = \text{Re}\{x + jy\} = x$. The notation for the *imaginary part* (imaginary part operator) is $\text{Im}\{z\} = \text{Im}\{x + jy\} = y$. As a simple example,
$$z + z^* = (x + jy) + (x - jy) = 2x \stackrel{also}{=} 2\text{Re}\{z\}.$$

Next, work through some calculations that utilize the basic complex math operations. Getting familiar with these types of calculations helps you prepare for more complicated topics in signals and systems.

Example 2-9: Let $z_1 = 1 + j7$ and $z_2 = -4 - j9$. The operations of Table 2-2 occur frequently, so it's time to run some numbers for $z_1 + z_2$, $z_1 z_2$, and z_1 / z_2. Here's how to work through the four calculations in sequence:

1. **Find the sum, $z_1 + z_2$, following immediately from the formula of Table 2-2.**

   The real parts add and the imaginary parts add: $z_1 + z_2 = [1 + (-4)] + j[7 + (-9)] = -3 - j2$.

2. **Work the multiplication result by using the polar form formula given in the multiplication row of Table 2-2.**

Begin by first converting z_1 and z_2 to polar form (see Examples 2-6 and 2-7), and then multiply the magnitudes and add the angles to find the polar form of the answer. Finally, convert back to rectangular form:

$$z_1 = 1 + j7 = \sqrt{1^2 + 7^2}\, e^{j\arctan(7/1)} = 7.0711 e^{j1.4289}$$

$$z_2 = -4 - j9 = \sqrt{(-4)^2 + (-9)^2}\; \underbrace{e^{j[\pi + \arctan(-9/-7)]}}_{\text{or } e^{j[\arctan(-9/-7)-\pi]}} = 9.8489 e^{-j1.9890}$$

$$z_1 z_2 = 7.0711 \cdot 9.8489 e^{j1.4289 + (-1.9890)} = 69.6419 e^{-j0.5601}$$

$$= 69.6419\left[\cos(-0.5601) + j\sin(-0.5601)\right] = 59 - j37$$

3. **Work the division, using the polar form formula of the division row of Table 2-2.**

Start with the polar form answers from Step 2, divide the magnitudes and subtract the angles, and finally convert back to rectangular form:

$$z_1/z_2 = \frac{7.0711 e^{j1.4289}}{9.8489 e^{-j1.9890}} = \frac{7.0711}{9.8489} \cdot e^{j[1.4289 - (-1.9890)]} = 0.7180 e^{-j2.8653}$$

$$= 0.7180\left[\cos(-2.8653) + j\sin(-2.8653)\right] = -0.6907 - j0.1959$$

4. **Using Python with PyLab as a numerical tool, check your results:**

```
In [57]: z_1 = 1 + 7j   # define z1
In [58]: z_2 = -4 - 9j   # define z2
In [59]: z_1 + z_2       # form the sum
Out[59]: (-3-2j)
In [60]: z_1*z_2         # form the product
Out[60]: (59-37j)
In [61]: z_1/z_2         # form the quotient
Out[61]: (-0.6907-0.1959j)
```

The Python results agree with the hand calculation. So pick up your calculator now and see whether you can get matching results.

Applying Euler's identities

Euler's formula is a useful mathematical result in signals and system analysis:

$$e^{j\theta} = \cos\theta + j\sin\theta$$

The formula is especially helpful when particular complex exponentials, where θ is a function of time, are used to develop the phasor addition formula (described in the next section). Given $z = re^{j\theta}$ and Euler's formula, the polar-to-rectangular conversion formula follows immediately, because $z = re^{j\theta} = r(\cos\theta + j\sin\theta) = r\cos\theta + jr\sin\theta$.

The inverse formula is perhaps even more useful:

$$\cos\theta = \frac{e^{j\theta} + e^{-j\theta}}{2} \text{ and } \sin\theta = \frac{e^{j\theta} - e^{-j\theta}}{2j}$$

In Part III of this book, I use both inverse formulas extensively, particularly to simplify expressions, demonstrating the importance of Euler's identity in real-world signals and systems work. As a quick example of the formula's practical application, consider factoring $e^{-j\omega_0 t/2}$ in the following to form a cosine:

$$1 + e^{-j\omega_0 t} = \underbrace{\left(e^{j\omega_0 t/2} + e^{-j\omega_0 t/2}\right)}_{2\times \text{ a cosine term}} e^{-j\omega_0 t/2} = 2\cos\left(\omega_0 t/2\right)\cdot e^{-j\omega_0 t/2}$$

Applying the phasor addition formula

When sinusoidal signals of like frequency ω_0 are added together *(superimposed)*, the result is a single sinusoidal signal having *composite* amplitude A and phase ϕ. In mathematical terms, given N sinusoidal functions (signals) $A_k \cos(\omega_0 t + \phi_k)$, $k = 1, 2, \ldots N$ each of identical frequency ω_0, you can show that the sum

$$\sum_{k=1}^{N} A_k \cos\left(\omega_0 t + \phi_k\right) = A\cos\left(\omega_0 t + \phi\right) \overset{\text{also}}{=} A\text{Re}\left\{e^{j(\omega_0 t + \phi)}\right\}$$

where, on the right side, the composite amplitude A and phase ϕ is related to the amplitude A_k and phase ϕ_k of each term through the summation.

$$Ae^{j\phi} = \sum_{k=1}^{N} A_k e^{j\phi_k}$$

Example 2-10: Given $x(t) = 4\cos\left(\omega_0 t + \pi/4\right) + 15\sin\left(\omega_0 t\right)$, find the composite signal $A\cos\left(\omega_0 t + \phi\right)$. Follow these steps to get A and ϕ:

1. **Rewrite the sine term as a cosine, so all terms in the sum fit the phasor addition formula mathematical form.**

   Using the Row 1/Column 1 trig identity of Table 2-1, $15\sin\left(\omega_0 t\right) = 15\cos\left(\omega_0 t - \pi/2\right)$.

2. **Identify the magnitude and angles $A_k e^{j\phi k}$, $k = 1, 2$, two terms in the problem statement:**

$$A_1 e^{j\phi_1} = 4e^{j\pi/4} \quad \text{and} \quad A_2 e^{j\phi_2} = 15e^{-j\pi/2}$$

3. **Add the two complex numbers that are in polar form.**

4. **Check the calculation on your calculator and then work the calculation in Python, as shown here:**

```
In [115]: A_phi=4*exp(1j*pi/4.) + 15*exp(-1j*pi/2.)
In [117]: (abs(A_phi), angle(A_phi)) # polar form
Out[117]: (12.4959, -1.3425)
```

5. **You find $Ae^{j\phi}$ in Step 3; write out the final form by putting the magnitude in front of the cosine and entering the angle inside the argument of the cosine:**

$$x(t) = 12.4959\cos(\omega_0 t - 1.3425)$$

Seeing the proof for the phasor addition formula

Here's the proof of the phasor addition formula:

1. Using Euler's formula and the real part operator, replace each term $\cos(\omega_0 t + \phi_k)$ with $\text{Re}\{e^{j(\omega_0 t + \phi_k)}\} = \text{Re}\{e^{j\phi_k} \cdot e^{j\omega_0 t}\}$ and recognize that the sum of real parts is identical to the real part of the sum:

$$\sum_{k=1}^{N} A_k \cos(\omega_0 t + \phi_k) = \sum_{k=1}^{N} A_k \text{Re}\{e^{j(\omega_0 t + \phi_k)}\} = \text{Re}\left\{\sum_{k=1}^{N} A_k e^{j\phi_k} \cdot e^{j\omega_0 t}\right\}$$

2. Inside the real-part operator on the far right, factor $e^{j\omega_0 t}$ outside the sum. What remains is the sum of N complex numbers in polar form, which finally produces a single complex number in polar form:

$$\sum_{k=1}^{N} A_k e^{j\phi_k} = Ae^{j\phi}$$

3. Make the substitution of Step 2 into Step 1, which results in

$$\sum_{k=1}^{N} A_k \cos(\omega_0 t + \phi_k) = \text{Re}\{Ae^{j\phi} \cdot e^{j\omega_0 t}\} = \text{Re}\{Ae^{j(\omega_0 t + \phi)}\}$$
$$= A\cos(\omega_0 t + \phi)$$

The expected result!

Catching Up with Calculus

Single variable integration and differentiation are main calculus topics at play in signals and systems work. In this section, I provide a table of differentiation properties and a short table of indefinite and definite integrals. The tables are for quick reference as the need arises. I also point out simple optimization theory, including min and max and efficient numerical techniques.

And lest you think I may have completely spaced the importance of geometric series formulas, don't worry! Formulas for finite and infinite geometric series are available in this section, too, so you can check back here if needed when you read Chapters 6, 11, and 14 on discrete-time signals and systems. I also describe tools for polynomial root finding in this section.

Differentiation

Finding the derivative of function is part of the core material in a calculus course. And being able to differentiate simple functions is essential to signals and systems analysis. This section reviews the fundamentals of differentiation and provides a table of key differentiation formulas.

The derivative of an equation is the slope at any point where the equation is evaluated. Formally, the function $f(x)$ has derivative $f'(x)$, at the point x if the limit $\dfrac{df(x)}{dx} = f'(x) = \lim\limits_{\Delta x \to 0} \dfrac{f(x+\Delta x) - f(x)}{\Delta x}$ exists.

If you need help when working problems where differentiation is involved, reference the differentiation formulas in Table 2-3. In this table, assume a is a constant and both u and v are possibly functions of x.

Table 2-3	Differentiation Formulas
$\dfrac{d}{dx}(ax^n) = nax^{n-1}$	$\dfrac{d}{dx}(uv) = u\dfrac{dv}{dx} + v\dfrac{du}{dx}$
$\dfrac{d}{dx}\left(\dfrac{u}{v}\right) = \dfrac{v(du/dx) - u(dv/dx)}{v^2}$	$\dfrac{dy(u)}{dx} = \dfrac{dy}{du}\cdot\dfrac{du}{dx}$ (chain rule)
$\dfrac{d}{dx}\cos(u) = -\sin(u)\dfrac{du}{dx}$	$\dfrac{d}{dx}\sin(u) = \cos(u)\dfrac{du}{dx}$
$\dfrac{d}{dx}\arctan(u) = \dfrac{1}{1+u^2}\dfrac{du}{dx}, -\dfrac{\pi}{2} < \arctan(u) < \dfrac{\pi}{2}$	$\dfrac{d}{dx}\ln(u) = \dfrac{1}{u}\dfrac{du}{dx}$
$\dfrac{d}{dx}e^{cx} = ce^{cx}$	$\dfrac{d}{dx}e^u = e^u\dfrac{du}{dx}$

Example 2-11: Given $x(t) = e^{-2t} \cdot \cos(\omega_0 t)$, $t \geq 0$, find $x'(t)$. To solve, use the formulas in Table 2-3:

$$x'(t) = \frac{d}{dt}\left[e^{-2t} \cdot \cos(\omega_0 t)\right]$$
$$= -2e^{-2t} \cdot \cos(\omega_0 t) + e^{-2t} \cdot (-\omega_0)\sin(\omega_0 t)$$
$$= -e^{-2t} \cdot \left[2\cos(\omega_0 t) + \omega_0 \sin(\omega_0 t)\right]$$

You can also use the CAS Maxima to check this differentiation.

Integration

What differentiation does, integration undoes. In mathematical terms, the integral of the function $f(x)$ is known as the *antiderivative.* The good news about signals and systems work is that although integration is present in the associated problem solving, you can use integration tables for both indefinite and definite integrals.

Indefinite integrals are the antiderivatives. So when I write

$$F(x) = \int f(x)\,dx$$

I'm saying that

$$f(x) = \frac{dF(x)}{dx}$$

In Table 2-4, I provide a collection of indefinite integrals that are useful for signals and systems work. In this table, assume a is a constant.

Table 2-4	Indefinite Integrals
$\int \sin(ax)\,dx = -\frac{1}{a}\cos(ax)$	$\int \cos(ax)\,dx = \frac{1}{a}\sin(ax)$
$\int e^{ax}\,dx = \frac{e^{ax}}{a}$	$\int xe^{ax}\,dx = \frac{e^{ax}}{a^2}(ax-1)$

Example 2-12: To see how numerical integration can be put into motion on a problem with known solution, find the integral of te^{-5t} on the interval $[0, 4]$ by using the Row 2/Column 2 entry of Table 2-4:

$$\int_0^4 te^{-5t}\, dt = \frac{e^{-5t}}{(-5)^2}(-5t-1)\Big|_0^4 = \frac{1-21e^{-20}}{25}$$

You can also use the CAS Maxima to check this integration.

Definite integrals, also known as *integrals with fixed limits,* are important in the study of signals and systems. Table 2-5 is a collection of useful definite integrals. In this table, assume a is a constant.

Table 2-5	Definite Integrals
$\int_0^\infty \frac{\sin(ax)}{x}\, dx = \frac{\pi}{2},\ a>0$	$\int_0^\infty x^n e^{-ax}\, dx = \frac{n!}{a^{n+1}},\ n$ integer, $a>0$
$\int_0^\infty e^{-ax}\cos(bx)\, dx = \frac{a}{a^2+b^2},\ a>0$	$\int_0^\infty e^{-ax}\sin(bx)\, dx = \frac{b}{a^2+b^2},\ a>0$

When *closed-form* integration isn't workable — that is, if the integral isn't available in a table — you need to use numerical integration. By *closed-form,* I mean the solution can be written in terms of a small number of well-known functions. Tools such as Python and the `integrate` module of SciPy provide a collection of numerical integration routines. I use the function `quad` in Example 2-13.

Numerical integration saves you the trouble of an extensive search for a closed-form integral, and it may be your only option in some cases. But use numerical integration with care to ensure convergence to the proper solution.

Example 2-13: To gain confidence in the use of numerical integration, I repeat the integration of Example 2-12, using Python and the function `quad` from the SciPy module `integrate`:

```
In [90]: from scipy.integrate import quad
In [91]: def my_inte(t): # integrand function
    ...: return t*exp(-5*t)
In [92]: quad(my_inte,0,4) # integrate from 0 to 4
Out[92]: (0.03999999826863095, 7.91215216153553e-14)
In [93]: (1 - 21*exp(-20))/25.
Out[93]: 0.03999999826863096
```

The calculation requires a Python function, Line [91], to first represent the integrand. Second, you call quad with the integrand function name followed by the lower and upper limits. The function returns the integral and the absolute error (see Line [92]). Comparing Line [92] with the known closed-form solution of Line [93] reveals very close agreement.

System performance

When you design a system, you need to maximize, minimize, or find critical values of performance functions. A *performance function,* such as a time-domain or frequency-domain response, provides a means to demonstrate that your design meets requirements set by your customer. Calculus and numerical methods based on calculus can help.

Elementary calculus reveals that function $f(x)$ is a maximum where the derivative is 0, as long as the function isn't passing through an inflection point. So if the solution to $f'(x) = 0$ is x_0, does x_0 correspond to a maximum or minimum? To find out, check the sign of the second derivative at x_0. If $f''(x_0) > 0$, then x_0 corresponds to a minimum. If $f''(x_0) < 0$, then x_0 corresponds to a maximum.

Example 2-14: To minimize the quadratic function $f(x) = x^2 + 10x + 20$, use this two-step process:

1. **Take the derivative of $f(x)$ and set the result equal to 0:**

$$f'(x) = 2x + 10 = 0 \Rightarrow x_0 = \frac{-10}{2} = -5$$

2. **See whether the point $f'(x_0) = 0$ corresponds to a maximum or minimum. Take the second derivative and check the sign at x_0:**

$$f''(x) = 2 > 0 \Rightarrow x_0 = -5 \text{ corresponds to a minimum!}$$

Example 2-15: More often than not, the performance function is nontrivial, meaning a pure calculus-based solution involves some work. Consider the step response waveform $y(t) = 1 - e^{-t/\sqrt{2}} \left[\sin\left(t/\sqrt{2}\right) + \cos\left(t/\sqrt{2}\right) \right]$, $t \geq 0$ (see Chapter 13). A plot of this function is given in Figure 2-2.

This function rises up from 0 to 1 as t increases from 0 to infinity but with a small amount of *overshoot,* which means the function exceeds 1 before settling back to 1.

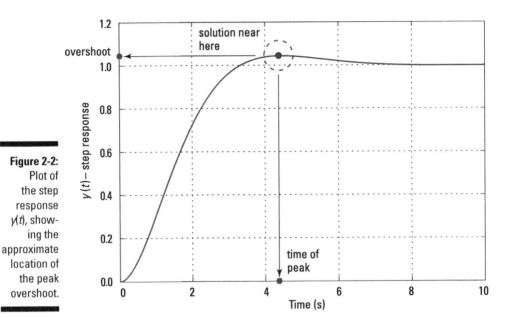

Figure 2-2:
Plot of
the step
response
y(t), show-
ing the
approximate
location of
the peak
overshoot.

Suppose the overshoot value and the corresponding time are both needed
for system performance assessment. You can find a numerical solution
quickly by first filling an array with closely spaced values of y(t) over 2 to 6 s.
Use max to get the maximum value on this interval, and then use find to find
the array index where the maximum occurs. Finally, plug the index into the t
array to reveal the time of the peak.

```
In [120]: t = arange(2,6,.001) # small time step
In [121]: y = 1-(exp(-t/sqrt(2))
                 *(sin(t/sqrt(2))+cos(t/sqrt(2))))
In [122]: max_y = max(y) # obtain peak numerically
In [123]: find(y == max_y) # find peak index
Out[123]: array([2443], dtype=int64) # at index 2443
In [124]: t_max = t[2443] # plug index into t array
In [125]: (max_y,t_max) # display results
Out[125]: (1.0432, 4.4429) # (overshoot, time in s)
```

The peak overshoot in this case is 1.0432, or about 4.3 percent, and it's
located at 4.44 s.

Geometric series

In the study of discrete-time signals and systems (covered in Chapters 6, 7,
11, 12, and 14), you frequently encounter geometric series — finite and
infinite. For example, in Chapter 6, both forms arise in impulse response

calculations, and in Chapters 11 and 14, where the discrete Fourier transform and z-transforms are studied, both forms are again essential for developing transform pairs.

The finite geometric series takes the following form, where a is a constant:

$$S_N = 1 + a + a^2 + \cdots + a^N = \sum_{n=0}^{N} a^n = \frac{1-a^N}{1-a}$$

The infinite series takes this form:

$$S = \sum_{n=0}^{\infty} a^n = \frac{1}{1-a}, |a| < 1$$

Table 2-6 provides a short table of geometric series sums.

Table 2-6	Series Sum Formulas
Sum	*Condition*
$\sum_{n=N_1}^{N_2} a^n = \dfrac{a^{N_1} - a^{N_2+1}}{1-a}$	none
$\sum_{k=0}^{N} a^k = \dfrac{1-a^{N+1}}{1-a}$	none
$\sum_{k=0}^{N} ka^k = \dfrac{a\left[1-(N+1)a^N + Na^{N+1}\right]}{(1-a)^2}$	none
$\sum_{n=0}^{\infty} a^k = \dfrac{1}{1-a}$	$|a| < 1$
$\sum_{n=0}^{\infty} ka^n = \dfrac{a}{(1-a)^2}$	$|a| < 1$

Example 2-16: To sum the series $S = 10 \sum_{n=5}^{\infty} \left(\frac{1}{2}\right)^n$, follow this three-step process.

Note: Table 2-6 doesn't contain an exact fit for this summation, so you need to do some improvising.

1. **Introduce a variable change to transform the sum to the standard form.**

   Let $m = n - 5$. Then, in the present sum, replace n with $m + 5$.

2. **Transform the sum limits, too:**

   $$n = 5 \Rightarrow m = 0 \text{ and } n = \infty \Rightarrow m = \infty$$

3. **Put Steps 1 and 2 together in the original sum:**

$$S = 10\sum_{m=0}^{\infty}\left(\tfrac{1}{2}\right)^{m+5} = 10\cdot\left(\tfrac{1}{2}\right)^{5}\cdot\sum_{m=0}^{\infty}\left(\tfrac{1}{2}\right)^{m} = \frac{10/2^{5}}{1-1/2} = 0.625$$

Finding Polynomial Roots

To perform stability analysis of systems (covered in Chapters 13 and 14), you need to be able to find the roots — values of the independent variable where the polynomial is 0 — of a polynomial of degree 2 and higher. The quadratic formula works for the second-order case, but numerical root finding is the best option for higher-order polynomials.

Numerical solving routines within Python and PyLab make calculating numerical roots a snap. Similar routines are available in other tool sets.

The problem here is to find the roots of the Nth-degree polynomial:
$P(s) = a_N s^N + a_{N-1}s^{N-1} + \cdots + a_1 s + a_0.$

The function pN = roots(AN) returns the polynomial roots into array pN given the coefficients a_N, a_{N-1}, $\dots$, a_1, a_0 are contained in the array AN.

A special Nth-degree polynomial that occurs in discrete-time systems modeling is $P(s) = s^N - 1$. Yes, you can use roots for this polynomial, but an analytical solution, known as the *roots of unity,* is readily available. To solve, set $P(s) = 0$, so $s^N - 1 = 0 \Rightarrow s^N = 1$. The roots are

$$p_k = \exp\left[\,j\frac{2\pi}{N}k\,\right], k = 0, 1, 2, \dots, N-1$$

because

$$p_k^N = \left(e^{j2\pi/N\cdot k}\right)^N = e^{j2\pi N/N\cdot k} = e^{j2\pi k} = 1$$

The roots are located in the complex plane equally spaced around the unit circle, at separation angle $2\pi/N$.

A variation on the original polynomial is $P_1(s) = s^N - a^N$. The N roots in this case take the same angle values but have magnitude a.

Chapter 3

Continuous-Time Signals and Systems

. . . .

In This Chapter

▶ Defining signal types

▶ Classifying specific signals

▶ Modifying signals

▶ Looking at linear and time-invariant systems

▶ Checking out a real-world example system

. . . .

*I*n signals and systems, the distinction *continuous time* refers to the independent variable, time *t,* being continuous (see Chapter 1). In this chapter, I provide an inventory of signal types and classifications that relate to electrical engineering and cover the process of figuring out the proper description for a particular signal. Like their discrete-time counterparts described in Chapter 4, continuous-time signals may be classified as deterministic or random, periodic or aperiodic, power or energy, and even or odd. Signals hold multiple classifications.

Also in this chapter, I describe the process of moving signals around on the time axis. Modeling the placement of signals on the time axis affects system functionality and relates to the convolution operation that's described in Chapter 5. Time alignment of signals entering and leaving a system is akin to composing a piece of music.

Don't worry; I include a drill down on system types and classifications here, too, focusing on five property definitions: linear, time-invariant, causal, memory-less, and stability. Find a system-level look at the signals and systems model of a karaoke machine at www.dummies.com/extras/signalsandsystems. The signal flow through this system consists of two paths: one for the recorded music and the other for the singer's voice that enters the micro-phone. The subsystems of the karaoke machine act upon the two input signal types — in this case, both random signals — to finally end up at the speak-ers, which convert the electrical signals to sound pressure waves that your ears can interpret.

Considering Signal Types

Knowing the different types of signals makes it possible for you to provide appropriate stimuli for the systems in your product designs and to characterize the environment in which a system must operate. Both desired and undesired signals are present in a typical system, and models make it easier to create products that can operate in a range of different conditions.

Some signals types are a means to the end — the signal is designed to fulfill a purpose, such as carrying information without wires. Other signal types are useful for characterizing system performance during the design phase. Still other signal types are intended to make things happen in an orderly fashion, such as timing events while a person arms and disarms a burglar alarm.

Signal types described in this section include sinusoids, exponentials, and various singularity signals, such as step, impulse, rectangle pulse, and triangle pulse. *Real signals* are typically composed of one or more signal types.

Exponential and sinusoidal signals

In this section, I introduce you to two of the most fundamental and important signal types:

- **Complex exponential:** This signal occurs naturally as the response (output) of linear time-invariant systems (see the section "Checking Out System Properties," later in this chapter) to arbitrary inputs.

- **Real and complex sinusoids:** These signals function inside electronic devices, such as wireless communications, and form the basis for the Fourier analysis (frequency spectra), which is described in Part III of this book.

A general complex exponential signal, which also includes exponentials and real complex sinusoidal signals, is $x(t) = Ae^{\gamma t}, -\infty < t < \infty$, where, in general, $\gamma = -\alpha + j\omega_0$ and $A = |A|e^{j\phi}$. This is a lot to handle when you think about it. The signal contains two parameters but, because each parameter is generally *complex* — having a real and imaginary part (for γ) or magnitude and phase (for $\tilde{A}$) — four parameters are associated with this signal. To make this concept manageable to study, I break down the signal into several special cases and discuss them in the following sections.

Real exponential

For γ real, ($\gamma = -\alpha$) and $\tilde{A} = A$, meaning $\phi = 0$. You arrive at the real exponential signal $x(t) = Ae^{-\alpha t}$, $-\infty < t < \infty$. Of the two parameters that remain, A controls the amplitude of the exponential and α controls the decay rate of the signal.

In practice, the real exponential signal also contains the function $u(t)$, known as the unit step function (see the later section "Unit step function" for details). For now, all you need to know about the step function is that it acts as a switch for a function of time. The switch $u(t)$ is 0 for $t < 0$ and 1 for $t > 0$. Using $u(t)$, limit $x(t)$ to turning on at $t > 0$ by writing $x(t) \cdot u(t) = Ae^{-\alpha t}u(t)$.

Example 3-1: To plot the real exponential signal $2e^{-\alpha t}u(t)$ for various values of α, use Python with PyLab. See the results in Figure 3-1.

```
In [689]: t = arange(-2,10,.01)
In [690]: plot(t,2*exp(-.2*t)*ssd.step(t))
In [691]: plot(t,2*exp(0*t)*ssd.step(t),'b--')
In [692]: plot(t,2*exp(.2*t)*ssd.step(t),'b-.')
```

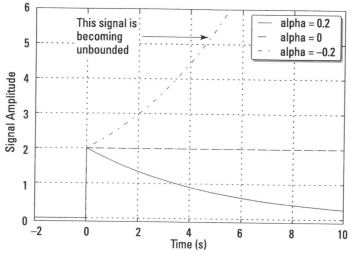

Figure 3-1: A real exponential signal for α positive, zero, and negative.

Complex and real sinusoids

For γ and A imaginary ($\gamma = j\omega_0$) and $\tilde{A}$ complex ($\tilde{A} = Ae^{j\phi}$), the signal becomes a complex sinusoid with real and imaginary parts being the cosine and sine signal, respectively. Take a closer look:

$$x(t) = \tilde{A} \cdot e^{\gamma t} = A \cdot \underbrace{e^{j\phi} \cdot e^{j\omega_0 t}}_{e^{j(\omega_0 t + \phi)}} = Ae^{j(\omega_0 t + \phi)}$$

$$= A\left[\cos(\omega_0 t + \phi) + j\sin(\omega_0 t + \phi)\right]$$

The last line comes from applying Euler's formula.

By breaking out the real and imaginary parts as $x(t) = x_{Re}(t) + jx_{Im}(t)$, you get the real cosine and sine signals: $x_{Re}(t) = |A|\cos(\omega_0 t + \phi)$, $-\infty < t < \infty$ and $x_{Im}(t) = A\sin(\omega_0 t + \phi)$, $-\infty < t < \infty$. In this signal model, A is the signal amplitude, ω_0 is the signal frequency in rad/s or when using the substitution $\omega_0 = 2\pi f_0$, f_0 is the frequency in hertz (cycles per second), and ϕ is the phase. See the later section "Deterministic and random" for more information on these parameters.

The real cosine signal is one of the most frequently used signal models in all of signals and systems. When you key in a phone call, you're using (and hearing) a combination of two sinusoidal signals to dial the phone number. The wireless LAN system you use at work or home also relies on the real sinusoid signal and the complex sinusoid. A real sinusoid signal with $\omega_0 \approx 377$ rad/s is the signal that enters your home to deliver energy (power) to run your appliances and lights. Modern life is filled with the power of the sinusoidal signal.

The case of K real sinusoids is also useful in general signal processing and communications applications. You simply add together the sinusoidal signals:

$$x(t) = \sum_{k=1}^{K} |A_k| \cos(2\pi f_k t + \phi_k), \text{ with the substitution } \omega_k = 2\pi f_k$$

This model can represent the result of pressing one or more keys on a piano keyboard.

Damped complex and real sinusoids

As a special case of the general complex exponential, assume that γ is complex $(\gamma = -\alpha + j\omega_0)$, A is complex, and $u(t)$ is included, so the signal turns on for $t > 0$. The signal becomes a damped complex sinusoid with real and imaginary parts being damped cosine and sine signals, respectively:

$$x(t) = \tilde{A}e^{\gamma t} = Ae^{j\phi} \cdot e^{(-\alpha + j\omega_0)t} = Ae^{-\alpha t} \cdot e^{j(\omega_0 t + \phi)}$$
$$= Ae^{-\alpha t}\left[\cos(\omega_0 t + \phi) + j\sin(\omega_0 t + \phi)\right]$$

The term *damped* means that $e^{-\alpha t}$ drives the signal amplitude to 0 as t becomes large. The sinusoid sine or cosine is the other time-varying component of the signal. Breaking out the real and imaginary parts,

$$x_{Re}(t) = |A|e^{-\alpha t}\cos(\omega_0 t + \phi)u(t)$$
$$x_{Im}(t) = |A|e^{-\alpha t}\sin(\omega_0 t + \phi)u(t)$$

Have you ever experienced a damped sinusoid? Physical examples include a non-ideal swinging pendulum or your car's suspension after you hit a pothole. In both cases, the oscillations eventually stop due to friction. With an appropriate measuring device, the physical motion can convert to an electrical signal. In the car example, the signal may be fed into an *active ride control* system and other inputs to modify the suspension system.

Singularity and other special signal types

As preposterous as it may sound, sinusoidal signals don't rule the world. You need singularity signals to model other important signal scenarios, but this type of signal is only *piecewise* continuous, meaning a signal that has a distinct mathematical description over contiguous-time intervals spanning the entire time axis $(-\infty,\infty)$. The piecewise character means that a formal derivative doesn't exist everywhere. A singularity signal may also contain jumps.

In this section, I describe a few of the most common singularity functions, including rectangle pulse, triangle pulse, unit impulse, and unit step. With these functions, you can put together many special waveforms, such as the one shown in Figure 3-8, later in this chapter.

Rectangle and triangle pulse

Two useful singularity signal types used in signals and systems analysis and modeling are the *rectangle pulse*, $\Pi(t/\tau)$, and *triangle pulse*, $\Lambda(t/\tau)$.

Here's the piecewise definition of the rectangle pulse:

$$\Pi\left(\tfrac{t}{\tau}\right) = \begin{cases} 1, & |t| \leq \tau/2 \\ 0, & \text{otherwise} \end{cases}$$

This signal contains a jump up and a jump down. **Note:** *Otherwise* is math terminology for all the remaining time intervals on $(-\infty,\infty)$ not specified in the definition. For the rectangle pulse, that would be $|t| > \tau/2$. And also assume that τ is positive.

Here is the definition for the triangle pulse:

$$\Lambda\left(\tfrac{t}{\tau}\right) = \begin{cases} 1 - \dfrac{|t|}{\tau}, & |t| \leq \tau \\ 0, & \text{otherwise} \end{cases}$$

See plots of these two pulse signals in Figure 3-2.

Figure 3-2:
The
rectangle
pulse (a) **(a)**
and the
triangle
pulse (b).

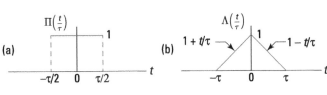

The full base width of the rectangle pulse is τ; for the triangle pulse, it's 2τ. This is no accident. Check out Chapter 5 to find out how to get a triangle pulse of base width 2τ by convolving two rectangle pulses with a width of τ.

Here are the Python functions for these two signal types:

```
def rect(t,tau):
    x = np.zeros(len(t))
    for k,tk in enumerate(t):
        if np.abs(tk) > tau/2.:
            x[k] = 0
        else:
            x[k] = 1
    return x

def tri(t,tau):
    x = np.zeros(len(t))
    for k,tk in enumerate(t):
        if np.abs(tk) > tau/1.:
            x[k] = 0
        else:
            x[k] = 1 - np.abs(tk)/tau
    return x
```

The functions are designed to accept `ndarray` variables when using PyLab. Later, in Figure 3-8, these functions create a quiz problem. Note that I use Maxima for computer algebra–oriented calculations. For numerical calculations, particularly those that benefit from working with arrays or vectors, I use Python.

Unit impulse

You use the *unit impulse,* or Dirac delta function (δ-function) signal type, $\delta(t)$ as a test waveform to find the *impulse response* of systems in Chapter 4. It's a fundamental but rather mysterious signal.

Tooling up

The functions `rect` and `tri` are contained in the Python code module `ssd.py` found at `www.dummies.com/extras/signals andsystems`. To access the functions from IPython, import the module, using `import ssd` at the IPython command prompt.

To access the functions `rect`, qualify them with the module name. In this case, that means something like `ssd.rect(t,2)` for a rectangle of width 2 centered at $t = 0$. If you want to create a rectangle of width 2 on a time axis that runs from [–5, 5] and plot the results, use this code:

```
In [6]: import ssd
In [7]: t = arange(-5,5,.01)
         # a vector of time samples
In [8]: x = ssd.rect(t,2) #
         create the signal vector
In [9]: plot(t,x)
```

If you make changes to the module `ssd.py`, such as adding another function, you need to reload the module (not `import` it).

You can define this signal only in an operational sense, meaning how it behaves. The signal appears as a spike, but the spike has zero width and unity area. This is confusing at first. But think of the cue ball striking another ball in a game of billiards. The momentum transfers instantly (impulsively), and the struck ball begins to roll. In an electrical circuit, think of a battery momentarily (very momentarily) making contact with the circuit input terminals. The circuit responds with its impulse response.

Operationally, the delta function has the following key properties:

$$\int_{t_1}^{t_2} \delta(t-t_0)dt = 1, \quad t_1 < t_0 < t_2$$

$$\delta(t-t_0) = 0, \quad t \neq t_0$$

The signal $\delta(t-t_0)$ appears as a function with unit area located at $t-t_0 = 0$ or $t = t_0$. You can sift out a single value of the function $x(t)$, which is assumed to be continuous at $t = t_0$, by bringing it inside the integral:

$$\int_{-\infty}^{\infty} x(t)\delta(t-t_0)dt = x(t_0)$$

This result is known as the *sifting property* of the unit impulse function. To actually put your hands on something that closely resembles the true unit impulse function, define the test function $\delta_\varepsilon(t)$:

$$\delta_\varepsilon(t) = \frac{1}{2\varepsilon}\Pi\left(\frac{t}{2\varepsilon}\right)$$

Notice that this function has unit area, like $\delta(t)$, and is focused at $t = 0$. In fact, as desired, the signal behaves like $\delta(t)$ as $\varepsilon \to 0$, or $\lim_{\varepsilon \to 0} \delta_\varepsilon(t) \to \delta(t)$. To illustrate the action of the sifting property, I first sketch the integrand by using $\delta_\varepsilon(t)$ in Figure 3-3a and then using $\delta(t)$ in Figure 3-3b.

The proper way of plotting $\delta(t)$ is to draw it as a vertical line with an arrow at the top. The location on the axis is where the argument is 0, as in t_0, and height corresponds to the area. That means $\delta(t)$ has location $t = 0$ and a height of 1, and $A\delta(t - t_0)$ has location $t = t_0$ and height A. See this in Figure 3-3b.

Figure 3-3: A graphical depiction of the sifting property, using $\delta_\varepsilon(t)$ (a) and then $\delta(t)$ (b).

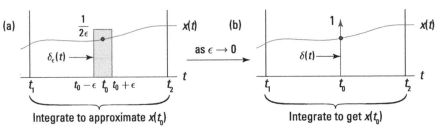

Integrate to approximate $x(t_0)$ Integrate to get $x(t_0)$

Unit step function

The unit step function, $u(t)$, is also a singularity function. A popular use is modeling signals with on and off gates. In Chapter 5, I describe the step response of a system. Start from the unit impulse:

$$u(t) = \int_{-\infty}^{t} \delta(\lambda) d\lambda = \begin{cases} 0, & t < 0 \\ 1, & t > 0 \\ \text{undefined}, & t = 0 \end{cases}$$

Thinking of $\delta_\varepsilon(t)$, you can say $u(0) = 1/2$. In the limit, $u(t)$ contains a jump at $t = 0$, so it's not defined.

You can program the unit step in Python:

```
def step(t):
    x = np.zeros(len(t))
    for k,tt in enumerate(t):
        if tt >= 0: # the jump occurs at t = 0
            x[k] = 1.0
    return x
```

Grasping the reality of step and impulse signals

A real-world step function doesn't suddenly jump from 0 to 1. The waveform smoothly transitions from 0 to 1 over a period of time. When you zoom in on the real signal, the gradual transition is visible. When you zoom way out, the signal appears as a true mathematical step function, apparently jumping from 0 to 1 in no time. When you differentiate the real-world step, the smooth transition from 0 to 1 has a derivative everywhere. The result is a pulse-like signal that results from the nonzero derivative over the step transition. When viewed from a distance, the signal looks like an impulse function.

The step function is also available in `ssd.py`. To return to unit impulse function, just differentiate $u(t)$ to get

$$\delta(t) = \frac{du(t)}{dt}$$

Don't sweat the details about the existence of the derivative at $t = 0$. Just replace the derivative at the discontinuity with a unit impulse function of height equal to the jump height. Note that this appears to violate what you probably learned in calculus about derivatives, but the unit impulse function fits perfectly in an engineering math sense.

Example 3-2: To practice working with the derivatives of signals that incorporate singularity functions, consider the derivative of the rectangle and triangle pulse functions.

- ✔ The derivative of $\Pi(t/\tau)$ contains two unit impulse functions because it contains two jumps. A jump by 1 at $t = -\tau/2$, and a jump down by 1 at $t = \tau/2$:

$$\frac{d\Pi(t/\tau)}{dt} = \delta(t + \tau/2) - \delta(t - \tau/2)$$

- ✔ The derivative of $\Lambda(t/\tau)$ doesn't contain jumps, but it does contain three points of discontinuity, $-\tau$, 0, and τ. When taking the derivative, focus on the derivative away from these points:

$$\frac{d\Lambda(t/\tau)}{dt} = \begin{cases} 1/\tau, & -\tau < t < 0 \\ -1/\tau, & 0 < t < \tau \\ 0, & \text{otherwise} \end{cases} \overset{\text{also}}{=} \frac{1}{\tau}\Pi\left(\frac{t+\tau/2}{\tau}\right) - \frac{1}{\tau}\Pi\left(\frac{t-\tau/2}{\tau}\right)$$

Figure 3-4 shows both signal derivatives.

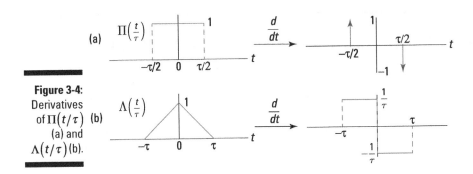

Figure 3-4:
Derivatives
of $\Pi(t/\tau)$ (b)
(a) and
$\Lambda(t/\tau)$ (b).

Getting Hip to Signal Classifications

Signals are classified in a number of ways based on properties that the signals possess. In this section, I describe the major classifications and point out how to verify and classify a given signal. Classifications aren't mutually exclusive. A periodic signal, for example, is usually a power signal, too. And a signal may be even and aperiodic.

Deterministic and random

A signal is classified as *deterministic* if it's a completely specified function of time. A good example of a deterministic signal is a signal composed of a single sinusoid, such as $x(t) = A\cos(2\pi f_0 t + \phi)$ with (A, f_0, ϕ) being the signal parameters. A is the amplitude, f_0 is the frequency (oscillation rate) in cycles per second (or hertz), and ϕ is the phase in radians. Depending on your background, you may be more familiar with radian frequency, $\omega_0 = 2\pi f_0$, which has units of radians/sample. In any case, $x(t)$ is deterministic because the signal parameters are constants.

Hertz (Hz) represents the *cycles per second* unit of measurement in honor of Heinrich Hertz, who first demonstrated the existence of radio waves.

A signal is classified as *random* if it takes on values by chance according to some probabilistic model. You can extend the deterministic sinusoid model $x(t) = A\cos(2\pi f_0 t + \phi)$ to a random model by making one ore more of the parameters A, f_0, or ϕ random. By introducing random parameters, you can more realistically model real-world signals.

To see how a random signal can be constructed, write $x(t, \zeta_i) = A_i \cos(2\pi f_i t + \phi_i)$, where ζ_i corresponds to the drawing of a particular set of (A_i, f_i, ϕ_i) values from a set of possible outcomes. Relax; incorporating random parameters in your signal models is a topic left to more advanced courses. I simply want you to know they exist because you may bump into them at some point.

To visualize the concepts in this section, including randomness, you can use the IPython environment with PyLab to create a plot of deterministic and random waveform examples:

```
In [234]: t = linspace(0,5,200)
In [235]: x1 = 1.5*cos(2*pi*1*t + pi/3)
In [237]: plot(t,x1)
In [242]: for k in range(0,5): # loop with k=0,1,...,4
     ...: x2 = (1.5+rand(1)-0.5))*cos(2*pi*1*t +
          pi/2*rand(1)) # rand()= is uniform on (0,1)
     ...: plot(t,x2,'b')
     ...:
```

See the results in Figure 3-5, which uses a 2×1 PyLab `subplot` to stack plots.

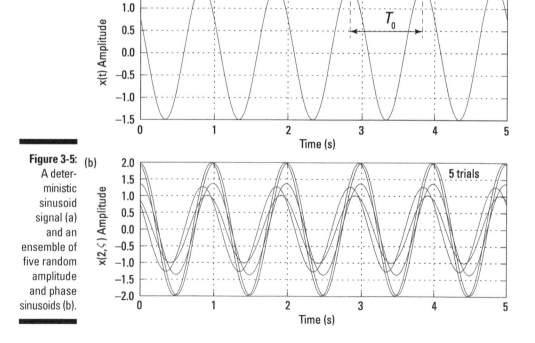

Figure 3-5: A deterministic sinusoid signal (a) and an ensemble of five random amplitude and phase sinusoids (b).

Generate the deterministic sinusoid by creating a vector of time samples, t, running from zero to five seconds. To create the signal, x1 in this case, I chose values for the waveform parameters $(A, f_0, \phi) = (1.5, 1.0, \pi/3)$.

For the random signal case, A is nominally 1.5, but I added a random number uniform over $(-0.5, 0.5)$ to A, making the composite sinusoid amplitude random. The frequency is fixed at 1.0, and the phase is uniform over $(0, \pi/2)$. I create five realizations of $x(t, \zeta)$ using a `for` loop.

Periodic and aperiodic

Another type of signal classification is *periodic* versus *aperiodic*. A signal is periodic if $x(t) = x(t + T_0)$, where T_0, the period, is the largest value satisfying the equality. If a signal isn't periodic, it's aperiodic.

When checking for periodicity, you're checking in a graphical sense to see whether you can copy a period from the center of the waveform, shift it left or right by an integer multiple of T_0, and if it perfectly matches the signal T_0 seconds away. The single sinusoid signal is always periodic, and the proof, which relies on simple trigonometry (flip to Chapter 2 for a trig refresher), allows you to determine what the period is.

Example 3-3: To establish that a single sinusoid signal is periodic and to determine the period, follow these steps:

1. **Ask yourself: What does it take to make the equality hold?**

   Making the arguments equal seems to be the only option.

   $$A\cos(2\pi f_0 t + \phi) \overset{?}{=} A\cos\left[2\pi f_0(t + T_0) + \phi\right]$$

2. **Expand the argument of the cosine on the right side to see the impact of the added T_0:**

   $$A\cos\left[2\pi f_0(t + T_0) + \phi\right] = A\cos(2\pi f_0 t + 2\pi f_0 T_0 + \phi)$$

   Cosine is a *modulo* 2π function; that is, the functional values it produces don't change when the argument is shifted by integer multiples of 2π, which makes the single sinusoid a periodic signal from the get-go.

3. **To establish the period, observe that forcing $2\pi f_0 T_0 = 2\pi$ means**

   $$T_0 \equiv \frac{1}{f_0} = \text{ the (fundamental) period.}$$

   Because T_0 is the fundamental period, f_0 is the fundamental frequency; they're reciprocals.

Example 3-4: Figure 3-6a shows a periodic signal known as a *periodic pulse train* because it has an infinite train of pulses. Each pulse has width τ, and the ellipsis indicate that the pulse train continues in both directions. The pulses that follow each other don't have to be periodic, though. In Figure 3-6b, a waveform with a single isolated pulse or just a few pulses makes the signal aperiodic.

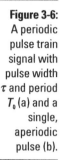

Figure 3-6: A periodic pulse train signal with pulse width τ and period T_0 (a) and a single, aperiodic pulse (b).

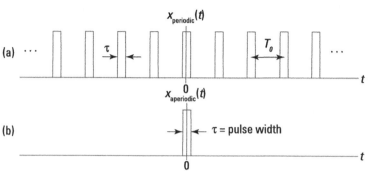

Considering power and energy

To classify a signal $x(t)$ according to its power and energy properties, you need to determine whether the energy is finite or infinite and whether the power is zero, finite, or infinite. The measurement unit for power and energy are *watts* (W) and *joules* (J).

In circuit theory, watts delivered to a resistor of R ohms is represented as $P = V \cdot I = \dfrac{V^2}{R} = I^2 \cdot R$, where V is voltage in volts and I is current in amps.

Figure 3-7 shows that a circuit is composed of resistor and voltage sources, demonstrating the interconnection of signals and systems with physics and circuit concepts. A simple power calculation in circuit analysis becomes instantaneous power with time dependence, and the resistance is normalized to 1 ohm as used in signals and systems terminology.

Figure 3-7: Relating power in a circuit to signals and systems 1-ohm normalization.

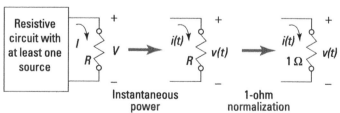

With the circuit calculation of power, you can introduce time dependence and define a signal's instantaneous power as

$$p(t)=v(t)\cdot i(t)=\frac{v^2(t)}{R}=i^2(t)\cdot R$$

Computer and electrical engineers use the abstraction available through mathematics to work in a convenient 1-ohm environment.

The significance of the 1-ohm impedance normalization is that instantaneous power is simply $v^2(t)$ or $i^2(t)$. It's convenient then to use $x(t)$ to represent the signal — voltage or current. Keep in mind that, unless told otherwise, $R=1\,\Omega$ in all signals. When modeling results need to be coupled to physical, real-world measurements, you can add back in the resistance, or impedance level.

Average power, P (watts), and average energy, E (joules), are defined as follows:

$$P=\lim_{T\to\infty}\frac{1}{2T}\int_{-T}^{T}\left|x(t)\right|^2 dt \qquad\qquad W\ (1\,\Omega\ \text{system})$$

$$E=\lim_{T\to\infty}\int_{-T}^{T}\left|x(t)\right|^2 dt=\int_{-\infty}^{\infty}\left|x(t)\right|^2 dt \qquad J\ (1\,\Omega\ \text{system})$$

The $|\ |^2$ is used when the signal happens to be complex. For $x(t)$ periodic, you can simplify the average power formula to

$$P^{\text{periodic}}=\frac{1}{T_0}\int_{0}^{T_0}\left|x(t)\right|^2 dt=\frac{1}{T_0}\int_{T_0}\left|x(t)\right|^2 dt$$

where T_0 is the period. Note that the single limit on the far right integral means that you can use any T_0 interval for the calculation. The limit is gone, and the integration now covers just one period.

For $x(t)$ to be a power signal, $0<P<\infty$ and $E\to\infty$. To understand why this is, think about a signal that has nonzero but finite power. Integrating the power over all time gives you energy. When you integrate nonzero finite power over infinite time, you get infinite energy.

An energy signal requires $0<E<\infty$ and $P\to0$. Yet some signals are neither power nor energy types because they have unbounded power and energy. For these, $P\to\infty$.

Mathematically, a signal can have infinite power, but that's not a practical reality. Infinite energy indicates that the signal duration is likely infinite, so it doesn't make sense to deal with energy. Power, on the other hand, is energy per unit time averaged over all time, making power more meaningful than energy.

Proving the simplified power calculation formula

Create the proof of the simplified power calculation formula for periodic signals in three steps:

1. Let $T \to NT_0$, which allows the entire time axis to be used for the average:

$$\lim_{T \to \infty} \frac{1}{2T} \int_{-T}^{T} |x(t)|^2 dt = \lim_{N \to \infty} \frac{1}{2NT_0} \int_{-NT_0}^{NT_0} |x(t)|^2 dt$$

2. Because $x(t) = x(t + T_0) = x(t \pm NT_0)$, N any integer, it follows that

$$\int_{-NT_0}^{NT_0} |x(t)|^2 dt = 2N \int_{0}^{T_0} |x(t)|^2 dt$$

3. Finally, you get

$$P = \lim_{N \to \infty} \frac{1}{2NT_0} \cdot 2N \int_{0}^{T_0} |x(t)|^2 dt = \frac{1}{T_0} \int_{0}^{T_0} |x(t)|^2 dt$$

Here are a few examples of power and energy calculations.

Example 3-5: To classify a signal with one or two cosines as a power or energy function, follow this process:

✔ For a single real sinusoid $x(t) = A\cos(2\pi f_0 t + \phi)$, $-\infty < t < \infty$, you can take advantage of the fact that $x(t)$ is periodic with a period $T_0 = 1/f_0$ in these calculations:

$$\lim_{N \to \infty} \int_{-NT_0}^{NT_0} |x(t)|^2 dt = \lim_{N \to \infty} 2N \int_{0}^{T_0} |x(t)|^2 dt$$

$$E = \lim_{N \to \infty} 2N \int_{0}^{T_0} A^2 \underbrace{\cos^2(2\pi f_0 t + \phi)}_{\text{use } \cos^2(x) = \frac{1}{2}[1 + \cos(2x)]} dt$$

$$= \lim_{N \to \infty} 2N \frac{A^2}{2} \int_{0}^{T_0} \left[1 + \underbrace{\cos\left[2\pi(2f_0)t + 2\phi\right]}_{\text{integrates to zero as 2 full periods}} \right] dt$$

$$= \lim_{N \to \infty} 2N \frac{A^2}{2} \cdot T_0 = \infty$$

$$P = \frac{1}{T_0} \cdot \frac{A^2}{2} \underbrace{\int_{0}^{T_0} \left[1 + \cos\left[2\pi(2f_0)t + 2\phi\right] \right] dt}_{\text{evaluates to } T_0} = \frac{A^2}{2}$$

This is a power signal because E is infinite and P is finite.

✔ For a signal with two real sinusoids, such as $x(t) = A_1 \cos(2\pi f_1 t + \phi_1) + A_2 \cos(2\pi f_2 t + \phi_2)$, as long as $f_1 \neq f_2$, you can use this equation to solve for P:

$$P = \frac{A_1^2}{2} + \frac{A_2^2}{2}$$

This is the sum of powers for single sinusoids, which is all you need. Periodicity of $x(t)$ isn't a requirement.

Example 3-6: Consider a signal with two sinusoids:

$$x(t) = |A_1| \cos(2\pi f_1 t + \phi_1) + |A_2| \cos(2\pi f_2 t + \phi_2), f_1 \neq f_2$$

Can this signal be periodic? The individual sinusoid periods are $T_1 = 1/f_1$ and $T_2 = 1/f_2$. For the composite signal to be periodic, T_1 and T_2 must be *commensurate*, which means you seek the smallest values N_1 and N_2 so that N_1 periods of duration T_1 are equal to N_2 periods of duration T_2. Here's how to set it up in an equation:

$$N_1 \cdot T_1 = N_2 \cdot T_2 \Rightarrow \frac{T_1}{T_2} = \frac{f_2}{f_1} \overset{\text{must}}{=} \frac{N_2}{N_1} = \text{Rational Number}$$

The ratio of the two periods and, likewise, the frequencies must be a rational number. The fundamental period is $T_0 = N_1 T_1 = N_2 T_2$. In algebraic terms, you can state it as the least common multiple (LCM) of the periods or the greatest common divisor (GCD) of the frequencies:

$$T_0 = \text{LCM}(T_1, T_2) \text{ or } f_0 = \text{GCD}(f_1, f_2)$$

If $f_1 = 20\text{Hz}$ and $f_2 = 45\text{Hz}$, $\text{LCM}(20, 45) = 5\text{Hz}$, so $4 \times T_1 = 9 \times T_2$.

If you have a calculator that incorporates a computer algebra system (CAS), you may be able to use the LCM and GCD functions to check your hand calculations. If not, consider using Maxima (see Chapter 1 for details).

The LCM and GCD functions in Maxima confirm the hand calculations presented earlier:

```
(%i1)  lcm(1/20,1/45);

(%o1)  1
       -
       5
```

```
(%i2)  gcd(20,45);
(%o2)  5
```

Periodicity among multiple sinusoids is essential to Fourier series (covered in Chapter 8).

Example 3-7: To classify $x(t) = Ae^{-\alpha t}u(t)$ as a power or energy (or neither) signal with respect to α, consider the following three cases. To properly classify $x(t)$, you need to find both the energy (E) and the power (P) values. Using the definition given in the section "Considering power and energy," find out whether E is finite or infinite and whether P is zero, finite, or infinite.

✔ **Case 1:** $\alpha > 0$

$$E = \int_{-\infty}^{\infty} \left| Ae^{-\alpha t} \right|^2 u(t)dt = A^2 \int_0^\infty e^{-2\alpha t}\, dt = A^2 \cdot \frac{e^{-\alpha t}}{-2\alpha} \bigg|_0^\infty = \frac{A^2}{2\alpha} < \infty$$

$$P = \lim_{T \to \infty} \frac{1}{2T} A^2 \int_0^T e^{-2\alpha t}\, dt = \lim_{T \to \infty} \frac{1}{2T} \cdot A^2 \cdot \frac{e^{-\alpha t}}{-2\alpha} \bigg|_0^T$$

$$= \lim_{T \to \infty} \frac{1}{2T} \cdot \frac{A^2}{2\alpha} \left[1 - e^{-\alpha T} \right] = 0$$

This is an energy signal because E is finite and P is zero.

✔ **Case 2:** $\alpha = 0$

$$E = \int_{-\infty}^{\infty} A^2 u(t)dt = A^2 \int_0^\infty dt = A^2 \cdot t \big|_0^\infty = \infty$$

$$P = \lim_{T \to \infty} \frac{1}{2T} A^2 \int_0^T dt = \lim_{T \to \infty} \frac{1}{2T} \cdot A^2 \cdot t \big|_0^T = \frac{A^2}{2} < \infty$$

This is a power signal because E is infinite and P is finite.

✔ **Case 3:** $\alpha < 0$:

$$E = A^2 \int_0^\infty e^{-2\alpha t}\, dt = A^2 \cdot \frac{e^{-\alpha t}}{-\alpha} \bigg|_0^\infty = \infty$$

$$P = \lim_{T \to \infty} \frac{1}{2T} \cdot A^2 \cdot \frac{e^{-\alpha t}}{-\alpha} \bigg|_0^T = \lim_{T \to \infty} \frac{1}{2T} \cdot \frac{A^2}{\alpha} \left[1 - e^{-\alpha T} \right] = \infty$$

This signal is neither power nor energy because P is infinite. The signal amplitude becomes unbounded (refer to Figure 3-1). The term *unbounded* means magnitude approaching infinity.

Sinusoidal signals are power signals. For a single sinusoid, the power is just $A^2/2$. For K sinusoids at distinct frequencies, the signal power is $P = \frac{1}{2} \sum_{k=1}^{K} A_k^2$.

If sinusoids have the same frequency, you need to combine these terms by using the phasor addition formula (described in Chapter 2). To figure out the power of each like frequency, you just need the equivalent amplitude.

Even and odd signals

Signals are sometimes classified by their symmetry along the time axis relative to the origin, $t = 0$. *Even* signals fold about $t = 0$, and *odd* signals fold about $t = 0$ but with a sign change. Simply put,

$$\text{Even} \Rightarrow x(-t) = x(t)$$
$$\text{Odd} \Rightarrow x(-t) = -x(t)$$

To check the even and odd signal classification, I use the Python `rect()` and `tri()` pulse functions to generate six aperiodic signals. Here's the code for generating two of the signals:

```
In [759]:  t = arange(-5,5,.005) # time axis for plots
In [760]:  x1 = ssd.rect(t+2.5,3)+ssd.rect(t-2.5,3)
In [763]:  x4 = ssd.rect(t+3,2)-ssd.tri(t,1)
                +ssd.rect(t-3,2)
```

Check out the six signals, including the classification, in Figure 3-8.

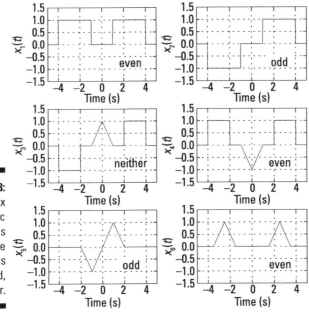

Figure 3-8: Six aperiodic waveforms that are classified as even, odd, or neither.

To discern even or odd, observe the waveform symmetry with respect to $t = 0$. Signals $x_1(t)$, $x_4(t)$, and $x_6(t)$ are even; they fold nicely about $t = 0$. Signals $x_2(t)$ and $x_5(t)$ fold about $t = 0$ but with odd symmetry because the waveform on the negative time axis has the opposite sign of the positive time axis signal. Signal $x_3(t)$ is neither even nor odd because a portion of the waveform, the triangle, is even about 0, while the rectangles are odd about 0. Taken in combination, the signals are neither even nor odd.

A single sinusoid in cosine form, without any phase shift, is even, because it's symmetric with respect to $t = 0$, or rather it's a mirror image of itself about $t = 0$. Mathematically, this is shown by the property of being an even signal:

$$A\cos(-2\pi f_0 t) = A\cos(2\pi f_0 t)$$

Similarly, a single sinusoid in sine form, without any phase shift, is odd, because it has negative symmetry about $t = 0$. Instead of an exact mirror image of itself, values to the left of $t = 0$ are opposite in sign of the values to the right of $t = 0$. This is mathematically an odd signal:

$$A\sin(-2\pi f_0 t) = -A\sin(2\pi f_0 t)$$

If a nonzero phase shift is included, the even or odd properties are destroyed (except for π or $\pi/2$).

Transforming Simple Signals

Signal transformations, such as *time shifting* and *signal flipping,* occur as part of routine signal modeling and analysis. In this section, I cover these signal manipulation tasks as well as superimposing of signals to help you get more comfortable with these processes as they relate to continuous-time signals. (For details on these tasks for discrete-time signals, flip to Chapter 6.)

Time shifting

Time shifting signals is a practical matter. By design, signals arrive at a system at different times. A signal sent from a cellphone arrives at the base station after a time delay due to the distance between the transmitter and receiver. Mathematically, modeling this delay allows you, the designer, to consider the impact of time on system performance.

Given a signal $x(t)$, consider $x(t - t_0)$, where t_0 may be positive, zero, or negative.

For the shifted or transformed sequence $x(t - t_0)$, $x(0)$ occurs when $t - t_0 = 0$ or $t = t_0$, which implies that the signal is shifted by t_0 seconds. If t_0 is positive, the shift is to the right; if t_0 is negative, the shift is to the left. For the unshifted signal $x(t)$, $x(0)$ occurs when $t = 0$.

When working with the rectangle or triangle pulse function, think of moving the active region, or *support interval,* of the pulse. For example, the support interval of $\Lambda(t/\tau)$ is $-\tau \le t \le \tau$. The shifted pulse $\Lambda\left[(t-t_0)/\tau\right]$ has support interval $-\tau \le t - t_0 \le \tau$. If you isolate t between the two inequalities by adding t_0 to both sides, then the support interval is $t_0 - \tau \le t \le t_0 + \tau$. The support interval shifts by t_0 seconds. See Figure 3-9 for a graphical depiction of time-axis shifting.

Figure 3-9:
Time
shifting
depicted for
the triangle
pulse.

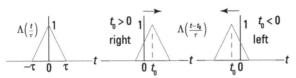

Flipping the time axis

Processing signals in the time domain by linear time-invariant systems requires a firm handle on the concept of flipping signals (see Chapter 5). Flipping, or *time reversal,* of the axis corresponds to $x(t) \to x(-t)$. The term *flipping* describes this process well because the waveform literally flips over the point $t = 0$. Everything reverses. What was at $t = 5$ is now located at $t = -5$, for example. You can apply this to all signals, including rectangular pulses, sinusoidal signals, or generic signals. Check out Figure 3-10 to see the concept of flipping for a generic pulse signal.

Figure 3-10:
Axis flipping
(reversing)
for a generic
pulse signal.

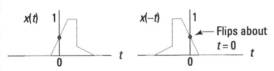

Putting it together: Shift and flip

Shifting and flipping a signal over the time axis corresponds to $x(t) \to x(t_0 - t)$. Think of t in this axis transformation as the axis for plotting the signal and t_0 as a parameter you can vary. You can visualize the transformed signal as a two-step process.

1. **Manipulate the signal argument so you can see the problem as shift and then flip:** $x(t_0 - t) = x[-(t - t_0)]$.

   If, for a moment, you ignore the minus sign between the bracket and parentheses on the right, $x[\times(t - t_0)]$, it looks like $x(t)$ is simply shifted to the right by t_0 (assume $t_0 > 0$).

2. **Bring the minus sign back in, realizing that this flips the signal.**

   The minus sign surrounds $(t - t_0)$, so the signal flips over the point $t - t_0 = 0$ or $t = t_0$.

Figure 3-11 shows a shifting and flipping operation.

Figure 3-11:
Combined signal transformation of shifting and flipping.

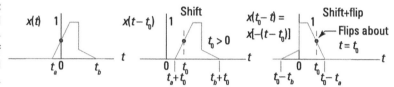

Confirm that the leading and trailing edges of $x(t)$, denoted t_a and t_b, respectively, transformed as expected. In the plot of $x(t_0 - t)$, does the point $t = t_0 - t_a$ correspond to the original leading edge t_a? To find out, plug $t = t_0 - t_a$ into $x(t_0 - t)$:

$$x(t_0 - t)\big|_{t \to t_0 - t_a} = x[t_0 - (t_0 - t_a)] = x(t_a)$$

To check the trailing edge location, consider the time location of the far left side of the signal. The leading edge is at the far right side of the signal.

Flip to Chapter 5 for more information on shifting and flipping signals when working with the convolution integral.

Superimposing signals

Think about a time when you've been in a loud public space and can clearly hear the voice of the person to whom you're talking; the other sounds register to you as background noise. The concept of superimposing signals is similar.

In mathematical terms, superimposing signals is a matter of signal addition:

$$x(t) = x_1(t) + x_2(t) + \cdots + x_K(t) = \sum_{k=1}^{K} x_k(t)$$

Each component signal, $x_k(t)$, may actually be an amplitude that's a scaled and time-shifted and/or flipped version of some waveform primitive.

The signals with multiple sinusoids described in the earlier section "Exponential and sinusoidal signals" represent superimposed signals, and the example signals in Figure 3-8 were generated by using time-shifted and added rectangle and triangle pulses. In real-world scenarios, you have environments that contain superimposed signals — whether by design or as a result of unwanted interference.

Example 3-8: Signal $x_4(t)$ from Figure 3-8 is a sum of three pulse signals. I used the Python `rect()` and `tri()` pulse functions to create the signal, but I was thinking about this mathematical description when I created it:

$$x_4(t) = \Pi\left(\tfrac{t+3}{2}\right) - \Lambda(t) + \Pi\left(\tfrac{t-3}{2}\right)$$

This involves the use of signal time shifting and superposition. Note also that the Python code for generating the signal as a vector (actually a PyLab `ndarray`) of signal samples, x4, is an almost verbatim match with the mathematical equation for $x_4(t)$.

Checking Out System Properties

Generally, all continuous-time systems modify signals to benefit the objectives of an engineering design (see Chapter 1). Consider the system input/output block diagram of Figure 3-12. The input signal has *fuzz* on it, and the output is clean, suggesting that the system operator $T\{\ \}$ is acting as a fuzz-removing filter. *Fuzz* to you is *noise* to me.

Figure 3-12:
System
$T\{\ \}$ input/
output block
diagram.

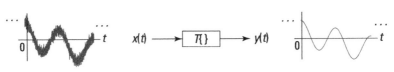

All systems have specific jobs. If you're a system designer, you look at your design requirements and create a system accordingly. In the noise-removing filter example, the filter design takes into account the very nature of the desired and undesired signals entering the filter. If you don't do this, your filter may pass the noise and block the signal of interest, and no one wants a coffee filter that passes the grounds and blocks the coffee.

In this section, I describe various properties of systems based on their mathematical characteristics.

Linear and nonlinear

Simply put, a system is linear if *superposition* holds. Superposition refers to the ability of a system to process signals individually and then sum them up to process all the signals simultaneously. For example, suppose that two signals $x_1(t)$ and $x_2(t)$ are present at the system input. When applied to the system individually, they produce

$$y_1(t) = T\{x_1(t)\}$$
$$y_2(t) = T\{x_2(t)\}$$

If superposition holds, you can declare that, for arbitrary constants a and b,

$$T\{ax_1(t) + bx_2(t)\} = aT\{x_1(t)\} + bT\{x_2(t)\}$$
$$= ay_1(t) + by_2(t)$$

The generalization for K signals superimposing at the system input is

$$y(t) = T\left\{\sum_{k=1}^{K} a_k x_k(t)\right\} = \sum_{k=1}^{K} a_k y_k(t)$$

If superposition doesn't hold, the system is *nonlinear*.

In practical terms, think about a karaoke system. You want the audio amplifier that drives the speakers in this kind of system to be linear so the music and singer's voice in the microphone can merge without causing distortion, which happens with a nonlinear amplifier. On the other hand, many hard rock guitar players send their signal through a nonlinear amplifier to get some distortion.

Time-invariant and time varying

A system is *time-invariant* if its properties or characteristics don't change with time. A mathematical statement of this is that given $y(t) = T\{x(t)\}$ and any time offset t_0, the time-shifted input $x_1(t) = x(t - t_0)$ must produce system output $y_1(t) = y(t - t_0)$.

Here, $y(t)$ is the system output to the present input $x(t)$. For time invariance to hold, the output of a system is unchanged (except for the time offset by t_0) when you apply the same input at any arbitrary offset t_0.

A system that doesn't obey the condition established for time invariance is said to be *time varying*. Creating a system with a time-varying property is as easy as twisting the volume control on your car stereo. Specifically, the *gain* of the system is time varying.

A noise-removing filter is typically designed to be time-invariant. Assuming the noise signal characteristics and the desired signal characteristics are fixed, the filter design should be time-invariant. A time-varying filter, known as an *adaptive filter*, is needed when the noise signal characteristics change over time. Think of noise-canceling headphones that give you relative peace and quiet riding in an airplane or on the flight deck of an aircraft carrier. These headphones are a time-varying system.

Causal and non-causal

A system that is causal is *nonanticipative;* that is, the system can't anticipate the arrival of a signal at the input. Sounds crazy, I know, but a non-causal system can predict the future (in a signals sense); it anticipates the signal input. Mathematically, you can define such a system, but building a physical system is impossible.

A system is causal if all output values, $y(t_0)$, depend *only* on input values $x(t)$ for $t \le t_0$ — or the present output depends *only* on past and present input values. A non-causal system is more of a mathematical concept than a practical reality. A system that can use future values of the input to form the present output can predict the future.

With discrete-time signals and systems (described in Chapter 4), it's possible to store a signal in memory and then process it later by using a non-causal system. The catch is that the processing is all being done by using past values of the input, because you're working with a recorded signal. The math of a non-causal system is still at work, because the system doesn't realize that the signal was prerecorded. For continuous-time systems, making this work is harder; you can perform non-causal processing on continuous-time signals with records and tapes of music recordings.

Memory and memoryless

Very simply, a system is memoryless if each output $y(t)$ depends only on the present input $x(t)$. Can a memoryless system be non-causal? If the output depends only on the present input, then no way can the future values of the input be used to form the present output. Yet causal systems aren't necessarily memoryless. A causal system can utilize past values of the input in forming the present output.

A system that filters a system generally does so by using the present and past values of the input to form the present output. A system described by a linear constant coefficient (LCC) differential equation is one such example. An electronic circuit that's composed of resistors, capacitors, and inductors is another example. The capacitors and inductors are the memory elements. A system with only resistors has no memory.

Bounded-input bounded-output

A system is bounded-input bounded-output (BIBO) stable only if every bounded input produces a bounded output. What's this bounded stuff? *Bounded* is a mathematical term that means a signal has magnitude less than infinity over all time. The signal $x(t)$ — which may be an input or an output — is bounded if some constant $B_x > 0$ exists such that $|x(t)| \le B_x < \infty$ for all values of t.

To show that the property holds for *any* bounded input is the fundamental challenge of this scenario. *Any* represents quite a lot of cases; testing them all can be prohibitive. Therefore, some proof-writing skills are required here (see Example 3-10).

Choosing Linear and Time-Invariant Systems

From a design and analysis standpoint, engineers are typically most interested in working with systems that are both linear and time-invariant because such systems can meet demanding real-world requirements and allow for smoother analysis in the time, frequency, and s-domains. The ability to analyze system performance is critical; you want to be confident that your design meets requirements before committing to expensive prototypes.

Example 3-9: Consider the system input/output relationship:

$$y(t) = T\{x(t)\} = 2x(t) + 5$$

Here's how to classify this system according to the five system properties: linear, time-invariant, causal, memoryless, and BIBO stable.

- ✔ Check linearity by verifying $T\{ax_1(t) + bx_2(t)\} \stackrel{?}{=} aT\{x_1(t)\} + bT\{x_2(t)\}$.

  Insert the system operator $2x(t) + 5$ into the left and right sides to see whether equality holds:

  $$2(ax_1(t) + bx_2(t)) + 5 \stackrel{?}{=} a(2x_1(t) + 5) + b(2x_2(t) + 5)$$
  $$= 2(ax_1(t) + bx_2(t)) + \underbrace{(a+b)5}_{\ne 5}$$

  The equality doesn't hold, so the system is nonlinear. If the 5 is set to 0, then linearity holds.

- ✔ Check time invariance by observing that nothing about the system is time varying. In particular, the coefficients — 2 and 5 in this case — are constants. The system is time-invariant.

✔ Check causality by noting that the present output depends only on the present input. No future values of the input are used for the present output. The system is causal.

✔ The present output depends only on the present input, so the system is memoryless.

✔ Check BIBO stability by observing that $|y(t)| = |2x(t) + 5| \leq |2x(t)| + 5 \leq 2B_x + 5 < \infty$ because $x(t)$ is a bounded input with $|x(t)| \leq B_x < \infty$. I used the triangle inequality ($|x + y| \leq |x| + |y|$) to complete this proof. The system is BIBO stable.

Example 3-10: Consider the system $y(t) = T\{x(t)\} = \sin(2\pi t) \cdot x(t) + u(t - 2)$, and check for the five system properties:

✔ Check linearity by verifying $T\{ax_1(t) + bx_2(t)\} \stackrel{?}{=} aT\{x_1(t)\} + bT\{x_2(t)\}$.

Insert the system operator $y(t) = T\{x(t)\} = \sin(2\pi t) \cdot x(t) + u(t - 2)$ into the left and right sides to see whether equality holds:

$$\sin(2\pi t)(ax_1(t) + bx_2(t)) + u(t - 2)$$
$$\stackrel{?}{=} a(\sin(2\pi t)x_1(t) + u(t - 2)) + b(\sin(2\pi t)x_2(t) + u(t - 2))$$
$$= \sin(2\pi t)(ax_1(t) + bx_2(t)) + \underbrace{(a + b)u(t - 2)}_{\neq u(t - 2)}$$

The equality doesn't hold, so the system is nonlinear. If $t < 0$, then the step function is turned off and linearity holds conditionally.

✔ Check time invariance by observing that the system contains two time-varying coefficients, $\sin(2\pi t)$ and $u(t - 2)$. The system isn't time-invariant.

✔ Check causality by noting that the present output depends only on the present input. No future values of the input are used for the present output. The time-varying coefficients have no impact on causality, so the system is causal.

✔ The present output depends only on the present input, so the system is memoryless.

✔ Check BIBO stability by observing that

$$|y(t)| = |\sin(2\pi t)x(t) + u(t - 2)| \leq |\sin(2\pi t)x(t)| + |u(t - 2)|$$
$$\leq \underbrace{|\sin(2\pi t)|}_{\leq 1} B_x + \underbrace{|u(t - 2)|}_{\leq 1} < \infty$$

Because $x(t)$ is a bounded input with $|x(t)| \leq B_x < \infty$ and both $\sin(2\pi t)$ and $u(t - 2)$ are *upper bounded* by 1, which means that the largest values these signals take on is one, the system is BIBO stable.

Chapter 4

Discrete-Time Signals and Systems

*W*e live in a continuous-time/analog world; but people increasingly use computers to process continuous signals in the discrete-time domain. From a math standpoint, discrete-time signals and systems can stand alone — independent of their continuous-time counterparts — but that isn't the intent of this chapter. Instead, I want to show you how discrete-time signals and systems get the job done in a way that's parallel to the continuous-time description in Chapter 3.

A great deal of innovation takes place in discrete-time signals and systems. High-performance computer hardware combined with sophisticated algorithms offer a lot of flexibility and capability in manipulating discrete-time signals. Computers can carry out realistic simulations, which is a huge help when studying and designing discrete-time signals and systems.

Continuous- and discrete-time signals have a lot in common. In this chapter, I describe how to classify signals and systems, move signals around the time axis, and plot signals by using software. Also, because discrete-time signals often begin in the continuous-time domain, I point out some details of converting signals late in the chapter.

Exploring Signal Types

Like the continuous-time signals, discrete-time signals can be exponential or sinusoidal and have special sequences. The unit sample sequence and the unit step sequence are special signals of interest in discrete-time. All these signals have continuous-time counterparts, but singularity signals (covered in Chapter 3) appear in continuous-time only.

Note: Discrete-time signals are really just sequences. The independent variable is an integer, so no in-between values exist. A bracket surrounds the time variable for discrete-time signals and systems — as in $x[n]$ versus the $x(t)$ used for continuous-time. The values that discrete-time signals take on are concrete; you don't need to worry about limits.

Exponential and sinusoidal signals

Exponential signals and real and complex sinusoids are important types of signals in the discrete-time world. Sinusoids, both real and complex, are firmly rooted in discrete-time signal model and applications. Signals composed of sinusoids can represent communication waveforms and the basis for Fourier spectral analysis, both the discrete-time Fourier transform (described in Chapter 11) and the discrete Fourier transform and fast Fourier transform (Chapter 12).

The exponential sequence $x[n] = \tilde{A}\gamma^n$, $-\infty < n < \infty$ is a versatile signal. By letting $\tilde{A}$ and γ be complex quantities in general, this signal alone can represent a real exponential, a complex sinusoid, a real sinusoid, and exponentially damped complex and real sinusoids. For definitions of these terms, flip to Chapter 3.

Figure 4-1 shows stem plots of three variations of the general exponential sequence.

In the remaining material, let $\gamma = \alpha e^{j\hat{\omega}_0}$ and $\tilde{A} = Ae^{j\phi}$.

The real exponential is formed when $\phi = 0$ and $\hat{\omega}_0 = 0$. A unit step sequence (described later in this section) is frequently included. The result is $x[n] = A\alpha^n u[n]$.

If $|\alpha| < 1$, the sequence is decreasing. Without any assumptions about $\tilde{A}$ and γ, you get a full complex exponential sequence:

$$x[n] = A\alpha^n e^{j(\hat{\omega}_0 n + \phi)}$$
$$= A\alpha^n \left[\cos(\hat{\omega}_0 n + \phi) + j\sin(\hat{\omega}_0 n + \phi) \right], \quad -\infty < n < \infty$$

The envelope may be growing, constant, or decaying, depending on $|\alpha| > 1, |\alpha| = 1$, or $|\alpha| < 1$, respectively. For $\alpha = 1$, you have a complex sinusoidal sequence:

$$x[n] = \tilde{A}e^{j\hat{\omega}_0 n} = Ae^{j(\hat{\omega}_0 n + \phi)}, \quad -\infty < n < \infty$$

Figure 4-1: Stem plots of a real exponential $\gamma = \alpha = e^{-0.25} = 0.78$, an exponentially damped cosine $\gamma = e^{-0.25} + j2\pi/10$, and a cosine sequence $\gamma = j2\pi/10$; $\tilde{A} = 1$ in all three plots.

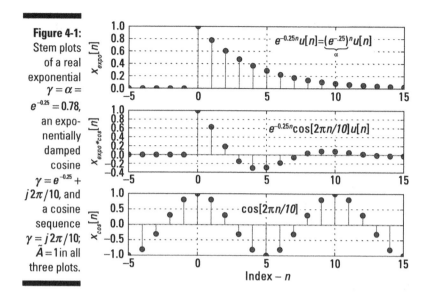

In the discrete-time domain, complex sinusoids are common and practical, because computers can process complex signals by just adding a second signal path in software/firmware. The continuous-time complex sinusoid found in Chapter 3 requires a second wired path, increasing the complexity considerably.

Finally, if you consider just the real part of $x[n]$, with $\alpha = 1$, you get the real sinusoid $x[n] = A\cos(\hat{\omega}_0 n + \phi)$, $-\infty < n < \infty$.

Just as in the case of the continuous-time sinusoid, the definition has three parameters: The amplitude A and phase ϕ carry over from the continuous-time case; the frequency parameter doesn't. What gives?

In the continuous-time case, the cosine argument (less the phase) is $\omega_0 t = 2\pi f_0 t$. Tracking the units is important here. The radian frequency ω_0 has units radians/second (f_0 has units of hertz), so the cosine argument has units of radians, which is expected. But in the discrete-time case, the cosine argument (again less the phase) is $\hat{\omega}_0 n$. The time axis n has units of *samples*, so it must be that $\hat{\omega}_0$ has units of radians/sample. It's now clear that $\hat{\omega}_0$ and ω_0 aren't the same quantity!

But there's more to the story. In practice, discrete-time signals come about by uniform sampling along the continuous-time axis t. Uniform sampling means $t \rightarrow nT$, where T is the sample spacing and $f_s = 1/T$ is the sampling rate in hertz (actually, samples per second). When a continuous-time sinusoid is sampled at rate f_s, $x[n]=A\cos(\omega_0 t+\phi)\big|_{t\to nT}=A\cos(\omega_0 nT+\phi)\overset{?}{=}A\cos(\hat{\omega}_0 n+\phi)$ so it must be that $\hat{\omega}_0=\omega_0 T$.

REMEMBER

A fundamental result of uniform sampling is that discrete-time frequencies and continuous-time frequencies are related by this equation:

$$\hat{\omega}_0=\omega_0 T=\frac{\omega_0}{f_s}=2\pi\cdot\frac{f_0}{f_s}$$

This result holds so long as $-f_s/2<f_0<f_s/2$, a condition that follows from low-pass sampling theory. When the condition isn't met, aliasing occurs. The continuous-time frequency won't be properly represented in the discrete-time domain; it will be *aliased* — or moved to a new frequency location that's related to the original frequency and the sampling frequency. See Chapter 10 for more details on sampling theory and aliasing.

Special signals

The signals I consider in this section are defined piecewise, meaning they take on different functional values depending on a specified time or sequence interval. The first signal I consider, the unit impulse, has only one nonzero value. Most of the special signals are defined over an interval of values (more than just a point).

The unit impulse sequence

The *unit impulse,* or *unit sample, sequence* is defined as

$$\delta[n]=\begin{cases}1, & n=0\\0, & \text{otherwise}\end{cases}$$

This definition is clean, as opposed to the continuous-time unit impulse, $\delta(t)$, defined in Chapter 3. Any sequence can be expressed as a linear combination of time-shifted impulses:

$$x[n]=\sum_{k=-\infty}^{\infty}x[k]\delta[n-k]$$

This representation of $x[n]$ is important in the development of the convolution sum formula in Chapter 6. Note that time or sequence shifting of $\delta[n]$— that is, $\delta[n-k]$— moves the impulse location to $n=k$. Why? The function *turns on* when $n-k=0$ because $n=k$ is the only time that $\delta[n-k]\neq 0$.

The unit step sequence

The *unit step sequence* is defined as

$$u[n] = \begin{cases} 1, & n \ge 0 \\ 0, & \text{otherwise} \end{cases}$$

Note that unlike the continuous-time version, $u[n]$ is precisely defined at $n = 0$. The unit step and unit impulse sequences are related through these relationships:

$$u[n] = \sum_{k=-\infty}^{n} \delta[k] \overset{\text{also}}{=} \sum_{k=0}^{\infty} \delta[n-k]$$

$$\delta[n] = u[n] - u[n-1]$$

As you can see, the unit step sequence is simply a sum of shifted unit impulses that repeats infinitely to the right, starting with $n = 0$. This is just one example of how any sequence can be written as a linear combination of shifted impulses — pretty awesome!

Example 4-1: Use the unit step sequence to create a rectangular pulse sequence of length L samples. The pulse is to *turn on* at $n = 0$ and stay on for L total samples.

The solution is quite simple, $x_L[n] = u[n] - u[n-L]$, but watch the details. When $n = L$, the second unit step turns on and begins to subtract +1s from the first step. The first nonzero point occurs at $n = 0$, and the last nonzero point occurs at $n = L-1$. So how many nonzero points are there?

The total number of points in the pulse is $\left[n_{\text{Stop}} - n_{\text{Start}} \right] + 1 = \left[(L-1) - 0 \right] + 1 = L$, as desired. Figure 4-2 shows the formation of the rectangular pulse sequence for $L = 10$.

The Python support function `dstep()`, found in `ssd.py`, generates a unit step sequence that's used to create a ten sample pulse sequence:

```
In [10]: import ssd # at the start of the session
In [11]: n = arange(-5,15+1) # create time axis n
In [13]: stem(n,ssd.dstep(n))# plot u[n]
In [18]: stem(n,ssd.dstep(n-10))# plot u[n-10]
In [23]: stem(n,ssd.dstep(n)-ssd.dstep(n-10))# difference
```

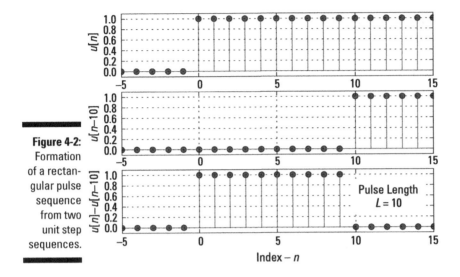

Figure 4-2: Formation of a rectangular pulse sequence from two unit step sequences.

Window functions

Signals pass through windows just as light passes through the glass in your house or office; how illuminating! You may want to think of windows as the rectangular pulses of discrete-time. Just keep in mind that windows may not always be rectangular; sometimes they're tapered. Nevertheless, these discrete-time signals are more than academic; they're shapes used in real applications.

In one application, you use a window function $w[n]$ to multiply or weight a signal of interest, thereby forming a *windowed sequence: $y[n] = x[n] \cdot w[n]$*.

Window functions are also used in digital filter design. The L sample rectangular pulse of Example 4-1 shows the natural windowing that occurs when capturing an L sample chunk of signal $x[n]$ by multiplying $x[n]$ by $w[n]$.

In general, the sample values of a window function smoothly taper from unity near the center of the window to zero at each end. The rectangular window, just like the rectangle pulse, is a constant value from end to end and thus provides no tapering. In spectral analysis, a tapered window allows a weak signal to be discerned spectrally from a strong one. Windowing in a digital filter makes it harder for unwanted signals to leak through the filter.

To fully appreciate windowing, you need to understand frequency-domain analysis, the subject of Part III. Specifically, I describe spectral analysis of discrete-time signals in Chapter 12.

Example 4-2: The `signal` package, specifically, `signal.windows`, of the Python library SciPy contains more than ten window functions! As a specific example, the Hann (also known as the Hanning) window is defined as follows:

$$w[n] = \begin{cases} 0.5\big(1 - \cos\big[2\pi n/(L-1)\big]\big), & 0 \le n \le L-1 \\ 0, & \text{otherwise} \end{cases}$$

The IPython command line input, which calls the `hann(L)` window function to generate a 50 sample window is

```
In [138]: import scipy.signal as signal
In [139]: w_hann = signal.windows.hann(50)
```

Pulse shapes

In digital communication, a pulse shape, $p[n]$, creates a digital communication waveform that's bandwidth or spectrally efficient. A mathematical representation of a digital communication signal/waveform is the sequence

$$x[n] = \sum_k d_k p\big[n - kN_s\big]$$

where the d_k are ± 1 data bits (think 0s and 1s), and N_s is the duration of each data bit in samples.

Popular pulse shapes used for digital communications include the rectangle (RECT), half-sine (HC), the raised cosine (RC), and the square-root raised cosine (SRC or RRC). You can find the definitions of these pulse shapes and more detailed information on their use in real systems in a digital communications case study at `www.dummies.com/extras/signalsandsystems`.

The Python function `NRZ_bits()` generates the digital communications signal $x[n]$. The function returns x, p, and `data`, which are the communications waveform, the pulse shape, and the data bits encoded into the waveform. Find details on the capabilities of this and supporting functions at `ssd.py`.

Surveying Signal Classifications in the Discrete-Time World

Like their continuous-time counterparts, discrete-time signals may be deterministic or random, periodic or aperiodic, power or energy, and even or odd. Discrete-time also has special sequences as well as exponential and sinusoidal signals. I explain the details of these signal types and point out how they compare to continuous-time signals in the following sections.

Deterministic and random signals

Discrete-time signals may be deterministic or random. Discrete signal $x[n]$ is *deterministic* if it's a completely specified function of time, n. As a simple example, consider the following finite sequence:

$$x[n] = \{..., 0, 2, 2, 2, \underset{\uparrow}{1}, 0, -1, -1, -3, 1, 1, 0, ...\}$$

The symbol $\uparrow$ is a timing marker denoting where $n = 0$. Outside the interval shown, I assume the sequence is 0.

The customary way of plotting a discrete-time signal is by using a *stem plot,* which locates a vertical line at each n value from zero to the sequence value $x[n]$; a stem plot also includes a marker such as a filled circle. PyLab and other similar software tools provide support for such plots.

TIP

To create the stem plot, using the sequence definition for $x[n]$, create a vector x (NumPy ndarray) to hold the signal values and a vector n to hold the time index values. The time index values consist of values from $-4 \leq n \leq 7$. These values correspond to each value of the sequence $x[n]$, with reference taken to the timing marker.

```
In [466]: n = arange(-4,7+1) # creates -4<=n<=7
In [467]: x = array([0,2,2,2,1,0,-1,-1,-3,1,1,0])
In [468]: stem(n,x) # create the stem plot
```

Figure 4-3 illustrates this stem plot.

The stem plot is ideal here because connecting the points isn't really appropriate for a signal that's defined only at integer values. PyLab's stem() function, which creates the stem plot, is quite flexible by allowing for different colors and stem head symbols (to see how, type stem? at the IPython command prompt).

TECHNICAL STUFF

A *random* signal, $w[n]$, takes on unpredictable values according to some probability distribution. A computer or *state machine* (a mathematical model of computation) is typically nearby in the case of discrete-time signals, so you can generate $w[n]$ by using a software-based random number generator. Technically speaking, the sequence is only pseudo-random, because computer-generated random numbers eventually repeat. Still, you can get a good approximation of a random signal in practice. Python, in particular NumPy, provides an excellent random number library.

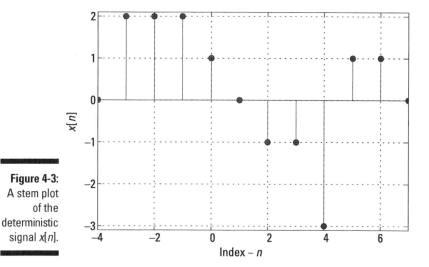

Figure 4-3:
A stem plot
of the
deterministic
signal x[n].

At the command prompt, type In [5]: random? for a helpful listing of the distribution types available.

Gaussian and uniform distributed random sequences are common choices in signal modeling; for example:

```
In [6]: randn(4) # 4 numbers, mean = 0, variance = 1
Out[6]:
array([-1.509427, -0.779072, 0.643483, -1.020021])
In [7]: uniform(0,1,4) # 4 numbers uniform on (0, 1)
Out[7]:
array([0.114319, 0.415064, 0.330576, 0.266975])
```

Periodic and aperiodic

Sinusoidal sequences behave differently in discrete-time situations; they're not always periodic here as they are in the world of continuous-time.

A discrete-time signal, $x[n]$, is *periodic* with period N_0, for the smallest integer N_0 resulting in $x[n] = x[n+N_0]$. If N_0 can't be found, the signal is *aperiodic*.

Example 4-3: Using IPython, you can create a short sequence, $x_p[n]$, of ten samples and then embed this sequence in a larger one, using the mod() function. Define $x_p[n] = \{\underset{\uparrow}{1}, 1, -1, 0, 2, 0, 0, 0, 0, 0\}$. A second sequence, $\cos(6n/10)$ is plotted.

Here is the Python code for plotting:

```
In [474]: n = arange(0,31) # n-axis for plotting
In [475]: x_p = array([1,1,-1,0,2,0,0,0,0,0]) # one period
In [476]: x_10 = x_p[mod(n,10)] # 3+ periods
In [477]: subplot(211); stem(n,x_10) # plot per. 10 seq.
In [478]: subplot(212); stem(n,cos(6*n/10.) # aper. cos
```

The two sequences are shown in Figure 4-4 as subplots.

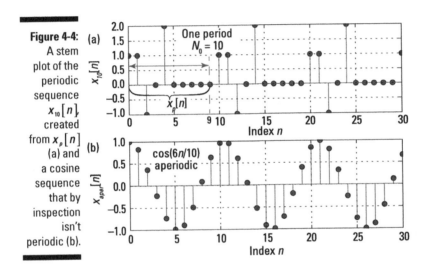

Figure 4-4: A stem plot of the periodic sequence $x_{10}[n]$, created from $x_p[n]$ (a) and a cosine sequence that by inspection isn't periodic (b).

Note the signal $x_p[n]$ by itself is aperiodic, because it's an isolated sequence of nonzero values. The cosine sequence appears to be periodic.

To make a sinusoidal sequence periodic, the following expressions must be equal:

$$A\cos[\hat{\omega}_0 n] \overset{?}{=} A\cos[\hat{\omega}_0(n+N_0)]$$

Cosine is a $\mathrm{mod}\,2\pi$ function, so for the equality to hold, the following must be true:

$$\hat{\omega}_0 N_0 \overset{\text{must}}{=} 2\pi k, \ k \text{ an integer}$$

Rearranging, you have

$$\hat{\omega}_0 = \frac{2\pi k}{N_0}, \ k \text{ an integer}$$

The conclusion is that a sinusoidal sequence can be periodic only if $\hat{\omega}_0$ is a rational number multiple of 2π. The smallest N_0 satisfying the equation is the period of the sinusoid.

For multiple sinusoids to be periodic, you need to find a common N_0 that works for all the sinusoids together.

In Example 4-6, $\hat{\omega}_0 = 6/10 = 2\pi{\cdot}3/(10\pi)$. This means that $3/(10\pi)$ is irrational by virtue of the π in the denominator.

Also, because sine/cosine are $\bmod 2\pi$ functions, the frequencies $\hat{\omega}_0$ and $\hat{\omega}_0 + 2\pi m$, $m = \pm 1, \pm 2, \ldots$ are *indistinguishable,* meaning they produce the same functional values. To be clear for $m = 1$, $\cos\left[(\hat{\omega}_0 + 2\pi)n\right] = \cos\left[\hat{\omega}_0 n + 2\pi n\right] = \cos\left[\hat{\omega}_0 n\right]$ because cosine is a $\bmod 2\pi$ function. This result is independent of the sinusoidal sequence being periodic. The *fundamental interval* is often taken as $\hat{\omega}_0 \in [0, 2\pi)$.

For the special case of a sinusoid having period N_0, the *distinguishable* frequencies are the N_0 values:

$$\hat{\omega}_k = \frac{2\pi k}{N_0}, \; k = 0, 1, \ldots, N_0 - 1$$

Distinguishable frequencies are distinct from all other frequencies that result in a sinusoidal sequence having period N_0 and lie on the fundamental frequency interval.

Example 4-4: Consider the signal $x[n] = \cos(2\pi n/10) + 3\sin(5\pi n/21)$. Find the period of $x[n]$, and calculate the power, P_x. The sinusoids have frequency $\hat{\omega}_1 = 2\pi{\cdot}1/10$ and $\hat{\omega}_2 = 2\pi{\cdot}5/21$. They're both periodic with periods of 10 and 21, respectively. The least common multiple is 210, so the overall period is $N_0 = 210$.

For multiple sinusoidal sequences, the following is true:

$$P_x = \frac{A_1^2 + A_2^2}{2} = \frac{1^2 + 3^2}{2} = \frac{10}{2} = 5$$

For continuous-time sinusoid $x(t) = A\cos(\omega_0 t + \phi)$, you can show that increasing ω_0 increases the *oscillation rate* (number of cycles per second). For $x[n] = A\cos(\hat{\omega}_0 n + \phi)$, the oscillation increases while $\hat{\omega}_0 \in [0, \pi]$, and then it decreases back to 0 for $\hat{\omega}_0 \in [\pi, 2\pi]$.

The oscillation rate increase and decrease property for the discrete-time sinusoid is verified in Figure 4-5.

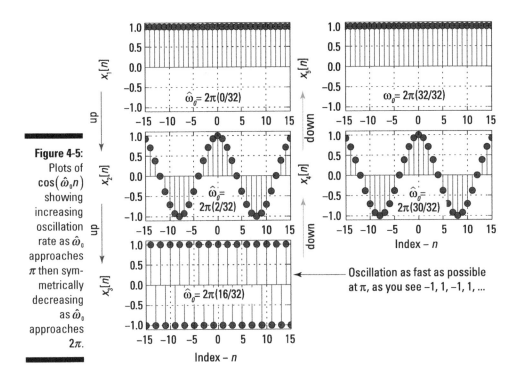

Figure 4-5:
Plots of $\cos(\hat{\omega}_0 n)$ showing increasing oscillation rate as $\hat{\omega}_0$ approaches π then symmetrically decreasing as $\hat{\omega}_0$ approaches 2π.

Recognizing energy and power signals

In discrete-time, a signal can be classified as being *energy*, *power*, or neither. Here are the expressions for discrete-signal energy, E_x, and power, P_x:

$$E_x = \lim_{N \to \infty} \sum_{n=-N}^{N} |x[n]|^2$$

$$P_x = \lim_{N \to \infty} \frac{1}{2N+1} \sum_{n=-N}^{N} |x[n]|^2$$

In the case of a continuous-time signal, the units for power and energy are watts (W) and joules (J). For the discrete-time case, the units don't formally apply unless you find $x[n]$ by sampling a continuous-time signal — that is, $x[n] = x(t)\big|_{t \to nT} = x(nT)$. The signal $x[n]$ is

✔ An energy signal if $E_x < \infty$ and $P_x = 0$

✔ A power signal if $P_x < \infty$ and $E_x \to \infty$

✔ Neither power nor energy if both E_x and P_x go to infinity

An *aperiodic signal* of finite duration is a good example of an energy signal, and a *sinusoidal signal* (sequence) is a good example of a power signal. See Chapter 3 for the definition of aperiodic signal.

Example 4-5: To find the energy in the signal $x_p[n] = \{1,1,-1,0,2,0,0,0,0\}$, plug the values into the energy formula:

$$E_{x_p} = \{1^2 + 1^2 + (-1)^2 + 2^2\} = 7 \text{ J}$$

Example 4-6: Find the power in the sinusoidal signal $x[n] = A\cos[\hat{\omega}_0 n + \phi]$, where $\hat{\omega}_0$ is the sinusoidal discrete-time frequency in radians per sample and $\phi \in (-\pi, \pi]$.

If $x[n]$ is periodic, you can take a shortcut and apply the definition over just one periodic: $P_x = \sum_{n=0}^{N_0-1} |x[n]|^2 / N_0$. You must use the full definition when $x[n]$ is aperiodic, so you calculate the power as follows:

$$P_x = \lim_{N \to \infty} \frac{1}{2N+1} \sum_{n=-N}^{N} A^2 \cos^2(\hat{\omega}_0 n + \phi)$$

$$= \lim_{N \to \infty} \frac{1}{2N+1} \cdot A^2 \sum_{n=-N}^{N} \tfrac{1}{2}\left[1 + \cos(2\hat{\omega}_0 n + 2\phi)\right]$$

$$= \frac{A^2}{2} + \underbrace{\lim_{N \to \infty} \frac{1}{2N+1} \cdot \frac{A^2}{2} \sum_{n=-N}^{N} \cos(2\hat{\omega}_0 n + 2\phi)}_{\to 0 \text{ as } N \to \infty}$$

Intuitively, the second term of the third line reduces to 0 for large N. The formula for power in a discrete-time sinusoid then becomes $A^2/2$. (The same result occurs in Chapter 3 for the continuous-time sinusoid.) The path you take to get the discrete-time results is different because of the summation instead of the integral, but the results are consistent with the continuous-time case.

Computer Processing: Capturing Real Signals in Discrete-Time

Example 4-7: For speech analysis, you can use a tool, such as Python with PyLab, to analyze real signals. By *real,* I mean a signal captured from the continuous-time domain. In this example, I assume you connect a microphone to the audio input of your computer and capture in wav file format the utterance of the two-syllable word *zero* twice. Perhaps you need to design a speech recognition system for "zero-zero." Figure 4-6 shows the system block diagram.

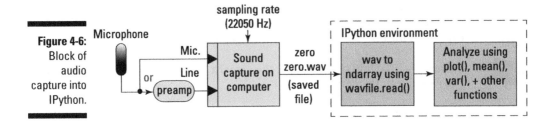

Figure 4-6:
Block of
audio
capture into
IPython.

Depending on the audio inputs on your computer, you may need to add a microphone preamp. In the following sections, I explain the wav file you capture and show how you can use Python to find the signal energy and average power in discrete-time.

Capturing and reading a wav file

Using readily available recording software, set the sampling rate to $f_s = 22.050$ kHz, which is 22,050 samples per second. The samples are spaced by the sampling period $T = 1/f_s = 45.35 \ \mu s$. Save the recorded sound in the file zerozero.wav. The module ssd.py contains functions for reading and writing wav files. The command for reading a wav file is:

```
In [481]: fs, zerozero_wav = ssd.wave_read('zerozero.wav')
```

Note that wavfile.read() returns the sampling rate fs and the NumPy ndarray zerozero_wav, ready for plotting and further analysis.

You can discover a little about the capture by using some of the properties associated with fs and zerozero_wav:

```
In [400]: zerozero_wav.dtype
Out[480]: dtype('float64')
In [480]: zerozero_wav.shape
Out[480]: (76024, 2)
In [481]: fs
Out[481]: 22050
```

The function scales the sample values from 16-bit signed integers to 64-bit floats (see Line [480]). Line [481] tells you that the input array is composed of two columns and 76,024 rows (samples). There are two columns because the recording was made in stereo. I used a single channel microphone so identical signal values are found in each column. The syntax zerozero_wav[:,0] extracts just column zero. If you divide the number of samples by the sampling rate, you get the duration of the recording: $N/f_s = 76,024/22,050 = 3.45$ s.

The signal is discrete-time, but a large number of points occur, so `plot()` is preferred over `stem()` in this case.

Figure 4-7 shows a `subplot()` array containing two views of the signal. The IPython command line code (abbreviated) is

```
In [488]: plot(zerozero_wav[:,0])
In [492]: plot(zerozero_wav[:,0])
In [495]: axis([12000, 22000, -0.3, 0.3]) # zoom axis
```

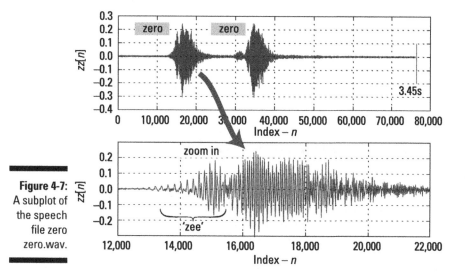

Figure 4-7: A subplot of the speech file zero zero.wav.

Figure 4-7 shows that the two *zero* utterances aren't all that similar. Surprised? The *z* in *zero* is distinctive in the zoom plot. Do you think a speech signal is deterministic or random? Random! In the context of this single data record, you can view the signal as a deterministic waveform. Because the signal has only two *segments,* you can also classify this as an energy signal.

If these two segments are part of a much longer speech record, you can use them to estimate the average power of the entire record.

Finding the signal energy

To calculate the signal energy of the two zeros recorded at 22,050 samples per second (sps), I form the sum of the squared sample values (Line [500]):

```
In [500]: E_zz = sum(zerozero_wav[:,0]**2) #sq & sum samps
In [541]: E_zz
Out[541]: 92.6137 # the signal energy
```

You can calculate the signal average value or mean by using mean() and the variance, essentially the signal power, by using var().

Want to know how indexing and slicing work in Python? Suppose x is a $N \times 1$ ndarray. In Python, *array indexes* start at 0 and end at N-1 (len(vec)-1). The first element is x[0], and the last element is x[-1] because a negative index means to count backward from the end; that is, x[-2] = x[len(x)-2]. *Slicing,* which involves the use of :, allows you to select a contiguous interval anywhere from 0 to N-1. So x[2:5] returns the elements 2 to 5-1, x[:5] returns elements 0 to 4, x[2:] returns elements 2 to N-1, and x[:] returns all the elements. In a two-dimensional array, indexing and slicing works the same, except you have two indexes to play with, as in x[i,j].

Classifying Systems in Discrete-Time

You can classify discrete-time systems based on their properties. Here, I point out how to check a discrete-time system for the following properties: linear/nonlinear, time-invariant/time-varying, causal/non-causal, memory/memoryless, and bounded-input bounded-output (BIBO) stability. (See Chapter 3 for a description of the mathematical properties of these classifications.)

Consider a generic discrete-time system $T\{\ \}$ that's defined as an *operator* that maps the input sequence to the output sequence. Figure 4-8 shows a block diagram representation of the system.

Figure 4-8:
Discrete-
time system
block
diagram
definition.

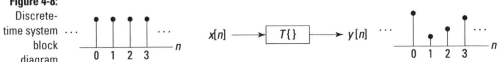

Checking linearity

A system is *linear* if superposition holds. So, if

$$y_1[n] = T\{x_1[n]\}$$
$$y_2[n] = T\{x_2[n]\}$$

and a and b arbitrary constants

$$T\{ax_1[n]+bx_2[n]\}=aT\{x_1[n]\}+bT\{x_2[n]\}$$
$$=ay_1[n]+by_2[n]$$

then the system is linear.

For K signals with $y_k[n]=T\{x_k[n]\}$, you can write:

$$y[n]=T\left\{\sum_{k=1}^{K}a_kx_k[n]\right\}=\sum_{k=1}^{K}a_ky_k[n]$$

If superposition doesn't hold, the system is nonlinear.

Investigating time invariance

A system is time-invariant if properties or characteristics of the system don't change with the time index. A mathematical statement of this is that given $y[n]=T\{x[n]\}$ and any sequence offset n_0, the time-shifted input $x_1[n]=x[n-n_0]$ must produce system output $y_1[n]=y[n-n_0]$.

A system that doesn't obey this property is said to be *time-varying*. In the continuous-time domain, time-varying behavior can be due to uncontrollables, such as environmental conditions like temperature and/or component aging. In the discrete-time domain, an adaptive filter with time-varying attributes can optimize overall system performance. Cellular phone systems make use of this today. The *channel* between you and the base station changes as you move, so your handset adapts to the environment by changing system attributes. Welcome to the world of *adaptive signal processing!*

Looking into causality

The mathematical definition of causality states that a system is causal if all output values, $y[n_0]$, depend *only* on input values $x[n]$ for $n \le n_0$. Another way of saying this is the present output depends *only* on past and present input values.

A system is *causal,* or *nonanticipative,* if it doesn't anticipate the arrival of the signal at the input. Non-causal systems can predict the future! For discrete-time systems, you're more likely to talk about non-causal systems. In discrete-time signal processing, a signal may be stored in a file and processed in such a way that non-causal operations are used.

What's the trick? Well, because the signal is stored in advance, the non-causal system has access to future values of the signal. The operations only appear non-causal. You can't predict the future here, but in discrete-time signal processing, you can pretend that you can.

Figuring out memory

A system is *memoryless* if each output $y[n]$ depends only on the present input $x[n]$. A memoryless system is always causal because there's no chance that values other than the present input will be used for the present output. A causal system, however, doesn't have to be memoryless. A causal system can use past values of the input to form the output. This puts memory into the system.

Checking system properties

To get a firmer grip on the discrete-time version of these five system properties, consider the moving average (MA) system, which is defined as

$$y[n]=\frac{1}{M_p+1+M_f}\sum_{k=-M_f}^{M_p}x[n-k]$$

With this system, the present output is the average (by virtue of the scale factor $1/(M_p+1+M_f)$ of the present input along with M_p past inputs and M_f future inputs. Investors use moving averages. Only past values can be used to guide your next stock purchase. A graphical depiction of the MA system for $M_f=2$ and $M_p=5$ is shown in the figure.

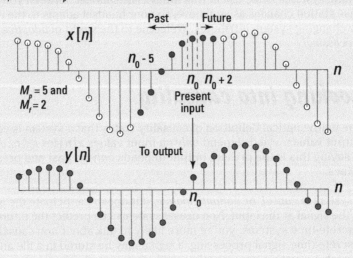

Run through the five properties to see where this systems stands:

✔ **Linearity:** The system is linear because the output is a linear combination of $M_p + 1 + M_f$ inputs.

✔ **Time invariance:** The system is time-invariant because the system *coefficients* (the scale factor applied to each $x[n-k]$ term) are constant.

✔ **Causality:** The system isn't causal as long as $M_f > 0$ (see the figure).

✔ **Memory:** The system isn't memoryless if $M_p > 0$ and/or $M_f > 0$.

✔ **BIBO stability:** The system is BIBO stable because the system coefficients are finite. To be more precise mathematically, I start with $|x[n]| < M_x < \infty$ and write

$$|y[n]| = \left| \frac{1}{M_p + 1 + M_f} \sum_{k=-M_f}^{M_p} x[n-k] \right| \le \frac{1}{M_p + 1 + M_f} \sum_{k=-M_f}^{M_p} |x[n-k]|$$

$$< \frac{1}{M_p + 1 + M_f} \sum_{k=-M_f}^{M_p} M_x = M_x < \infty. \ QED$$

The last term of the first line is due to the triangle inequality, which states that $|x+y| \le |x| + |y|$, for any x and y.

Testing for BIBO stability

A system is said to be *bounded-input bounded-output* (BIBO) stable if and only if every bounded input produces a bounded output. In other words, the signal $x[n]$ (which may be the input or output) is bounded if some constant $B_x > 0$ exists, such that $|x[n]| \le B_x < \infty$ for all values of n.

A challenge associated with this property is to show the property holds for any bounded input. *Any* is a lot of cases to test and requires some proof-writing skills.

Example 4-8: Consider the system $y[n] = 2x[n] + 5x[n-1] + u[n+2]$. Run through the five properties to see where this system stands.

✔ **Linearity:** The system isn't linear because

$$u[n+2] + 2(ax_1[n] + bx_2[n]) + 5(ax_1[n-1] + bx_2[n-1])$$
$$\ne u[n+2] + 2ax_1[n] + 5ax_1[n-1]$$
$$+ u[n+2] + 2bx_2[n] + 5bx_2[n-1]$$

Why? From left to right, $u[n+2] \ne 2u[n+2]$.

✔ **Time invariance:** The system isn't time-invariant because the system coefficient $u[n+2]$ is time-varying. For $n < -2$, it's 0, and for $n \ge -2$, it's 1.

✔ **Causality:** The system is causal. No way, you say! The system term $u[n+2]$ does turn on before $n=0$, but this is a system property, not a future value of the input. Only past and present values of the input form each and every output. The *time-varying bias* term is out of the picture.

✔ **Memory:** The system isn't memoryless because the past input $x[n-1]$ forms the present output.

✔ **BIBO stability:** The system is BIBO stable because the system coefficients are finite.

Interfacing to the analog world

The core building blocks for going from the analog (continuous-time) domain to the discrete-time domain and back are two *mixed-signal* subsystems. The analog-to-digital converter (ADC) and the digital-to-analog converter (DAC) perform these functions. The term *mixed signal* denotes that the electronic circuit design of these subsystems involves both analog and digital circuit design. I provide additional details in Chapter 10, where I describe sampling theory.

The top-level model of the ADC and DAC interfaces to the discrete-time domain are shown in the figure.

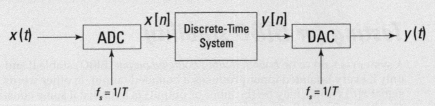

A sampling rate of $f_s = 1/T$ samples per second (sps) is shown. This means that $x[n] = x(nT)$ and $y(nT) \approx y[n]$, with interpolation of the $y[n]$ values is used to create values for $y(t)$ between the sample values. As long as the frequency content of $x(t)$ is less than $f_s/2$, you can also say that $y[n] = x[n]$. In practice, the discrete-time system modifies $x[n]$ to suit your needs, so $y(t)$ differs from $x(t)$, but the difference is a good thing.

The back half of the figure is similar to what goes on when you listen to MP3 or iPod music.

Stereo audio samples from a CD or other digital music source having a nominal sampling rate of 44.1 ksps, are compressed 11:1 using what is known as *perceptual coding*. The compression reduces the memory needed to store the music and also the bandwidth needed for streaming over the Internet. Following decompression, the samples are further manipulated in some discrete-time system. The sampling rate is likely increased to 48 ksps, 96 ksps, or as high as 192 ksps. The DAC steps in, and you're able to hear beautiful music.

Part II
Exploring the Time Domain

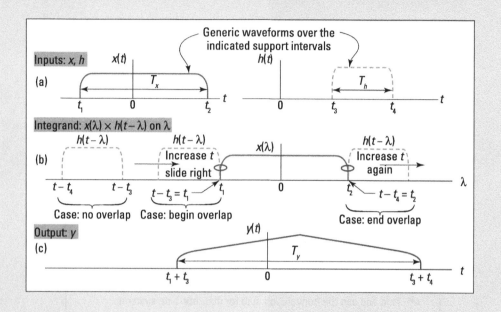

In this part . . .

✔ Explore system models related to linear constant coefficient (LCC) differential equations and their cousins, the LCC difference equations.

✔ Find out why engineers are so darn hot for the impulse response.

✔ Dig into the convolution integral for continuous-time systems.

✔ Find and use the convolution sum for discrete-time systems.

✔ Discover the sinusoidal steady-state response for differential and difference equations.

Chapter 5

Continuous-Time LTI Systems and the Convolution Integral

. . . .

In This Chapter

▶ Understanding the general input/output relationship

▶ Digging into the convolution integral and its properties

▶ Working with the step response, impulse response, and stability implications

. . . .

*R*eady to get pumped up on what you can do in the time domain with linear time-invariant (LTI) systems?

The starting point is the convolution integral, the main time-domain tool for relating the output of a system to its input in combination with the impulse response (IR). In other words, the IR enables you to calculate a system's output in terms of the input. You can use the IR in different ways; one way is convolution.

In this chapter, I describe the relationship between the impulse response and convolution integral and introduce properties related to these concepts. I also provide tips on how to manage the nitty-gritty details of convolution.

The mechanics of the convolution integral are tedious, no doubt, so I provide several worked-through examples and include information about computer tools that can help you evaluate the convolution integral and self-check your plotting work. I also scratch the surface of a powerful stability theorem for LTI systems that relies almost solely on the IR. This theorem grows even more powerful in Chapter 13 when it's combined with the Laplace transform.

Establishing a General Input/Output Relationship

A general continuous-time system input/output relationship is described mathematically as $y(t) = T\{x(t)\}$ where $T\{\ \}$ represents the system or operator of interest. Figure 5-1 depicts this relationship in block diagram form. The operator $T\{\ \}$ describes a mapping of the input sequence, $x(t)$, to the output sequence, $y(t)$.

Figure 5-1:
Block diagram depicting a general input/output relationship.

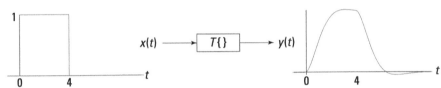

In Chapter 3, I define what it means for a system to be both linear and time-invariant (LTI). For LTI systems, the $T\{\ \}$ in Figure 5-1 is replaced by $L\{\ \}$ to signify that the systems of focus in this chapter are LTI.

The business of developing a general input/output relationship in this section begins with the definition of impulse response because, with the impulse response in hand, you can develop the convolution integral as a means to calculate the system output for any input and the impulse response.

Developing the convolution integral requires both the linearity and time invariance assumptions. I show you the steps in this important development. I also touch on useful properties of the convolution integral.

LTI systems and the impulse response

The impulse response is the beating heart of signals and systems work for LTI systems. Here's why: If you know the impulse response for any LTI system, you can use that information to figure out the system's response to any input. This is a really big super enormous deal.

The impulse response of a continuous-time LTI system, $h(t)$, for example, is defined as the output produced by an *at-rest* system, when given input $x(t) = \delta(t)$. The signal $\delta(t)$ is the unit impulse (see Chapter 3). The at-rest condition ensures that the resulting $h(t)$ is due solely to the input signal. Figure 5-2 shows how to find the impulse response.

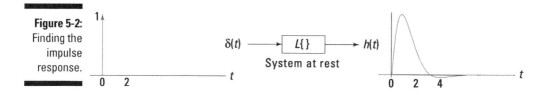

Figure 5-2: Finding the impulse response.

$\delta(t) \longrightarrow \boxed{L\{\}} \longrightarrow h(t)$

System at rest

Mathematically, Figure 5-2 tells you that $h(t) = L\{\delta(t)\}$. This information along with any input $x(t)$ allows you to calculate the system output $y(t)$.

Developing the convolution integral

To get $y(t)$ from $x(t)$ and $h(t)$, start with the sifting property (see Chapter 3) representation of $x(t)$ and then form a numerical approximation to the integral:

$$x(t) = \int_{-\infty}^{\infty} x(\lambda)\delta(t-\lambda)d\lambda \approx \lim_{K\to\infty} \sum_{k=-K}^{K} x(k\Delta t)\delta(t-k\Delta t)\Delta t$$

Figure 5-3 shows a rectangular partition approximation to the sifting property integral representation of $x(t)$, with λ replaced by $k\Delta t$ and $d\lambda$ replaced by Δt.

Figure 5-3: Time-shifted impulse approximation to the sifting integral for $x(t)$.

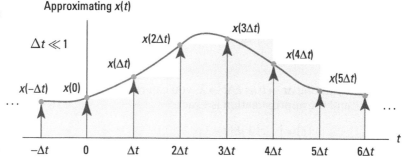

Approximating $x(t)$

Now you can operate on both sides of the $x(t)$ integral approximation with the LTI system, $L\{\ \}$, and in three steps establish the convolution integral:

1. **On the left side, $L\{x(t)\} = y(t)$; on the right side, take advantage of the linearity property:**

$$y(t) \approx \lim_{K\to\infty} L\left\{ \sum_{k=-K}^{K} x(k\Delta t)\delta(t-k\Delta t)\Delta t \right\} \ (\text{move } L\{\ \} \text{ inside the } sum)$$

$$\approx \lim_{K\to\infty} \sum_{k=-K}^{K} x(k\Delta t)L\{\delta(t-k\Delta t)\}\Delta t$$

When $L\{\ \}$ moves inside the sum, it finds $\delta(t - k\Delta t)$ as the only function of time t to operate on.

2. **Take advantage of time invariance to write** $y(t) \approx \lim\limits_{K \to \infty} \sum\limits_{k=-K}^{K} x(k\Delta t) h(t - k\Delta t) \Delta t.$

   Time-invariant systems always give the same response, to within a time offset, that's independent of time axis shifts.

3. **Notice that the rectangular partition approximation to the input is now a linear combination of time-shifted impulse responses,** $h(t - k\Delta t)$, **weighted by the sample values** $x(k\Delta t)$ **of the input.**

   Figure 5-4 shows the individual terms in the approximation, assuming $h(t)$ is an exponential pulse, $e^{-at}u(t)$, decaying to zero for $a > 0$.

   To formally get to the convolution integral, you let $K \to \infty$ and let $\Delta t \to d\lambda$ such that $k\Delta t \to \lambda$. The sum becomes the convolution integral with the limits running from $-\infty$ to ∞.

Figure 5-4: Approximating $y(t)$ as a sum of weighted and time-shifted impulse responses.

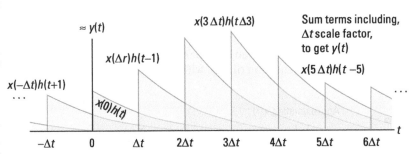

By letting $\Delta t \to 0$ as $k\Delta t \to \lambda$, you can replace the sum with an integral, and the approximation is exact:

$$y(t) = \int_{-\infty}^{\infty} x(\lambda) h(t - \lambda) d\lambda$$

The customary notation to indicate a convolution is $*$:

$$y(t) = x(t) * h(t) = \int_{-\infty}^{\infty} x(\lambda) h(t - \lambda) d\lambda$$

By the change of variables, $\alpha = t - \lambda$, you can also write the convolution integral as

$$y(t) = \int_{-\infty}^{\infty} x(\lambda) h(t - \lambda) d\lambda \bigg|_{\alpha = t - \lambda} = \int_{-\infty}^{\infty} x(t - \alpha) h(\alpha) d\alpha = h(t) * x(t)$$

Looking at useful convolution integral properties

The convolution integral obeys the standard commutative, associative, and distributive algebraic properties, as Table 5-1 shows:

Table 5-1	Convolution Integral Properties
Property	**Algebraic Representation**
Commutative	$x(t) * h(t) = h(t) * x(t)$
Associative	$\left[x(t) * h_1(t) \right] * h_2(t) = x(t) * \left[h_1(t) * h_2(t) \right]$
Distributive	$\left[x_1(t) + x_2(t) \right] * h(t) = x_1(t) * h(t) + x_2(t) * h(t)$
	$x(t) * h_1(t) + x(t) * h_2(t) = x(t) * \left[h_1(t) + h_2(t) \right]$

You can find information on these properties in Chapter 6 in context of discrete-time systems and the convolution sum.

The commutative and associative properties also lead to the series/cascade and parallel connection block diagrams that are shown in Figure 5-5. The series/cascade connection result states that

$$y(t) = \underbrace{\left[x(t) * h_1(t) \right]}_{w(t)} * h_2(t) \overset{\text{assoc. prop.}}{=} x(t) * \underbrace{\left[h_1(t) * h_2(t) \right]}_{h_{\text{cascade}}(t)}$$

The parallel connection results says

$$y(t) = \left[x(t) * h_1(t) \right] + \left[x(t) * h_2(t) \right] \overset{\text{distrib. prop.}}{=} x(t) * \underbrace{\left[h_1(t) + h_2(t) \right]}_{h_{\text{parallel}}(t)}$$

Additionally, linearity allows you to swap the order of the subsystems $h_1(t)$ and $h_2(t)$ in the series connection.

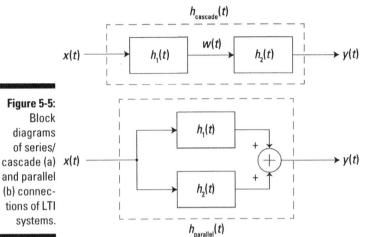

Figure 5-5:
Block
diagrams
of series/
cascade (a)
and parallel
(b) connec-
tions of LTI
systems.

Convolution involving the unit impulse has some memorable outcomes,
meaning you need to remember to take advantage of the simplicity of these
convolution forms; you won't regret it. The following identities hold:

$$\delta(t) * x(t - t_0) = x(t - t_0)$$
$$\delta(t - t_0) * x(t) = x(t - t_0)$$
$$\delta(t - t_1) * x(t - t_2) = x(t - t_1 - t_2)$$
$$\delta(t - t_1) * \delta(t - t_2) = \delta(t - t_1 - t_2)$$

Practical cascading and parallel connections

Being able to conveniently model series
and parallel connections of LTI systems is a
powerful tool, especially when you need to
interconnect continuous-time systems. Just be
sure to account for the impedance levels and
maximum input/output signal swings.

In control system applications, for example,
one system may be an electromechanical
actuator (a motor) or a mechanical sensor (a
tachometer), so you need to pay attention to
unit conversion between the motor's rotational
velocity, perhaps in revolutions per minute
(RPM), and the tachometer that may be

producing a voltage output proportional to a
rotational velocity in radians/s.

You don't need to worry about circuit level
interfacing and/or mechanical-to-electrical
interfacing with discrete-time systems because
the interaction happens between blocks of
code or digital logic. But in software, you do
need to worry about data types, such as signed
integer, unsigned integer, and float. Scaling is
also a concern, to avoid numerical saturation or
overflow. This is where you can really put your
programming skills to work.

Working with the Convolution Integral

Ready to jump in? In this section, I show you how to work through the details of convolution integral examples. And details, by the way, are especially important in convolution integrals, so be sure to follow each step. These examples rely heavily on signal flipping and shifting transformations. If you need a refresher on signal flipping and shifting, take a peek at Chapter 3.

Seeing the general solution first

To solve a convolution problem, you need to know one of the two forms of the integrand: $x(\lambda)h(t-\lambda)$ or $x(t-\alpha)h(\alpha)$. It all comes down to integrating the product of the two functions that form the integrand. As t increases from $-\infty$ to ∞, $h(t-\lambda)$ slides from left to right along the λ-axis. For each value of t, consider the product $x(\lambda)h(t-\lambda)$ and integrate the nonzero or *overlap* intervals along the λ-axis.

For each t phrase sounds daunting at first, but contiguous intervals of t values occur where the integration on the λ-axis uses the same integration setup.

For the alternative integrand form, $x(t-\alpha)h(\alpha)$, do the same thing, except $x(t-\alpha)$ slides as t changes.

The solution is *piecewise continuous*, meaning expressions for $y(t)$ are valid for some $t_i \le t < t_{i\prime}$ which corresponds to case i, $i = 1, \ldots, N$. When you put all the *pieces* together, you have a solution that's valid for $-\infty < t < \infty$.

The four steps in Figure 5-6 outline the general solution procedure.

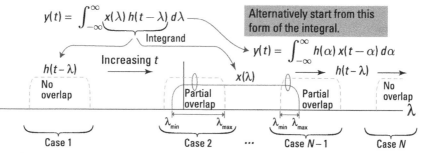

Figure 5-6: Solving a convolution integral problem in four steps with *cases* as $h(t-\lambda)$ slides from left to right.

① Sketch the two waveforms of the integrand on the integration variable axis, λ.
② Find values of t (*cases*) where overlap (partial or full) occurs in the integrand.
③ Establish the integration limits, λ_{min} to λ_{max}, from the waveform alignment.
④ Integrate, and then move on.

You don't want to mess up the sketching step because it provides a road map for all the analysis/calculations that follow. A bad sketch leads you along the wrong roads.

Example 5-1: To see the patterns of convolution problems for finite-duration signals, solve for the general *support* interval — the *t*-axis interval where $y(t) = x(t) * h(t) \neq 0$. Here are some things to notice in Figure 5-7:

- Figure 5-7a shows the signal and impulse response to be convolved.

- Figure 5-7b shows both signals on the λ-axis with *t* as a free variable.

- Also in Figure 5-7b, the flipped and shifted waveform $h(t - \lambda)$ is positioned by selection of *t* (dashed line) at the start of overlap $(t - t_3 = t_1)$ and at the end of overlap $(t - t_4 = t_2)$.

Figure 5-7:
Convolving two generic finite length waveforms: the convolution inputs (a), the integrand waveforms $x(\lambda)$ and $h(t - \lambda)$ along the λ-axis (b), and the output $y(t)$ (c).

Generic waveforms over the indicated support intervals

Inputs: *x, h*

(a)

Integrand: $x(\lambda) \times h(t - \lambda)$ on λ

(b)

Case: no overlap   Case: begin overlap   Case: end overlap

Output: *y*

(c)

To calculate the convolution integral, using the form $y(t) = \int_{-\infty}^{\infty} x(\lambda) h(t - \lambda) d\lambda$, follow the steps from Figure 5-6 but only in a general way because the exact waveform details aren't available for this example.

1. **Sketch the two waveforms of the integrand on the integration variable axis.**

   The example in Figure 5-7b sketches the two waveforms on the λ-axis.

2. **Find values of *t* (cases) where overlap (partial or full) occurs in the integrand.**

   Refer to the example in Figure 5-7: As *t* increases, the point of first overlap occurs when the *leading edge* of $h(t - \lambda)$ located at $t - t_3 = t_1 \Rightarrow t = t_1 + t_3$.

For $t < t_1 + t_3$, no overlap occurs, and the integrand is 0, meaning $y(t) = 0$ as well. As t continues to increase, eventually the trailing edge of $h(t - \lambda)$ is located at $t - t_4 = t_2 \Rightarrow t = t_2 + t_4$.

For $t > t_2 + t_4$, no overlap exists, and the integrand is again 0, so $y(t) = 0$.

For $t_1 + t_3 \le t \le t_2 + t_4$, you'll likely have more than one overlap case to consider, depending on the waveform shapes.

3. **Establish the integration limits from the waveform alignment.**

   Integration limits are set by the specific cases, so they don't apply in this example. In a gross sense, all you can say is that the fixed signal $x(\lambda)$ constrains the limits on λ to be $\left[t_1, t_2 \right]$.

4. **Integrate, and then move on.**

   You don't carry out the integration in this example.

A few things to note about the convolution result: The support interval for the output is at most $t_1 + t_3 \le t \le t_2 + t_4$. But you have no guarantee that $y(t) \neq 0$ for at least some values of t on this interval. The output starts at $t_{start} = t_1 + t_3$ and ends at $t_{end} = t_2 + t_4$. The duration or length of the support interval is $T_y = (t_2 + t_4) - (t_1 + t_3) = \underbrace{(t_2 - t_1)}_{T_x} + \underbrace{(t_4 - t_3)}_{T_h} = T_x + T_h$, which is the sum of the input signal durations.

Using the definitions established in Example 5-1, $T_{start} = t_1 + t_3$, $T_{stop} = t_2 + t_4$, and $T_y = T_x + T_h$.

Solving problems with finite extent signals

This section contains examples for working convolution problems with real numbers, where both the signal and the impulse response are of finite duration.

Example 5-2: Convolve the rectangular pulse signals $x(t) = 2\Pi(t - 3.5)$ and $h(t) = \Pi\left[(t - 1)/2\right]$ (these are real signals), using the same form of the convolution integral as Example 5-1: $y(t) = \int_{-\infty}^{\infty} x(\lambda) h(t - \lambda) d\lambda$.

1. **Sketch the two waveforms of the integrand on the integration variable axis.**

   Check out the waveforms sketched in Figure 5-8a and the λ-axis signal product cases sketched in Figures 5-8b through 5-8f.

2. **Find values of t (cases) where overlap (partial or full) occurs in the integrand.**

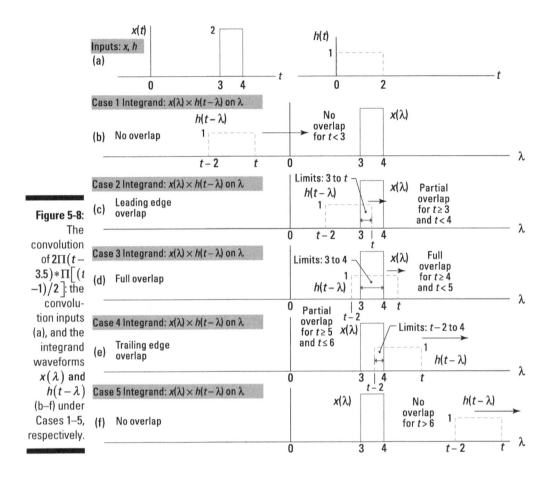

Figure 5-8: The convolution of $2\Pi(t-3.5)*\Pi\big[(t-1)/2\big]$: the convolution inputs (a), and the integrand waveforms $x(\lambda)$ and $h(t-\lambda)$ (b–f) under Cases 1–5, respectively.

For Example 5-2, you have five cases to consider:

- Case 1 and Case 5 occur when $t < 3$ and $t-2 > 4$ or $t > 6$, respectively. For Case 1, $y(t) = 0$ for $t < 3$, and for Case 5, $y(t) = 0$ for $t > 6$. As a check from the results of Example 5-1, the output starts at $3+0=3$ and ends at $4+2=6$.

- Case 2 is what I call *partial overlap leading edge*. You need $t \geq 3$ and $t < 4$ or $3 \leq t < 4$ for the active interval.

- Case 3 is *full overlap* or a straddle that occurs when $t-2 < 3$ and $t \geq 4$. The active interval is $4 \leq t < 5$.

- Case 4 is *partial overlap trailing edge*. This condition requires $t-2 \geq 3$ and $t-2 < 4$, so the active interval is $5 \leq t \leq 6$.

3. Establish the integration limits from the waveform alignment.

With the help of Figure 5-6, the integration limits for each case fall into place quite easily.

- Under Case 2, the limits on λ are 3 to t.
- Under Case 3, the limits on λ are set by $x(\lambda)$, so just 3 to 4.
- Under Case 4, the limits on λ are $t-2$ to 4.

4. **Integrate, and then move on.**

- The Case 2 integration is $y(t) = \int_3^t 2 \cdot 1 d\lambda = 2\lambda\big|_3^t = 2(t-3), \quad 3 \le t < 4.$
- The Case 3 integration is $y(t) = \int_3^4 2 \cdot 1 d\lambda = 2\lambda\big|_3^4 = 2, \quad 4 \le t < 5.$
- The Case 4 integration is $y(t) = \int_{t-2}^4 2 \cdot 1 d\lambda = 2\lambda\big|_{t-2}^4 = 2(6-t), \quad 5 \le t < 6.$

Pulling all the pieces together in one big case expression yields the following:

$$y(t) = \begin{cases} 2(t-3), & 3 \le t < 4 \\ 2, & 4 \le t < 5 \\ 2(6-t), & 5 \le t < 6 \\ 0, & \text{otherwise} \end{cases}$$

To plot $y(t)$, write a piecewise function in Python, using the IPython environment. First, create time axis vector, t, and pass it to the function example_5_2(t). You can see the results in Figure 5-9. The basic coding approach I use here is useful whenever you need to plot a piecewise function.

TIP

When faced with an unfamiliar waveform scenario, try visualizing your results as a troubleshooting aid. Assuming $x(t)$ and $h(t)$ are free of impulse functions, the plot of $y(t) = x(t) * h(t)$ should be *piecewise continuous,* meaning no jumps between the waveform pieces.

```
In [187]: t = arange(0,8,0.05) # t from 0 to 8s
In [188]: def example_5_2(t): # embedded IPython function
     ...:     y = zeros(len(t))
     ...:     for k, tt in enumerate(t):
     ...:         if tt >= 3 and tt < 4:
     ...:             y[k] = 2*(tt - 3)
     ...:         elif tt >= 4 and tt < 5:
     ...:             y[k] = 2
     ...:         elif tt >= 5 and tt < 6:
     ...:             y[k] = 2*(6 - tt)
     ...:     return y
In [189]: y = example_5_2(t) #call the function with t
In [190]: plot(t,y) # plot the results (you add labels)
```

When you first see the piecewise expression for $y(t)$, it may not be obvious that it describes a trapezoid-shaped waveform. The trapezoid height corresponds to the overlap area, $2 \cdot 1 = 2$. For identical rectangle pulse widths, such as $x(t) = h(t) = \Pi(t/T)$, the convolution is a triangle. In particular, you can show that $y(t) = \Pi\left(\frac{t}{T}\right) * \Pi\left(\frac{t}{T}\right) = T\Lambda\left(\frac{t}{T}\right)$, which is the triangle pulse that's defined in Chapter 3, scaled by T and of full base width of $2T$.

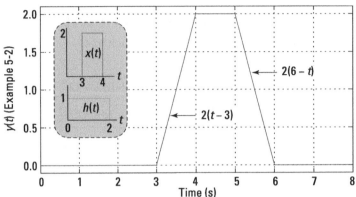

As a modification to the rectangle input, let $x(t) = 2(t - 3) \cdot \Pi(t - 3.5)$, which is a right triangle pulse shape. Because $x(t)$ has support over the interval $[0, 2]$, all work is reusable for Steps 1 through 3, so the five cases remain the same as well as the integration limits established for Cases 2 through 4. The changes come under Step 4 when you have to integrate. The integrand is now $x(\lambda) \cdot h(t - \lambda) = 2(\lambda - 3) \cdot 1$. With the integrand change, you get the following:

$$y(t) = \begin{cases} t^2 - 6t + 9, & 3 \le t < 4 \\ 1, & 4 \le t < 5 \\ -24 + 10t - t^2, & 5 \le t < 6 \\ 0, & \text{otherwise} \end{cases}$$

The waveform peak, which is the maximum overlap area of $x(\lambda) \cdot h(t - \lambda)$, is now the area of the triangle formed by $x(\lambda) \cdot 1$, which is $2 \cdot 1/2 = 1$. Figure 5-10 shows that making $x(t)$ triangular, smoother than the original rectangular pulse, results in a smoother $y(t)$.

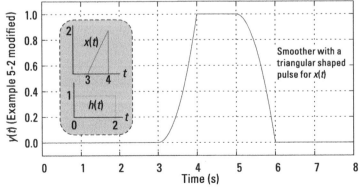

To check your analysis, use the Python function `conv_integral(x1, t1,x2,t2)` in the module `ssd.py` at www.dummies.com/extras/signals andsystems. With this function, you can numerically calculate the convolution integral by using simple rectangular integration. The time step when creating the `t` array needs to be small to achieve good numerical accuracy, as I did in Figure 5-10. Check out the code summary:

```
In [325]: t = arange(0,8,0.01) # create t axis, dt=0.01
In [326]: x = 2*(t - 3)*ssd.rect(t-3.5,1) # create x(t)
In [327]: h = ssd.rect(t-1,2) # create h(t)
In [328]: y,ty = ssd.conv_integral(x,t,h,t) # convolve
In [329]: plot(ty,y) # plot results
```

Dealing with semi-infinite limits

Distortion sometimes enters the picture when a simple system/filter has an exponential impulse response. The system *time constant* — the time it takes e^{-at} to decay from 1 to e^{-1}, $1/a$ (1 s in Example 5-12) — slows down the edges of the input pulse. For a fixed time constant, the pulse duration needs to be on the order of 10 times the time constant for the output pulse to resemble the input.

In the real world, a wired network (Ethernet) connection relies on a *twisted-pair* cable to transfer digital pulses from one end to the other. The cable acts as a filter similar to the exponential impulse response of this example. Trying to send pulses (symbols) at too high a rate, $R = 1/T$, results in pulse distortion. Too much pulse distortion increases the chance of symbols being received in error.

Example 5-3: Consider $x(t) = \Pi\left[(t - T/2)/T\right]$ and $h(t) = e^{-at}u(t)$. The convolution integral form is $y(t) = \int_{-\infty}^{\infty} h(\alpha)x(t - \alpha)d\alpha$.

I'd rather flip and shift the rectangular pulse signal and leave the exponential pulse fixed; it's a matter of comfort and confidence for me. For practice, I suggest reworking this example, using the other formulation.

Using the rules established in Example 5-1, the output support interval is [0 + 0, $T + \infty$] = $\left[0, \infty\right]$. From the get-go, you may expect to discover that $y(t) = 0$ for $t < 0$. Here are the four steps of the convolution integral:

1. **Sketch the two waveforms of the integrand on the integration variable axis, the α-axis for the convolution integral form chosen here.**

   See the waveforms sketched in Figure 5-11a and the α-axis sketched in Figure 5-11b.

2. **Find values of t (cases) where overlap (partial or full) occurs in the integrand.**

Figure 5-11:
Convolution
integral
waveform
sketches:
the convolu-
tion inputs
(a) and the
integrand
waveforms,
$x(t-\alpha)$
and $h(\alpha)$,
along the
α-axis (b),
under
Cases 1–3.

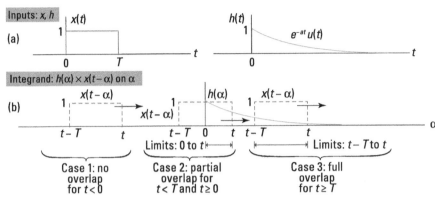

This problem breaks down into three cases.

- Case 1 is the no overlap condition, when $t < 0$. It follows that $y(t) = 0$ under this condition.

- Case 2 is partial overlap, which corresponds to $t \geq 0$ and $t - T < 0$. The active interval is thus $0 \leq t < T$.

- Case 3 is full overlap, which corresponds to $t - T \geq 0$ or $t \geq T$.

3. **Establish the integration limits from the waveform alignment.**

From Figure 5-11, the integration limits under Cases 1 and 2 include the following:

- For Case 2, α runs from 0 to t.

- For Case 3, α runs from $t - T$ to t.

4. **Integrate, and then move on.**

- The Case 2 integration is $y(t) = \int_0^t 1 \cdot e^{-a\alpha} \, d\alpha = \frac{e^{-a\alpha}}{-a}\Big|_0^t = \frac{1}{a}(1 - e^{-at})$, $0 \leq t < T$.

- The Case 3 integration is $y(t) = \int_{t-T}^t e^{-a\alpha} \, d\alpha = \frac{e^{-a\alpha}}{-a}\Big|_{t-T}^t = \frac{e^{-at}}{a}(e^{aT} - 1)$, $T \leq t < \infty$.

Pulling all the pieces together into a case expression yields the following:

$$y(t) = \begin{cases} 0, & t < 0 \\ \frac{1}{a}(1 - e^{-at}), & 0 \leq t < T \\ \frac{e^{-at}}{a}(e^{aT} - 1), & T \leq t < \infty \end{cases}$$

Figure 5-12 shows a family of output curves for $a = 1$ and T varying from 1 to 10 s.

Figure 5-12:
The result
of convolv-
ing a pulse
of duration
T seconds
with an
exponential
impulse
response
having time
constant
$1/a$.

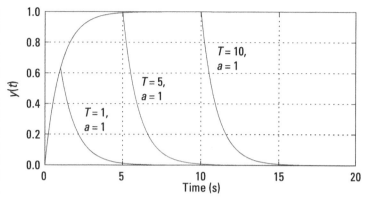

Example 5-4: Consider a 1-Hz sinusoid passing through a system having rectangular impulse response of amplitude $1/T$ and width T starting at $t = 0$. Start with the sinusoid turning on at $t = 0$ by incorporating a unit step function; later, you can remove the step to see what it's like to deal with an infinite duration signal.

The convolution setup is $y(t) = x(t) * h(t) = \int_{-\infty}^{\infty} x(\lambda)h(t-\lambda)d\lambda$ where $x(t) = A\sin(2\pi t)u(t)$ and $h(t) = \Pi\left[(t - T/2)/T\right]/T$. The output support interval is $[0+0, T+\infty] = [0, \infty]$. Now you're ready to work through the four steps:

1. **Sketch the two waveforms of the integrand on the integration variable axis, λ in this case.**

   See the waveforms sketched in Figure 5-13a and the two waveforms of the integrand sketched on the λ-axis in Figures 5-13b through 5-13d, as t takes on increasing values.

2. **Find values of t (cases) where overlap (partial or full) occurs in the integrand.**

   This problem breaks down into three cases, shown in Figures 5-13b through 5-13d. These cases are identical to Example 5-3, except now you're working on the λ-axis. See Figure 5-13 for details.

3. **Establish the integration limits from the waveform alignment.**

   From Figure 5-13, the integration limits under Cases 2 and 3 (the overlap cases) also follow from Example 5-3, but now on the λ-axis.

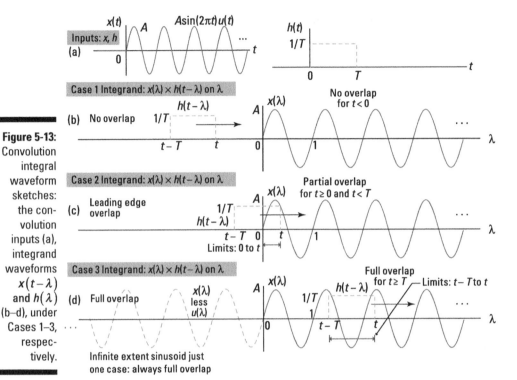

Figure 5-13:
Convolution integral waveform sketches: the convolution inputs (a), integrand waveforms $x(t-\lambda)$ and $h(\lambda)$ (b–d), under Cases 1–3, respectively.

4. **Integrate, and then move on.**

   The Case 2 integration is

   $$y(t)=\int_0^t \frac{A}{T}\sin(2\pi\lambda)\cdot 1\,d\lambda = \left.\frac{-A\cos(2\pi\lambda)}{2\pi T}\right|_0^t = A\frac{1-\cos(2\pi t)}{2\pi T}, \quad 0\le t<T$$

   The Case 3 integration is

   $$y(t)=\int_{t-T}^t \frac{A}{T}\sin(2\pi\lambda)\cdot 1\,d\lambda = \left.\frac{-A\cos(2\pi\lambda)}{2\pi T}\right|_{t-T}^t$$

   $$=A\frac{\cos\left[2\pi(t-T)\right]-\cos(2\pi t)}{2\pi T}, \quad T\le t<\infty$$

   Note: From the geometric series sum formula (see Chapter 2), this expression can also be written as

   $$y(t)=\frac{A\left|e^{-j2\pi T}-1\right|}{2\pi T}\cos\left[2\pi t+\angle\left(e^{-j2\pi T}-1\right)\right]$$

The complete case expression is

$$y(t) = \begin{cases} A\dfrac{1-\cos(2\pi t)}{2\pi T}, & 0 \leq t < T \\[2mm] A\dfrac{\cos\left[2\pi(t-T)\right]-\cos(2\pi t)}{2\pi T}, & T \leq t < \infty \\[2mm] 0, & \text{otherwise} \end{cases}$$

If you remove $u(t)$ from $x(t)$, the input sinusoid is now of infinite duration (see the dashing on the left side of Figure 5-13d), and you have only one case to consider. The output support interval is $\left[-\infty + 0, \infty + T\right] = \left[-\infty, \infty\right]$. The original Case 3 result provides the complete solution:

$$y(t) = A\frac{\cos\left[2\pi(t-T)\right]-\cos(2\pi t)}{2\pi T}, \quad -\infty < t < \infty$$

REMEMBER

No matter what value you choose for t in $h(t-\lambda)$, the signal $\sin(2\pi\lambda)$ is always nonzero on $\lambda \in \left[t-T, t\right]$.

The semi-infinite and infinite input signal results are compared in Figure 5-14.

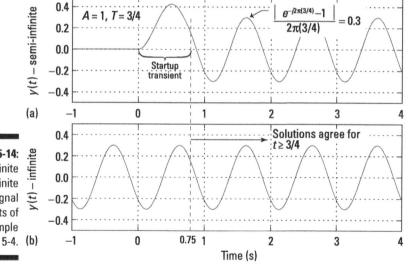

Figure 5-14: Semi-infinite and infinite input signal results of Example 5-4.

Stepping Out and More

The impulse response (IR) is closely related to the step response, which is another useful system characterization input signal. If you have the step response or the impulse response, you can find the other. Two for the price of one! You can establish this relationship through differentiation or integration, depending on the conversion direction.

The LTI specialization of this chapter also describes important stability and causality results. Stability and causality are directly related to the impulse response of a system, emphasizing the importance of a system's IR.

Step response from impulse response

The impulse response completely characterizes an LTI system. To determine the output of a system with a unit step response as the input, you simply convolve the unit step signal (see Chapter 3) with the impulse response $h(t)$. What more could you ask for?

The step response is defined as the system output, from an at-rest condition, when given a unit step input: $y_s(t) = u(t) * h(t)$. This is a popular characterizing waveform in control systems. Looks easy enough, right? Just another convolution. But wait! Maybe you want to investigate the details of the convolution integral.

Here, I expand the convolution integral for $y_s(t)$:

$$y_s(t) = h(t) * u(t) = \int_{-\infty}^{\infty} u(t-\alpha)h(\alpha)d\alpha = \int_{-\infty}^{t} h(\alpha)d\alpha$$

The final form follows by noting that $u(t-\alpha)$ turns off when $\alpha > t$, so I set the upper limit to t. The conclusion is that integrating the impulse response gives you the step response:

$$y_s(t) = \int_{-\infty}^{t} h(\alpha)d\alpha$$

In the opposite direction, differentiating the step response returns you to the impulse response:

$$h(t) = \frac{dy_s(t)}{dt}$$

Did you really just read that? Yes! The step response is the integration of the impulse response, which is the derivative of the step response. This means that if you have $h(t)$ or $y_s(t)$, you can compute the other.

BIBO stability implications

Bounded-input bounded-output (BIBO) stability (described in detail in Chapter 3) requires that the system output $y(t)$ remains bounded $\left[\left|y(t)\right| \le B_y < \infty\right]$ for any bounded-input $x(t)\left[\left|x(t)\right| \le B_x < \infty\right]$. When the system is also LTI, you can establish BIBO stability from the impulse response alone, by seeing whether $\int_{-\infty}^{\infty}\left|h(t)\right|dt \overset{?}{<} \infty$.

Example 5-5: Consider $h(t) = e^{-2t}\,u(t)$. According to the LTI stability theorem, you need $\int_{-\infty}^{\infty}\left|e^{-2t}u(t)\right|dt = \int_{0}^{\infty}e^{-2t}\,dt = \left.\dfrac{e^{-2t}}{-2}\right|_{0}^{\infty} = \dfrac{1}{2} < \infty$. Therefore, the system is BIBO stable. Note that the unit sets the integration lower limit to 0 in this calculation.

Causality and the impulse response

A system is causal if the present output relies only on past and present inputs. From an implementation perspective, *causal* means that the system can be implemented in real hardware. You want to be able to build what you design, right? In this section, I explore the connection between the system impulse response and causality, providing an easy way to spot causal systems.

For an LTI system, you have the convolution integral to establish the relationship between the input and output, along with the impulse response. How must $h(t)$ be constrained to ensure causality? Consider the following:

$$y(t) = \int_{-\infty}^{\infty}x(\lambda)\,h(t-\lambda)d\lambda$$

$$= \underbrace{\int_{-\infty}^{t}x(\lambda)\,h(t-\lambda)d\lambda}_{\text{uses past and present inputs}} + \underbrace{\int_{t}^{\infty}x(\lambda)\,h(t-\lambda)d\lambda}_{\text{uses future inputs}}$$

From the last integral in the second line, you see that the system can't use future inputs if $h(t-\lambda) = 0$ for $\lambda > t$. Rewriting, a causal system has $h(t) = 0$ for $t < 0$.

A direct consequence of the causality on $h(t)$ is that the $(-\infty, \infty)$ convolution limits are now constrained:

$$y(t) \overset{\text{if causal}}{=} \int_{-\infty}^{t}x(\lambda)h(t-\lambda)d\lambda = \int_{0}^{\infty}h(\alpha)x(t-\alpha)d\alpha$$

Example 5-6: Consider the moving-average system: $y(t) = \dfrac{1}{10}\int_{-5}^{5}x(t-\alpha)d\alpha$.

You can find the impulse response for this system by setting $x(t) = \delta(t)$:

$$h(t) = \frac{1}{10}\int_{-5}^{5} \delta(t-\alpha)d\alpha = \frac{1}{10}\Pi\left(\frac{t}{10}\right)$$

Does the integration make sense? When you integrate the unit impulse, you get 1 as long as the impulse lies on the integration interval. The integration interval on the α is [–5, 5] in this case, which holds so long as $-5 < t < 5$. This describes a rectangle pulse centered at $t = 0$.

The present system is non-causal because $h(t) \neq 0$ for $t < 0$. The system averages inputs five seconds on either side of the present input to form the present output. You can make the system causal simply by shifting the impulse response five seconds to the right; for example, $h(t) = \Pi\left[(t-5)/10\right]/10$. Now you form the average by using inputs from zero to ten seconds in the past.

Chapter 6

Discrete-Time LTI Systems and the Convolution Sum

· ·

In This Chapter

▶ Getting familiar with the impulse response and the convolution sum

▶ Understanding the convolution sum and its properties

▶ Figuring out convolution sum problems

· ·

*C*hances are good that sometime within the last 24 hours you've interfaced with an electronic device that uses a discrete-time linear timeinvariant (LTI) system. In fact, you've probably done so with more than one device! Surprised? Maybe not. The candidate devices I have in mind are popular ones, including digital music players, cellphones, HDTVs, and laptop computers.

Another name for a discrete-time LTI system is *digital filter,* and I cover the analysis and design of digital filters over multiple chapters in this book. The emphasis in this chapter is the time-domain view. (Check out Chapter 11 for details on the frequency domain and Chapter 14 for information on the z-domain view.)

The drill down in this chapter on the time-domain analysis, design, and behavior of discrete-time LTI systems leads to discrete-time convolution, or the *convolution sum.* The convolution sum is the mathematics of processing the input signal to the output of a digital filter. A related time-domain topic, linear constant coefficient (LCC) difference equations, is covered in Chapter 7, where I explain that the LCC difference equation implementation of a discrete-time LTI system produces an efficient processing algorithm. (Say *that* three times fast.)

Specializing the Input/Output Relationship

A general system input/output relationship is described mathematically as $y[n] = T\{x[n]\}$, where $T\{\}$ represents the system or operator of interest. Figure 6-1 shows this relationship in block diagram form. The operator $T\{\}$ describes a mapping of the input sequence, $x[n]$, to the output sequence, $y[n]$.

Figure 6-1:
Block
diagram
depicting
a general
input/output
relationship.

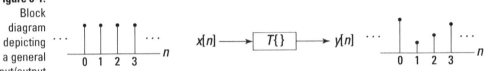

In Chapter 4, I define classifications of discrete-time systems. Here, I focus on systems that are both linear and time-invariant (LTI). I first describe the impulse response as a means to characterize an *at-rest* system (memory is set to zero). You can think of the impulse response as a form of system signature in the time domain. The fact that the system is time-invariant means that the impulse response is the same no matter when it's applied to the system, less the time shift due to the application time. I then show that time invariance combined with system linearity makes it possible to express the system output as a linear combination of the input signal and the impulse response. This linear combination is given the special name *convolution sum*.

In real-world applications, most discrete-time systems are linear and time-invariant because the analysis and design of LTI systems are tractable and, in the vast majority of cases, bring satisfying results. After all, engineers are problem solvers with a finite number of tools who tend to pursue effective processes and reliable outcomes.

Using LTI systems and the impulse response (sequence)

The impulse response of a discrete-time LTI system, $h[n]$, is defined as the output produced by an at-rest system, when given input $x[n] = \delta[n]$. (Check out Figure 6-2 to see a graphical depiction of the input/output relationship.)

The at-rest condition ensures that the resulting $h[n]$ is due solely to the input sequence. Keep in mind that other characterizations are possible, too, such as the step response, which I describe in the section "Step response from the impulse response," later in this chapter.

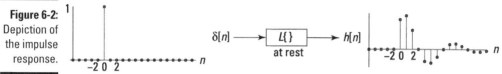

Figure 6-2: Depiction of the impulse response.

With information from the impulse response and the discrete-time LTI system input, you can calculate the output sequence via the convolution sum. The convolution sum for discrete-time systems is analogous to the convolution integral for continuous-time systems found in Chapter 5.

Getting to the convolution sum

To establish the convolution sum formula, any sequence $x[n]$ can be expressed as a superposition of time-shifted impulse functions (flip to Chapter 4 for details):

$$x[n] = \sum_{k=-\infty}^{\infty} x[k]\delta[n-k]$$

Combining the preceding impulse function expansion with the LTI assumption leads to the following conclusion:

$$y[n] = \underbrace{\sum_{k=-\infty}^{\infty} x[k]\,h[n-k]}_{x[n]*h[n]} \overset{\substack{\text{variable}\\\text{change}}}{=} \underbrace{\sum_{m=-\infty}^{\infty} h[m]\,x[n-m]}_{h[n]*x[n]}$$

The * is the customary shorthand notation to convey convolution between $x[n]$ and $h[n]$. This means — at last! — you've arrived at the famous convolution sum formula for computing the system output $y[n]$ given the impulse response $h[n]$ and input $x[n]$. Note that the two forms given are equivalent.

The doubly infinite sum is a necessary evil, because the *support interval* — the n-axis interval where $x[n]$ and $h[n]$ are nonzero — for both $x[n]$ and $h[n]$ could be $(-\infty, \infty)$.

Applying linearity and time invariance

If you're interested in seeing the full development of the convolution sum, here it is. In three steps, I take you through the process of applying linearity and time invariance to get the convolution sum formula. **Note:** I use the terms *smaller scale* to mean finite sum (where only two terms are considered) and *larger scale* to mean infinite sum or infinite sequence.

1. Apply $L\{\}$ to both sides of impulse function expansion of $x[n]$. On the left side, $y[n] = x[n]$; on the right, the operator can move inside the summation because $L\{\}$ is linear:

$$\underbrace{L\{x[n]\}}_{y[n]} = L\left\{\sum_{k=-\infty}^{\infty} x[k]\delta[n-k]\right\} \overset{\text{utilize linearity in part}}{=} \sum_{k=-\infty}^{\infty} L\{x[k]\delta[n-k]\}$$

On a smaller scale, consider a sum of just two terms, then $L\{(x[0]\delta[n-0] + x[1]\delta[n-1])\} = L\{x[0]\delta[n-0]\} + L\{x[1]\delta[n-1]\}$.

2. Apply linearity to further break down the result of Step 1 by first observing that $L\{x[0]\delta[n-0]\} = x[n]L\{\delta[n-0]\}$. So on a larger scale:

$$y[n] \overset{\text{utilize linearity completely}}{=} \sum_{k=-\infty}^{\infty} x[k]\, L\{\delta[n-k]\}$$

3. Utilize time invariance, which states $L\{\delta[n-k]\} = h[n-k]$. Again, on a larger scale:

$$y[n] \overset{\text{utilize time invariance}}{=} \sum_{k=-\infty}^{\infty} x[k]\, h[n-k]$$

The proof is complete!

You can get an alternative form for the convolution sum formula by making the change of variables $m = n - k$:

$$y[n] = \sum_{k=-\infty}^{\infty} h[m]x[n-m] = h[n] * x[n]$$

The two formulations of the convolution sum aren't in competition with each other; they provide options. Depending on the problem type and how your brain happens to be working, you can decide which approach is best for you at the time.

REMEMBER

The word *convolve* simply means "to roll or coil together; to entwine." From the first form of the convolution sum, you combine/sum time-shifted copies of the impulse response $h[n - k]$ weighted by the input sequence values $x[k]$ that correspond to the time shift:

$$y[n] = \cdots + x[-1]h[n-(-1)] + x[0]h[n-0] + x[1]h[n-1] + \cdots$$

Example 6-1: To practice solving a convolution sum problem, consider a two-sample moving average system with a three-sample input sequence:

$$x[n] = \delta[n] + 2\delta[n-1] + \delta[n-2] \overset{also}{=} \{\underset{\uparrow}{1}, 2, 1\}$$

$$h[n] = \frac{1}{2}\left(\delta[n] + \delta[n-1]\right) \overset{also}{=} \{\underset{\uparrow}{\tfrac{1}{2}}, \tfrac{1}{2}\}$$

The ↑ is a timing mark used to denote where $n = 0$ in the sequences. This sequence notation is particularly convenient for finite duration sequences. Outside the interval shown, you can assume that the sequences are zero.

When the problem you're working on is simple, you can get the answer quickly with a direct attack. In this case, take the second form of the convolution sum and insert $h[n]$:

$$y[n] = \sum_{k=-\infty}^{\infty} h[k]x[n-k] = \sum_{k=-\infty}^{\infty} \frac{1}{2}\left(\delta[k] + \delta[k-1]\right)x[n-k]$$

The doubly infinite sum can be intimidating, but notice that $h[0] = 1/2$, $h[1] = 1/2$, and $h[k] = 0$ for all other values of k. Only two terms of the sum survive, so you can write

$$y[n] = \frac{1}{2}\left(x[n] + x[n-1]\right)$$

In words, the output is the average of the present input, $x[n]$, and $x[n-1]$, which is the input itself, just one sample in the past. For the given input, this becomes

$$y[n] = \frac{1}{2}\{\underset{\uparrow}{1}, 2, 1\} + \frac{1}{2}\{\underset{\uparrow}{0}, 1, 2, 1\} = \frac{1}{2}\{\underset{\uparrow}{1}, 3, 3, 1\}$$

$$= \frac{1}{2}\delta[n] + \frac{3}{2}\delta[n-1] + \frac{3}{2}\delta[n-2] + \frac{1}{2}\delta[n-3]$$

Check to see that the output is indeed a two-sample moving average of the input. The first output (at $n = 0$) is 1/2. The present input is 1, and the previous input is 0, making the average $(0 + 1)/2 = 1/2$. The second output (at $n = 1$) is 3/2. The average of the present and past inputs is $(1 + 2)/2 = 3/2$, the expected result. Note the present input, 2, averaged with the past input, 1, is indeed 3/2. You can check the last two nonzero outputs.

From this example, you can now state that for systems having *finite impulse response* (FIR), $h[n]$ is of the form

$$h[n] = \begin{cases} b_k, & 0 \le k \le N \\ 0, & \text{otherwise} \end{cases}$$

It follows that $y[n] = \sum_{k=0}^{N} b_k \, x[n-k]$.

The FIR nature of $h[n]$ means that the limits in the convolution sum extend through $[0, N]$, rather than $(-\infty, \infty)$. FIR systems, or *filters,* are common in today's electronic applications. The action of the filter is to form the present output as a linear combination of N past inputs and the present input. Is this a causal (non-anticipatory) relationship? Yes, because future inputs aren't used to form the present output.

Simplifying with Convolution Sum Properties and Techniques

To simplify convolution work and avoid careless errors, knowing some fundamental properties and techniques of working with the convolution sum is important. These properties and techniques are similar to those developed for the convolution integral (covered in Chapter 5), but for continuous-time systems, you deal with integrals instead of summations.

Applying commutative, associative, and distributive properties

Do the words *commutative, associative,* and *distributive* properties ring any bells? I don't want to bring back terrifying memories from your past, but the convolution sum obeys these properties, which means you can write the properties given in Table 6-1.

Table 6-1	Convolution Sum Properties
Property	*Formula*
Commutative	$x[n] * h[n] = h[n] * x[n]$
Associative	$(x[n] * h_1[n]) * h_2[n] = x[n] * (h_1[n] * h_2[n])$
Distributive	$(x_1[n] + x_2[n]) * h[n] = x_1[n] * h[n] + x_2[n] * h[n]$
	$x[n] * h_1[n] + x[n] * h_2[n] = x[n] * (h_1[n] + h_2[n])$

The block diagram relationships of Figure 6-3 show the commutative, associative, and distributive properties. Figure 6-3a describes the series (cascade) connection of two systems and follows from the associative property. The impulse response of the cascade is given by $h_{cascade}[n] = h_1[n] * h_2[n]$.

Considering the proof for impulse response properties

Parallel and series connections of LTI systems are an important practical matter. So here's a detailed mathematical proof of these properties for impulse responses. Note that in the case of the cascade, linearity allows the order of the $h_1[n]$ and $h_2[n]$ systems to be freely interchanged.

✔ In the series connection that's rendered in Figure 6-3a, it follows that

$$y[n] = w[n] * h_2[n]$$
$$= \left(x[n] * h_1[n] \right) * h_2[n]$$
$$= x[n] * \underbrace{\left(h_1[n] * h_2[n] \right)}_{h_{cascade}[n]} \overset{also}{=} x[n] * \underbrace{\left(h_2[n] * h_1[n] \right)}_{h_{cascade}[n]}$$

✔ In the parallel connection (shown in Figure 6-3b), it follows that

$$y[n] = x[n] * h_1[n] + x[n] * h_2[n]$$
$$= x[n] * \underbrace{\left(h_1[n] + h_2[n] \right)}_{h_{parallel}[n]}$$

Figure 6-3b shows a parallel connection LTI system and follows from the second line of the distributive property. The impulse response of the parallel connection is given by $h_{parallel}[n] = h_1[n] + h_2[n]$.

Notice that these discrete-time cascade and parallel connections track the continuous-time case exactly (see Chapter 5). Here, the sequence notation indicates that you're working with a discrete-time signal instead of a continuous-time signal.

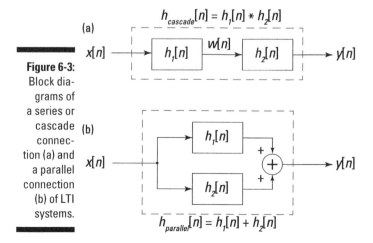

Figure 6-3: Block diagrams of a series or cascade connection (a) and a parallel connection (b) of LTI systems.

Convolving with the impulse function

In some situations, the convolution sum is just plain easy to work. One such situation is when you convolve *anything* (I mean any signal) with a time delayed (shifted) impulse function. You get that *anything* back, only time delayed by the delay factor of the impulse function.

Convolution with the impulse sequence, as in $\delta[n-n_0]*x[n]$, is a freebie — meaning you have no real convolution to do; just evaluate the function at the location of the impulse it's being convolved with and you're done. Check it out (n_0 is constant):

$$\delta[n-n_0]*x[n]= \underbrace{\sum_{k=-\infty}^{\infty} \delta[k-n_0]x[n-k]}_{\text{sifting property}}=x[n-n_0]$$

A simple extension result is $\delta[n-n_0]*x[n-n_1]=x[n-n_0-n_1]$.

Transforming a sequence

For discrete-time signals, you can perform time (sequence) shifting, axis flipping, and superimposing. These operations are useful for aligning multiple signals and for filtering applications, such as convolution. For details on these tasks as they apply to continuous-time signals, check out Chapter 3.

For example, the signal $x[n-n_0]$ is $x[n]$ *shifted* by n_0 samples. For $n_0 > 0$, the shift is to the right, and for $n_0 < 0$, the shift is to the left. The key observation is that by letting $n \rightarrow n-n_0$, you need to move forward in time by an additional n_0 steps to compensate for the $-n_0$ in the argument of $x[\ ]$.

Both forms of the convolution sum formula require *flipping* and *sliding* (shifting) one of the two sequences. Sounds fun, right? This technique is vital to problem solving because no guessing is allowed; you have to pay attention to every detail to get the correct answer.

Flipping a signal is just changing the sign of the argument; that is, $x[n] \rightarrow x[-n]$. The sequence value at $n=0$ is unchanged, but the values on the positive axis are now flipped to the negative axis, and those of the negative axis are flipped to the positive axis. In short, the flipping is with respect to $n=0$. You can shift the flipping point to $n=n_0$ by writing $x[-(n-n_0)]$.

With this information on shifting and flipping, consider the variable change $x[n] \rightarrow x[n-k]$, viewed on the k axis. A convenient way to view the shifted and flipped sequence is to write $n-k=-(k-n)$. Now with respect to the k axis, the sequence is flipped about the point $k=n$. Changing n moves the flipped sequence along the k axis.

Example 6-2: Consider the sequence

$$x[n] = \begin{cases} 8-n, & -2 \le n \le 5 \\ 0, & \text{otherwise} \end{cases}$$

Sketch $x[k]$, $x[-k]$, and $x[n-k]\big|_{n=2}$, using Python. The custom function `ex6_2(k)` contained in `ssd.py`, generates $x[k]$ given a time index array.

Then plot the two sequences, using a k-axis vector running over the interval [−10, 10]. Here's an abbreviated version of the command line code. The plots are given in Figure 6-4.

```
In [800]: import ssd # needed for current session
In [801]: k = arange(-10,10) # k-axis for plotting
In [802]: stem(k,ssd.ex6_2(k)) # plot x[k]
In [804]: stem(k,ssd.ex6_2(-k),'g','go') # plot x[-k]
In [806]: stem(k,ssd.ex6_2(2-k),'r','ro') # plot x[2-k]
```

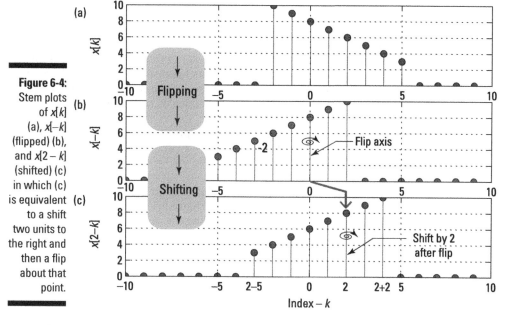

Figure 6-4: Stem plots of $x[k]$ (a), $x[-k]$ (flipped) (b), and $x[2-k]$ (shifted) (c) in which (c) is equivalent to a shift two units to the right and then a flip about that point.

To verify that all is well in the flip and shift, plug the edge k values into $x[n-k]$ to see that the sequence support interval is consistent with the original $x[n]$ sequence:

> ✔ The left or *trailing* edge of $x[k]$ is –2. With shifting and flipping, the index $k = -2 \rightarrow n - k = n + 2$. The flipping action means that the left edge becomes the right edge of $x[n-k]$. For the case $n = 2$, the right edge is thus sitting at $k = 2$. Agreement with Figure 6-4c!
>
> ✔ The right edge of $x[k]$ is +5. With shifting and flipping, $k = 5 \rightarrow n - k \big|_{n=2,\,k=5} = 2 - 5 = -3$. Agreement with Figure 6-4c!

Solving convolution of finite duration sequences

As with the convolution integral of Chapter 5, you solve a convolution sum problem by starting from one of two forms of the sum argument: $x[k]h[n-k]$ or $x[n-k]h[k]$. From there, you sum the product of the two sequences that form the sum argument.

As n increases from $-\infty$ to ∞, $h[n-k]$ slides from left to right along the k-axis. For each value of n, you consider the product $x[k]h[n-k]$ and sum the non-zero or *overlap* intervals along the k-axis. For some problems, you have to consider a lot of different n values, but this is where using a table method is helpful. (I examine this method in the later section "Using spreadsheets and a tabular approach." You can try your hand at the table method in Example 6-5.) For other problems, you find contiguous intervals of n values where you can use the geometric series sum formula (see Chapter 2). For the form $h[k]$ $x[n-k]$, do the same thing, except $x[n-k]$ slides as n changes.

The general solution is piecewise, meaning expressions for $y[n]$ are valid for some $n_i \leq n < n_{ib}$ which corresponds to case i, $i = 1, \ldots, N$. When you put all the pieces together, you have a solution valid for $-\infty < n < \infty$.

Unlike the convolution integral, the convolution sum may have a support interval of a single point. The four steps in Figure 6-5, which parallel the continuous-time scenario of Chapter 5, outline the general solution procedure. Find a table-based method for solving this problem in Example 6-5.

Example 6-3: Consider the convolution of $x[n] = \{1, \underset{\uparrow}{2}, 3\}$ and $h[n] = \{\underset{\uparrow}{2}, 1\}$, using the convolution sum form $y[n] = \displaystyle\sum_{k=-\infty}^{\infty} x[k]h[n-k]$.

Figure 6-6a shows the signal and impulse response to be convolved. Figure 6-6b shows both signals on the k-axis with n as a free variable. In Figures 6-6b through 6-6f, the flipped and shifted sequence $h[n-k]$ is additionally positioned to various overlap cases. And Figure 6g shows the output.

Figure 6-5:
Four steps
to solving
a convolu-
tion sum
problem and
a graphical
depiction of
the *cases*
as $h[n-k]$
slides
(shifts) from
left to right.

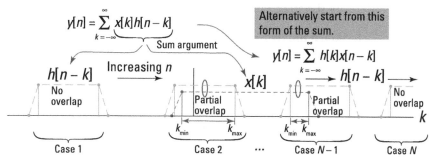

① Sketch the two sequences of the sum argument on the sum variable axis, k.
② Find values of N (cases) where overlap (partial or full) occurs in the sum argument.
③ Establish the sum limits, k_{min} to k_{max}, from the sequence alignment.
④ Sum products $x[k]h[n-k]$ over limits (perhaps use geometric series formulas).

Figure 6-6:
The convo-
lution of $x[n]$
with $h[n]$:
the convolu-
tion inputs
(a), the sum
argument
sequences
$x[k]$ and
$h[n-k]$ (b–f)
under Cases
1–6, respec-
tively, and
the output
$y[n]$ (g).

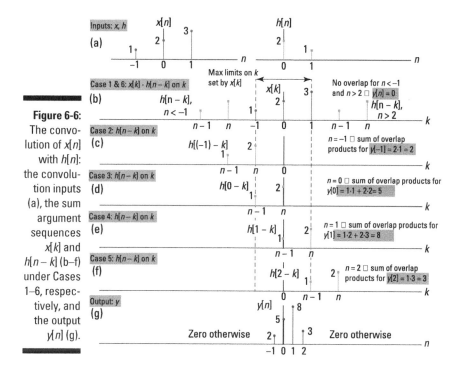

Here are the steps you take to get $y[n]$:

1. **Sketch the two sequences of the sum argument on the sum variable axis, k.**

   In Figure 6-6b, $x[k]$ and $h[n - k]$ are sketched for $n < -1$ and $n > 2$ along the k-axis. In Figures 6-6c through 6-6f, n is chosen to position $h[n - k]$ at all the overlap cases of interest.

2. **Find values of n (cases) where overlap occurs in the sum argument.**

   - Case 1 and Case 6 occur when $n < -1$ and $n - 1 > 1$ or $n > 2$, respectively. For Case 1, $y[n] = 0$ for $n < -1$, and for Case 6, $y[n] = 0$ for $n > 2$.

   - Case 2 is what I call *partial overlap leading edge*. This occurs when $n = -1$.

   - Case 3 is *full overlap* when $n = 0$.

   - Case 4 is *full overlap* when $n = 1$.

   - Case 5 is *partial overlap trailing edge* when $n = 2$.

3. **Establish the sum limits from the sequence alignment.**

   With the help of Figure 6-6, the sum limits lie at most on the interval $[-1, 1]$ but are less than this for the partial overlap cases.

   - Under Case 2, the limit on k is simply $k = -1$.

   - Under Case 3, the limits on k are -1 to 0.

   - Under Case 4, the limits on k are 0 to 1.

   - Under Case 5, the limit on k is simply 1.

4. **Form the sum of products between the two sequences, and then move on.**

   - The Case 2 sum of overlap products is $y[-1] = 2(1) = 2$.

   - The Case 3 sum of overlap products is $y[0] = 1(1) + 2(2) = 5$.

   - The Case 4 sum of overlap products is $y[-1] = 1(1) + 2(3) = 8$.

   - The Case 5 sum of overlap products is $y[-1] = 1(3) = 3$.

Example 6-4: In this example, I build on the pattern established by Example 6-2 to show you the support interval relationships for the convolution of two generic finite length sequences $x[n]$ and $h[n]$ (as shown in Figure 6-7).

In Figure 6-7, the support interval is displayed as a generic rectangle (dotted lines) for convenience. The reality is that any nonzero sequence value can appear on the closed intervals $[n_1, n_2]$ and $[n_3, n_4]$. Here, $x[n]$ has duration N_x and $h[n]$ has duration N_h.

Figure 6-7:
Convolving
two generic
finite length
sequences:
the convolu-
tion inputs
(a), the sum
argument
sequences
$x[k]$ and
$h[n-k]$
along the
k-axis (b),
and the out-
put $y(t)$ (c).

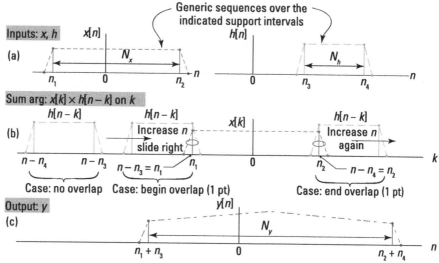

In general you can call the duration of $x[n]$ N_x and, similarly, for $h[n]$, the duration is N_h. The subscript notation is commonly used in signals and systems; N is used for sequence duration (N = number of samples in a sequence, sequence length, and so forth). Adding the subscript clarifies that you're talking about the duration of a specific sequence.

Now you can utilize your flipping and sliding skills to discover the support interval for $y[n]$ in this general problem, where

$$y[n] = \sum_{k=-\infty}^{\infty} x[k]h[n-k] \neq 0$$

You must evaluate the sum for each value of independent variable n. Nonzero results can occur only when the product $x[k]h[n-k] \neq 0$ and terms don't cancel. This analysis considers only the former situation, so the result is the largest possible support interval, barring any sum term cancellations to zero.

The following steps walk you through the process of determining the support interval of the convolution sum. Verifying the support interval *before* crunching the numbers of the convolution sum itself is important and helps you avoid having to do the whole thing over again. **Note:** The actual convolution sum isn't computed in this example.

1. **Sketch the two sequences of the sum argument on the sum variable axis, k.**

   Figure 6-7b shows a sketch of the two waveforms on the k-axis.

2. **Find values of n (cases) where overlap occurs in the sum argument.**

   Refer to the example in Figure 6-7: As $h[n-k]$ slides to the right for increasing n, overlap and the first nonzero value in the product occur when $n - n_3 = n_1$ or $n = n_1 + n_3$ — when the leading edge of $h[n-k]$ just touches the trailing edge of $x[k]$. The last nonzero product occurs when the trailing edge of $h[n-k]$ is just touching the leading edge of $x[k]$. Mathematically, the condition occurs when $n - n_4 = n_2$ or when $n = n_2 - n_4$.

3. **Establish the sum limits from the sequence alignment.**

   The sum limits are set by the specific cases, so they don't apply in this example. In a gross sense, all you can say is that the fixed sequence $x[k]$ constrains the limits on k to be at most $[n_1, n_2]$.

4. **Form the sum of products between the two sequences, and then move on.**

   You don't carry out the sum of products in this example.

This exercise reveals the following information about the properties or characteristics of the convolution of two finite duration sequences:

- ✔ The nonzero output sequence $y[n]$ begins at the sum of the $x[n]$ and $h[n]$ sequence starting points: $n = n_1 + n_3$.

- ✔ The output sequence $y[n]$ stops at the sum of the $x[n]$ and $h[n]$ sequence ending points: $n = n_2 + n_4$.

- ✔ The output sequence $y[n]$ has maximum support duration that's equal to the sum of the durations of $x[n]$ and $h[n]$ minus 1: $y[n]$ is nonzero on at most $n_1 + n_3 \le n \le n_2 + n_4$, so

$$N_y = \underbrace{(n_2 + n_4)}_{\text{stop}} - \underbrace{(n_1 + n_3)}_{\text{start}} + 1$$
$$= \underbrace{(n_2 - n_1 + 1)}_{N_x} + \underbrace{(n_4 - n_3 + 1)}_{N_h} - 1$$
$$= N_x + N_h - 1$$

As a bonus, the results for finite duration sequences easily extend to semi-infinite duration sequences (check out Examples 6-6 and 6-7).

- ✔ If $x[n]$ has support $[0, \infty)$ and $h[n]$ has support $[5, 20]$, then the output sequence $y[n]$ support interval begins at $0 + 5 = 5$ and ends at $\infty + 20 = \infty$. The support interval for $y[n]$ is $[5, \infty)$ and the duration is $\infty - 5 + 1 = \infty$.

- ✔ If $x[n]$ and $h[n]$ have support intervals $(-\infty, \infty)$ and $[0, 10]$ respectively, then the support interval for $y[n]$ is $(-\infty, \infty)$. Know why? Minus infinity plus 0 is still minus infinity, and infinity plus 10 is still infinity.

In terms of actual convolution, there is always overlap between $x[n]$ and some nonzero values of $h[n]$.

Working with the Convolution Sum

In this section, I help you develop practical hands-on skills for solving convolution sum problems of all types. The first approach is table-based and intuitive. The second is more of a direct analysis attack, which works well when the sum expressions under various *cases* are actually summable in closed form, as when using geometric series sum formulas (see Chapter 2).

Using spreadsheets and a tabular approach

Example 6-5: A tabular approach works well for $x[n]$ and $h[n]$ finite duration sequences. Consider these sequences:

$$x[n] = \{\underset{\uparrow}{1}, 2, 3, 1\}, \quad h[n] = \{1, \underset{\uparrow}{2}, 1, -1\}$$

Decisions decisions . . . which form of the convolution sum should you use? The choice really comes down to comfort factor. You need to flip and slide one or the other sequence and do it without error. Choosing one over the other offers some flexibility. For me, I generally choose to flip and slide the least complex of the two sequences. You may see things differently and choose always to flip and slide the input $x[n]$. Here, I choose

$$y[n] = \sum_{k=-1}^{2} h[k]x[n-k]$$

Notice that the sum limits have already been constrained to the support interval for $h[k]$: $(-\infty, \infty) \to [-1, 2]$.

Here's a step-by-step solution and spreadsheet check:

1. **Sketch the two sequences of the sum argument on the sum variable axis, k.**

   Get your bearings by making a quick sketch of $h[k]$ stacked over $x[n-k]$, such as the sketch in Figure 6-8.

2. **Find values of n (cases) where overlap occurs in the sum argument.**

   Case 1 in Figure 6-8 occurs when $n < -1$; there's no overlap. As $x[n-k]$ slides to the right for increasing n, overlap (and the first nonzero value in the product) occurs when $n = -1$ — when the leading edge of $x[n-k]$ just touches the trailing edge of $h[k]$ at a single point. The last nonzero product occurs when $n - 3 = 2$ or $n = 5$ — when the trailing edge of $h[n-k]$ is just touching the leading edge of $x[k]$.

Cases 2 through 8 correspond to n stepping from –1 to 5. The table described later in Step 4 manages the seven overlap cases.

3. **Establish the sum limits from the sequence alignment.**

   With the help of Figure 6-8, the sum limits are visible for each n (case) where overlap occurs. For example, $n = 2$ has full overlap, so you sum the $h[k]x[n-k]$ products from $k = -1$ to $k = 2$. The table described in Step 4 can help you visualize the limits for all the overlap cases.

4. **Form the sum of products between the two sequences, and then move on.**

   You set up a table (such as the one in Figure 6-9) to house the $x[n-k]$ and $h[k]$ values for each case, identify the sum limits, and perform the sum of products as n steps from –1 to 5 — truly the heart of the convolution sum. You use the sequence plots of $h[n]$ and $x[n-k]$ found in Figure 6-8 to make the table entries. You fill in entries for $h[k]$ in the top center of the table and fill in $x[n-k]$ in the rows below $h[k]$ at the needed shift values for n. Using a spreadsheet table is optional; the table is more of a "housekeeping" means, so I recommend at minimum a blank table paper worksheet (great for your next quiz). The spreadsheet just automates the math, so it isn't a requirement.

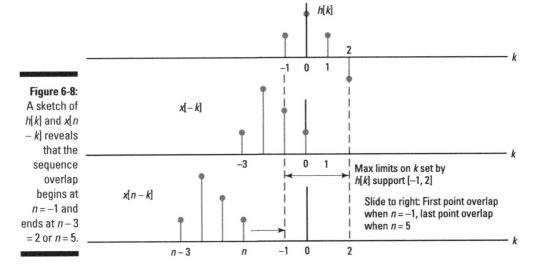

Figure 6-8: A sketch of $h[k]$ and $x[n-k]$ reveals that the sequence overlap begins at $n = -1$ and ends at $n - 3 = 2$ or $n = 5$.

Max limits on k set by $h[k]$ support [–1, 2]

Slide to right: First point overlap when $n = -1$, last point overlap when $n = 5$

Consider the case $n = 2$ as highlighted in Figure 6-9:
$$y[2] = 1\cdot1 + 2\cdot3 + 1\cdot2 + (-1)\cdot1 = 8.$$

Sum of products formula for case $n = 2$

O7 f_x =F$2*F7+G$2*G7+H$2*H7+I$2*I7

	A	B	C	D	E	F	G	H	I	J	K	L	M	N	O
1	n							h[k] for k on [-1, 2]							
2		x[n-k] as n increases				1	2	1	-1						y[n]
3	-2	1	3	2	1	0	0	0	0	0	0	0	0	0	0
4	-1	0	1	3	2	1	0	0	0	0	0	0	0	0	1
5	0	0	0	1	3	2	1	0	0	0	0	0	0	0	4
6	1	0	0	0	1	3	2	1	0	0	0	0	0	0	8
7	2	0	0	0	0	1	3	2	1	0	0	0	0	0	8
8	3	h[k]x[n − k] sum				0	1	3	2	1	0	0	0	0	3
9	4	of products for				0	0	1	3	2	1	0	0	0	-2
10	5	case n = 2				0	0	0	1	3	2	1	0	0	-1
11	6	0	0	0	0	0	0	0	1	3	2	1	0	0	0
12		k=-5	k=-4	k=-3	k=-2	k=-1	k=0	k=1	k=2	k=3	k=4	k=5	k=6	k=7	

n index   h[k]x[n − k] overlap region in   y output
                              convolution sum

Figure 6-9: Convolution sum calculation for finite length sequences fit into a spreadsheet table.

As a check, see whether your results here agree with the bullet items found in Step 4 of Example 6-3. In particular, check that the first nonzero output is at the sum of the sequence starting points to find that $n = -1 + 0 = -1$. Now check that the last nonzero point is the sum of the sequence ending points to find that $n = 2 + 3 = 5$. The start and stop points agree with the table results.

TIP

5. **As a final check, use IPython and the custom function `conv_sum()` to numerically convolve the two sequences and keep track of the time index n for plotting (find the function in the code module `ssd.py`).**

   The function makes use of the `scipy` subpackage `signal` for the underlying convolution function. The array `x` holds the $x[n]$ values, and the array `nx` holds the corresponding time index. The array `h` holds the $h[n]$ values, and the array `nh` holds the corresponding time index.

   Here are the IPython commands to produce a numerical output:

```
In [418]: y,ny = ssd.conv_sum([1,2,3,1],[0,1,2,3],[1,2,1,-
            1],[-1,0,1,2]) # input x, nx, h, nh
In [419]: ny
Out[419]: array([-1, 0, 1, 2, 3, 4, 5]) # index values
In [420]: y
Out[420]: array([ 1, 4, 8, 8, 3, -2, -1])# Output y checks
```

Attacking the sum directly with geometric series

Purely analytical solutions provide valuable insights in the design of LTI systems. In the context of the convolution sum, consider the examples in this section. They're all detailed in nature, and I use geometric series sum formulas to simplify solutions into a compact piecewise form. The goal in these examples, by the way, is to find $y[n] = x[n] * h[n]$.

Example 6-6: Consider a rectangular pulse sequence of duration N samples for both $x[n]$ and $h[n]$: $x[n] = h[n] = u[n] - u[n - N]$, using the convolution sum form $y[n] = \sum_{k=-\infty}^{\infty} x[k]h[n-k]$.

To avoid careless errors in your preliminary analysis, check the support interval for the output, using the starting and ending point rules established in Example 6-3. The output starts at $n = 0 + 0 = 0$ and concludes at $n = (N-1) + (N-1) = 2N - 2$. These values guide the remaining steps.

Now follow the steps from Figure 6-5 to get $y[n]$:

1. **Sketch the two sequences of the sum argument on the sum variable axis, k.**

   Figure 6-10a sketches $x[k]$ and Figures 6-10b through 6-10e sketch $h[n - k]$ for five different cases of overlap.

Figure 6-10: The convolution of $x[n]$ with $h[n]$: the convolution input $x[k]$ (a), the input $h[n-k]$ (b–e) flipped and shifted to illustrate Cases 1–4.

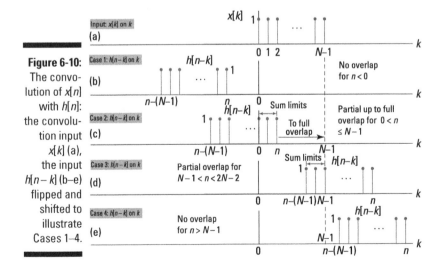

2. **Find values of n (cases) where overlap occurs in the sum argument.**

   - Case 1 and Case 4 occur when $n < 0$ and $n - (N-1) > N-1$ or $n < 2N-2$, respectively. For Case 1, $y[n] = 0$ for $n < 0$; for Case 4, $y[n] = 0$ for $n < 2N-2$.

   - Case 2 is partial overlap of the leading edge leading to full overlap. This occurs when $n \geq 0$ and $n - (N-1) < 0$ or $n < N-1$. An interval, as opposed to a single point, works here because the sequence product is simply $1 \times 1 = 1$.

   - Case 3 is partial overlap of the trailing edge that occurs when $0 < n - (N-1) \leq N-1$ or $N-1 < n \leq 2N-2$.

3. **Establish the sum limits from the sequence alignment.**

   With the help of Figure 6-9, you can identify the sum limits.

   - Under Case 2, the limits on k are 0 to n.

   - Under Case 3, the limits on k are $n - (N-1)$ to $N-1$.

4. **Form the sum of products between the two sequences, and then move on.**

   The Case 2 sum of overlap products is

   $$y[n] = \sum_{k=0}^{n} 1 \cdot 1 = \underbrace{n+1}_{\text{points in sum}}, \ 0 \leq n \leq N-1$$

   Note that when $n = N-1$ (full overlap) the output achieves the maximum output value of N.

   The Case 4 sum of overlap products is

   $$y[n] = \sum_{k=n-(N-1)}^{N-1} 1 \cdot 1 = \overbrace{\underbrace{(N-1)}_{\text{upper limit}} - \underbrace{[n-(N-1)]}_{\text{lower limit}} + 1}^{\text{total number of points}}$$

   $$= 2N - 1 - n, \ N-1 < n \leq 2N-2$$

   Find the final answer by counting the number of ones being added together. The sum is the upper limit minus the lower limit plus one.

   Finally, you can summarize (whew!) $y[n]$ into a compact piecewise form:

   $$y[n] = \begin{cases} n+1, & 0 \leq n \leq N-1 \\ 2N-1-n, & N-1 < n \leq 2N-2 \\ 0, & \text{otherwise} \end{cases}$$

In Chapter 5, an exercise in convolving two rectangle pulses (continuous-time) led to a triangle pulse. It may not be obvious from the piecewise form of $y[n]$, but you should expect to have a similar result here. To see for yourself, plot

$y[n]$ by hand or use a tool. I use the Python function `con_sum()` to convolve two rectangle sequences, using the step function `dstep()` (you can find both functions in the code module `ssd.py`, which you must `import` into the IPython environment). The result is shown in Figure 6-11 and is triangular in shape, just as expected!

The IPython commands to produce the plot of Figure 6-11 are

```
In [187]: import ssd # to access module functions
In [188]: n = arange(-5,20) # time axis
In [189]: x = ssd.dstep(n) - ssd.dstep(n-10)# N=10 pulse
In [190]: y, ny = ssd.conv_sum(x, n, x, n) # x conv x
In [191]: stem(ny,y) # plot output
```

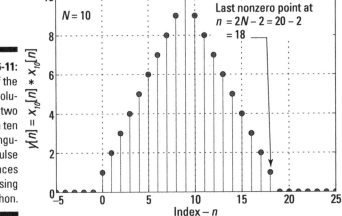

Figure 6-11: A plot of the convolution of two length ten rectangular pulse sequences using Python.

Example 6-7: Consider two semi-infinite duration sequences: a step function input $x[n] = u[n]$ and an exponential impulse response $h[n] = a^n u[n]$, where a is taken as a real constant with magnitude less than one.

I introduce these discrete-time signal types in Chapter 4 and explain that the function $u[n]$ is the unit step function, which is 1 for $n \geq 0$ and 0 for $n < 0$. From a systems-modeling standpoint, this example requires you to find the step response of the exponential decay system. (See the section "Connecting the step response and impulse response," later in this chapter, to explore the relationship between the impulse response and the step response.)

On to the solution! Use direct substitution into the convolution sum formula:

$$y[n] = \sum_{k=-\infty}^{\infty} h[k]x[n-k] = \sum_{k=-\infty}^{\infty} a^k u[k]u[n-k]$$

This form of the convolution sum, I think, makes it easier to flip and shift $u[n]$.

As preliminary analysis, the support interval for $y[n]$ is $[0+0, \infty+\infty] = [0, \infty]$.

With the preliminary analysis complete, you can jump right into the detailed step-by-step analysis:

1. **Sketch the two sequences of the sum argument on the sum variable axis, k.**

   Figure 6-12a sketches $h[k]$ and Figures 6-12b and 6-12c sketch $x[n-k]$ for two different cases of overlap.

Figure 6-12:
The convolution of $x[n]$ with $h[n]$: the convolution input $h[k]$ (a) and the input $x[n-k]$ (b and c) flipped and shifted to illustrate Cases 1 and 2.

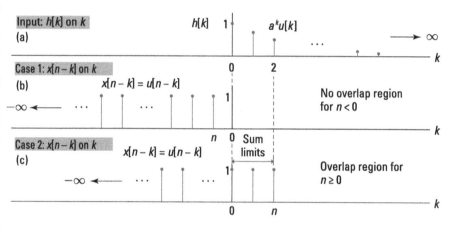

2. **Find values of n (cases) where overlap occurs in the sum argument.**

   • Case 1 when $n < 0$ is no overlap so $y[n] = 0$.

   • Case 2 occurs when $n \geq 0$.

3. **Establish the sum limits from the sequence alignment.**

   With the help of Figure 6-12, you can identify the sum limits. Under Case 2, the limits on k are 0 to n.

4. **Form the sum of products between the two sequences, and then move on.**

   Case 2 offers a bit of excitement — the limits and sum argument lead to a geometric series–based solution:

   $$y[n] = \sum_{k=0}^{n} 1 \cdot a^k \quad \overset{\text{finite geometric series}}{=} \quad \frac{1-a^{n+1}}{1-a}, \; n \geq 0$$

The summation form is picture perfect: a is the geometric series ratio. The series sums to 1 plus the ratio raised to one power higher than the upper sum limit all over the quantity 1 minus the ratio (find details on geometric series in Chapter 2).

Now you can combine the solution into a single expression by using the unit step function:

$$y[n] = \frac{1-a^{n+1}}{1-a}u[n]$$

Figure 6-13 illustrates the general form of $y[n]$. The form of the step response is a charge-up that starts from 1 at $n = 0$ and reaches a steady-state value of $1/(1-a)$. Why? For $|a| < 1$, large n has the quantity $a^{n+1} \to 0$. This response is similar to the continuous-time step response of $h(t) = e^{-at}u(t)$ that I describe in Chapter 5.

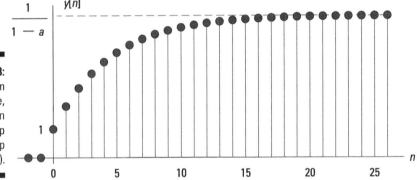

Figure 6-13: The system response, y[n], for an input step (the step response).

Example 6-8: Consider an N-sample rectangular pulse input of the form $x[n] = u[n] - u[n-N]$ and a system with exponential impulse response $h[n] = a^n u[n]$. This is a classic signal/impulse response pair-up.

You have options here! Brute-force plugging into the convolution sum works, but you can discover alternative methods from Example 6-7. As preliminary analysis, the support interval for $y[n]$ is $[0+0, N-1+\infty] = [0, \infty]$.

Here's an alternative method, which utilizes basic convolution sum properties:

1. **Use the distributive property of convolution to decompose $y[n]$ into two terms by first substituting $x[n] = u[n] - u[n - N]$:**

$$y[n] = x[n] * h[n]$$
$$= (u[n] - u[n-N]) * h[n]$$
$$= u[n] * h[n] - u[n-N] * h[n]$$
$$= u[n] * a^n u[n] - (u[n] * a^n u[n])\Big|_{n \to n-N}$$

The third line follows from the distributive property, and the last term of the last line follows from time invariance.

2. **Evaluate the convolution** $u[n] * a^n u[n]$ **that remains (both terms of the last line).**

This convolution is worked in Example 6-6, so now you just need to package up the pieces for this scenario:

$$y[n] = \frac{1-a^{n+1}}{1-a} u[n] - \frac{1-a^{n+1}}{1-a} u[n]\Big|_{n \to n-N}$$

$$= \frac{1-a^{n+1}}{1-a} u[n] - \frac{1-a^{n-N+1}}{1-a} u[n-N]$$

TIP

You may be satisfied with the mathematical form you achieve in Step 2, but you get a different form of the same solution if you use the head-on convolution approach. It's always good to see how additional manipulation can yield alternative (and sometimes more insightful) forms.

To find the alternative form, start by noting that for $0 \le n < N$, only the first term is active; for $n \ge N$, both terms are active, which means

$$y[n] = \frac{\cancel{1} - a^{n+1} - (\cancel{1} - a^{n-N+1})}{1-a} = \frac{a^{n-N+1}(1-a^N)}{1-a}, \, n \ge N$$

In summary, the piecewise solution for $y[n]$ is

$$y[n] = \begin{cases} 0, & n < 0 \\ \dfrac{1-a^{n+1}}{1-a}, & 0 \le n < N \\ \dfrac{a^{n-N+1}(1-a^N)}{1-a}, & n \ge N \end{cases}$$

The analysis reveals that as the rectangular pulse passes through the system, it goes through a charge-up phase (line 2) and then a discharge phase (line 3).

To see this graphically, write a Python function or simulate the output by using the convolution sum function used in Example 6-4 (be sure to import the ssd.py code module at command prompt when you start your IPython session). Using the simulation approach, first create the sequences $x[n]$ and $h[n]$. The input, impulse response, and output sequences are created directly at the IPython command prompt. In the following example, I use subplots to stack all three sequences, but to save space, I don't show the plot labeling commands.

```
In [211]: import ssd # to access module functions
In [212]: nx = arange(0,10) # generate x[n]
In [213]: x = ssd.dstep(n) - ssd.dstep(n-5)
In [214]: nh = arange(0,15) # generate h[n]
In [215]: h = exp(-0.8*nh)  # set a = 0.8
In [216]: y, ny = ssd.conv_sum(x,nx,h,nh) # convolve
```

See Figure 6-14 for a complete picture of all three sequences with the charge and discharge intervals evident.

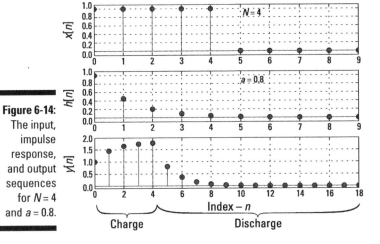

Figure 6-14:
The input,
impulse
response,
and output
sequences
for $N = 4$
and $a = 0.8$.

The action of the impulse response, which behaves as a low-pass filter when $0 < a < 1$, is to slow down the edges of the rectangular pulse (refer to Figure 6-14). By *slow down*, I mean that when compared to the input $x[n]$, which jumps up and down in one sample, the output $y[n]$ slowly rises up starting at 0 and slowly falls back down starting at $n = 4$. Depending on the application, the filtering action in this example may be exactly what the design requires.

Think of the pulse signal as a simplified digital signaling waveform, such as an activation signal for opening a garage door. The system parameter a can be chosen to act as a sequence smoothing filter by choosing $0 \leq a < 1$. Figure 6-14 illustrates this.

When you press the garage door button, it may bounce between zero and one before settling at one. With the filter, you need to hold the button down long enough (the pulse length N) for the filtered signal to rise above an amplitude threshold that toggles the door to either open or close. The filter reduces the chance of false triggering.

Example 6-9: Consider the cascade of two LTI systems, as shown in Figure 6-15.

Figure 6-15:
Finding the
impulse
response of
a cascade
of two LTI
systems.

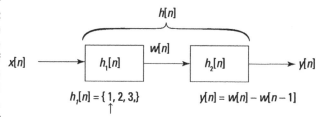

Next, I tell you how to find the impulse response of the cascade by using the table method, first shown in Example 6-5.

> ✔ The impulse response you seek, $h[n]$, is produced by $h_1[n] * h_2[n]$ (refer to Figure 6-3a).
>
> How can you find $h_2[n]$? Well, $h_2[n]$ is the impulse response of the second system. So if you force the input to the second system (called $w[n]$ in Figure 6-15) to be $\delta[n]$, the output $y[n]$ will be the impulse response, $h_2[n]$, by definition. Now, because you're given $\delta[n-n_0] * x[n]$, the resulting impulse response is $h_2[n] = \delta[n] - \delta[n-1]$. Now you have what you need to start crunching those numbers!
>
> *Note:* This impulse response is an example of a simple difference equation (see Chapter 7 for the lowdown on simple difference equations).
>
> ✔ The table in Figure 6-16 is set up to number crunch the answer, using the sum argument $h_1[k]h_2[n-k]$. From the support intervals of $h_1[n]$ and $h_2[n]$, the support interval for $h[n]$ is $[0 + 0, 2 + 1] = [0, 3]$. As you work with the table, keep these limits in mind.

Sum of products formula for case $n = 0$

N4 f_x =F$2*F4+G$2*G4+H$2*H4

Figure 6-16:
Spread-
sheet
table for
convolving
the impulse
responses
of the
cascade.

	A	B	C	D	E	F	G	H	I	J	K	L	M	N
1	n						h1[k] for k on [0, 2]							
2		h2[n-k] as n increases				1	2	3						h[n]
3	-1	0	0	-1	1	0	0	0	0	0	0	0	0	0
4	0	0	0	0	-1	1	0	0	0	0	0	0	0	1
5	1	0	0	0	0	-1	1	0	0	0	0	0	0	1
6	2	*h1[k]h2[n−k]*			0	0	-1	1	0	0	0	0	0	1
7	3	sum of products			0	0	0	-1	1	0	0	0	0	-3
8	4	for case $n = 0$			0	0	0	0	-1	1	0	0	0	0
9		k=-4	k=-3	k=-2	k=-1	k=0	k=1	k=2	k=3	k=4	k=5	k=6	k=7	

Index   *h[k]x[n − k]* overlap region in   Output
convolution sum

Connecting the step response and impulse response

The step response is the output of an LTI system when driven by input $x[n] = u[n]$. To visualize that statement, consider the graphical depiction of Figure 6-17.

Figure 6-17:
Graphical
depiction
of the step
response.

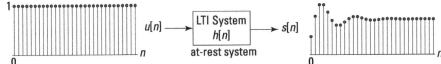

Figure 6-17:
Graphical
depiction
of the step
response.

The step response, denoted as $s[n]$, is the system output for an at-rest system with input $u[n]$. You can establish the connection between the impulse response and step response rather easily. If you set $n_0 = 0$ and $x[n] = u[n]$ in the formula $u[n] = \delta[n] * u[n]$ (introduced in the section "Convolving with the impulse function," earlier in this chapter), it follows that $s[n] = h[n] * u[n] = \sum_{k=-\infty}^{\infty} h[k]u[n-k] = \sum_{k=-\infty}^{n} h[k].$

This may seem like a lowly result, but if you consider $u[n]$ (the unit step function) as the impulse response of an LTI system, the result is significant. Check out the upper path of Figure 6-18, where a cascade of a $u[n]$ system with the $h[n]$ system produces the step response when the cascade input is an impulse function.

Figure 6-18:
Block
diagram
explana-
tion of the
impulse
response
to step
response
transforma-
tion.

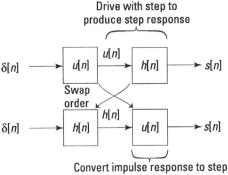

Drive with step to
produce step response

Convert impulse response to step
response

The $u[n]$ system output is a step function. The commutative property of convolution says that the order of the cascade can be swapped, so you arrive at the second pathway in Figure 6-18, which shows $s[n]$ is achieved from the impulse response by convolving $h[n]$ with $u[n]$.

Here's how to connect the impulse response to the step response and the inverse:

1. **Expand the convolution between the impulse response and the step response:**

$$s[n] = h[n]*u[n] = \sum_{k=-\infty}^{\infty} h[k]u[n-k] = \sum_{k=-\infty}^{n} h[k]$$

2. **Set the upper sum limit to n by flipping and shifting the unit step, $u[n-k]$.**

3. **Find the inverse relationship with a simple difference operation:**

$$h[n] = \sum_{k=-\infty}^{n} h[k] - \sum_{k=-\infty}^{n-1} h[k] = s[n] - s[n-1]$$

A system having $u[n]$ as the impulse response is known as an *accumulator*. In transforming the impulse response to the step response, you witness the accumulator in action. The accumulator output is the accumulation, or sum, of all past inputs up to the present.

Example 6-7 revealed the step response of a system having exponential decay $h[n] = a^n u[n]$ if $|a| < 1$. With that information, you can find the step response through this relationship between the impulse and step responses:

$$|y[n]| < \infty$$

It works! Flip back to Example 6-6 to see for yourself.

Checking the BIBO stability

Bounded-input bounded-output (BIBO) stability for discrete-time and continuous-time systems guarantees that the output remains bounded ($|x[n]| < \infty$) for any bounded ($\sum_{n=-\infty}^{\infty} |h[n]| < \infty$) input. A system is BIBO stable if, for any bounded input, the output is also bounded.

When the system is LTI, BIBO stability follows if

$$\left| y[n] \right| < \left| \sum_{k=-\infty}^{\infty} h[n]x[n-k] \right| \le \sum_{k=-\infty}^{\infty} \left| h[n] \right| \underbrace{\left| x[n-k] \right|}_{<B_x}$$

$$< B_x \sum_{k=-\infty}^{\infty} \left| h[n] \right| < \infty$$

The proof is straightforward:

$$B_x < \infty$$

Because $\sum_{n=-\infty}^{\infty} \left| h[n] \right| < \infty$ if the input is bounded, the output is bounded if $\sum_{n=-\infty}^{\infty} \left| h[n] \right| < \infty$. This parallels the integral formulation for BIBO stability in Chapter 5.

Example 6-10: Consider the exponential impulse response $h[n] = a^n u[n]$. The LTI test for BIBO stability yields the following:

$$S = \sum_{n=-\infty}^{\infty} \left| a^n u[n] \right| = \sum_{n=0}^{\infty} \left| a \right|^n \overset{?}{<} \infty$$

When a meets the following conditions, the sum is finite:

- Case $|a| < 1$: From the infinite geometric series sum formula (covered in Chapter 2), you find that $\sum_{n=0}^{\infty} r^n = 1/(1-|r|), |r| < 1$, so $S = 1/(1-|a|) < \infty$.

- Case $|a| \ge 1$: Under this condition, the sum diverges: $S \to \infty$.

  Therefore, $|a| \ge 1$ results in an unstable system.

This example also shows that $a = 1$, which implies $h[n] = u[n]$ isn't BIBO stable. But don't let this worry you too much. Accumulators, which have $h[n] = u[n]$, are still useful building blocks. The key is how they're utilized in a composite system.

Checking for system causality

The key concept of causality for discrete-time systems is that present output values can rely on only past and present input values. Future input values have no influence on the present output. (Flip to Chapter 4 for more info on causality related to discrete-time systems.) For LTI systems, causality requires that the impulse response $h[n]$ is 0 for $n < 0$. Figure 6-19 shows that only $h_b[n]$ is causal; both $h_a[n]$ and $h_c[n]$ are nonzero for $n < 0$.

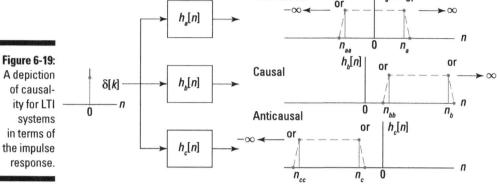

Example 6-11: Consider the following scenarios

✔ Suppose that $h_1[n] = u[n + 5] - u[n - 10]$. From the mathematical description, the impulse response turns on at $n = -5$ and turns off at $n = 9$. Five samples of the impulse response violate the LTI system causality requirement, so the system is non-causal.

✔ Say that $h_2[n] = 0.7^n u[n-4] + 2\delta[n+1]$. The first term works because the step function doesn't turn on until $n \geq 4$. The second term is an isolated sample sitting at $n = -1$, thus this system is non-causal.

✔ Think about $h_3[n] = h_2[n - 1]$. If you move $h_2[n]$ one sample to the right, all is well — $h_3[n] = 0$ for $n < 0$. This system is causal.

Seeing zero's importance

To see why $h[n] = 0$ for $n < 0$ follows for a causal system, I break the convolution sum into two terms. The first term involves future values of the input, and the second term uses only past and present values of the input:

$$y[n] = \sum_{k=-\infty}^{\infty} h[k]x[n-k] = \underbrace{\sum_{k=-\infty}^{-1} h[k]x[n-k]}_{\text{uses future inputs}} + \underbrace{\sum_{k=0}^{\infty} h[k]x[n-k]}_{\text{uses past and present inputs}}$$

For a causal system, the contribution from the first term must be zero. It follows from the sum limits of the first term and the presence of $h[k]$ in the sum argument that $h[k]$ must be zero for $k < 0$.

Chapter 7

LTI System Differential and Difference Equations in the Time Domain

- -

- -

A special class of LTI systems contains systems that have linear constant coefficient (LCC) differential or difference equation representations in the continuous- or discrete-time domains, respectively. (Find general information about LTI systems in the time domain in Chapters 5 and 6.)

One thing to know right off the bat here is that the abbreviation LCCDE often applies to both the continuous- and discrete-domain systems, but *differential equations* aren't the same as *difference equations*. To lessen the confusion, I use the acronym LCC to shorten *linear constant coefficient* in this chapter and then spell out whether I'm talking about a differential equation system or a difference equation. If in doubt when you're reading other engineering materials, it's probably safe to bet that LCCDE, if not otherwise defined, is referring to linear constant coefficient *difference* equations.

I'm guessing that you've taken a course in differential equations (check out *Differential Equations For Dummies* if you need a refresher on this topic), and I assume that you've had little to no reason to think much about difference equations. You can find both system types in your music player and cellphone, but here's the deal: The LCC differential equation system has been the mainstay of electrical engineers for a long time, and it's easy to build and manage; just wire up a circuit composed of resistors, capacitors, and inductors, and you have an LCC differential equation system that runs in real time without a computer.

The LCC difference equation is a different story. Although it's found in mathematics, science, certain branches of engineering, and business, LCC difference equation systems are relatively new to the signals and systems world. Plus you need a computer to implement the LCC difference equation for signal-processing applications.

In this chapter, I focus on the time-domain solution of LCC differential and difference equations under sinusoidal steady-state conditions. For the case of LCC difference equations, I also show you how the LCC difference equation implementation of a discrete-time LTI system allows for the instantiation of an efficient processing algorithm. (To see how a digital filter can be designed to remove interference in an audio stream being played back on a computer, check out the real-world case studies at www.dummies.com/extras/signalsandsystems.)

Getting Differential

The study of signals and systems requires exposure to differential equations because they allow you to mathematically model continuous-time systems and find out how they respond to certain signals. The natural setting for LCC differential equations is analog/continuous-time electronic circuits.

These systems also show up in mechanical, chemical, biological, and other sciences to model time-varying signals and LTI systems. The other science disciplines may not seem relevant for an electrical engineer, but electronic circuitry sometimes takes measurements and/or controls operations in other domains. For example, a hybrid electromechanical system, such as the cruise control on a car, may have an LCC differential equation governing its operation. The system designer needs to model the complete system to ensure that the resulting product will meet performance requirements, including safety standards.

Solving LCC differential equations in all but the simplest cases requires powerful mathematics. In this section, I restrict my attention to the sinusoidal steady-state solution. I analyze the LCC differential equation in more detail in the frequency/Fourier domain in Chapter 9 and show you how to find general time-domain solutions by using the Laplace transform in Chapter 13.

Introducing the general Nth-order system

An order N LCC differential equation has this mathematical representation:

$$\sum_{k=0}^{N} a_k \frac{d^k y(t)}{dt^k} = \sum_{k=0}^{M} b_k \frac{d^k x(t)}{dt^k}$$

where $x(t)$ is the input and $y(t)$ is the output. The model order N is the highest derivative present in the output $y(t)$.

The coefficient sets of $\{a_k\}$ and $\{b_k\}$ control the precise behavior of the model for given N and M values. The coefficients *completely* describe the system; you find N and M from the highest order derivatives of the output and input variables, respectively. Lower order coefficients could be zero.

Formally, because the input and its derivatives are present in this equation, it's called a *nonhomogeneous differential equation*. If $x(t)$ and its derivatives are 0, you have the corresponding homogeneous equation. The right side is 0 in the homogeneous equation.

Short-hand notation for the derivative terms are a *superscript* or a prime symbol:

$$y^{(k)}(t) = \frac{d^k y(t)}{dt^k} \quad \text{or} \quad y'(t) = \frac{dy(t)}{dt}$$

You can get the complete solution to the nonhomogeneous differential equation by using the Laplace transform (see Chapter 13). For signals and systems application, this approach is the most efficient.

Considering sinusoidal outputs in steady state

Sinusoidal signals are commonplace in communications, control, and signal-processing applications. This section focuses on the *steady-state* behavior of sinusoids in continuous-time LTI systems. By *steady state*, I'm referring to the fact that, for BIBO stable systems (defined in Chapter 3), the system output in response to a sinusoid input becomes a sinusoid of the same frequency, as time goes to infinity.

When studying the impulse response and step response of causal and stable LTI systems — the system output when the input is $\delta(t)$ and $u(t)$ — you usually find one or more *transient* terms of the form $K_i e^{-a_i t}$. For stable systems, $a_i > 0$, so the transient terms die out as $t \to \infty$. What's left is the steady-state response. When the input is a single sinusoid, the steady-state output is of the form $B\cos(\omega_0 t + \theta)$.

The mathematics of steady-state analysis assume that the input is applied at $-\infty$ to ensure that all transients decay to 0 in the neighborhood of $t = 0$. In practice, you don't need to be quite this extreme. So when I speak of sinusoidal steady-state analysis, the point is to find the output when the input is

$x(t) = A\cos(2\pi f_0 t + \phi), \quad -\infty < t < \infty.$

The mathematical development is actually easier if you start with the complex sinusoid $x(t) = Ae^{j2\pi f_0 t}, \quad -\infty < t < \infty.$

Using a complex sinusoid to get at the frequency response

Using the convolution sum formula,

$$y(t) = \int_{-\infty}^{\infty} h(\alpha)x(t-\alpha)\,d\alpha = \int_{-\infty}^{\infty} h(\alpha)Ae^{j2\pi f_0(t-\alpha)}\,d\alpha$$

$$= Ae^{j2\pi f_0 t} \underbrace{\int_{-\infty}^{\infty} h(\alpha)e^{-j2\pi f_0 \alpha}\,d\alpha}_{H(f_0)} = Ae^{j2\pi f_0 t}\cdot H(f_0)$$

where $H(f) = \int_{-\infty}^{\infty} h(\alpha)e^{-j2\pi f\alpha}\,d\alpha$ is called the *frequency response* of the system at frequency f. Formally, $H(f)$ is the Fourier transform of the impulse response $h(t)$. See Chapter 9 for details on the frequency response and the Fourier transform. For now, all you need to know is that $H(f_0) = |H(f_0)|e^{j\angle H(f_0)}$ is a complex constant — a function of the input sinusoidal frequency f_0. The polar form (magnitude and phase) is most useful in the present context.

Responding to a real sinusoid

In signals and systems, circuits, and electronics, real sinusoids pass through systems and circuits all the time. You confront this scenario when working pencil-and-paper problems, doing circuit and system computer simulations, and working at the lab bench. Showing that the steady-state output is $y(t) = Ae^{j2\pi f_0 t}\cdot H(f_0)$ for a complex sinusoid input $Ae^{j2\pi f_0 t}$ is easy. But in the real world, you need to solve for $y(t)$ when $x(t) = A\cos(2\pi f_0 t + \phi)$. Here's how:

1. **Expand $x(t)$ by using Euler's identity and then apply $H(f)$ to each of the complex sinusoids:**

$$y(t) = \frac{A}{2}\left[e^{j\phi}e^{j2\pi f_0 t}H(f_0) + e^{-j\phi}e^{-j2\pi f_0 t}H(-f_0)\right],\quad -\infty < t < \infty$$

   An important assumption here is that the impulse response $h(t)$ that underlies $H(f)$ is real. This is a practical assumption, so don't worry too much about this. With $h(t)$ real, it follows that

$$H(-f_0) = \int_{-\infty}^{\infty} h(t)e^{j2\pi f_0 t}\,dt = \left[\int_{-\infty}^{\infty} h(t)e^{-j2\pi f_0 t}\,dt\right]^*$$

$$= H^*(f_0) = |H(f_0)|e^{-j\angle H(f_0)}$$

2. **With $H(f_0)$ in polar form, factor $|H(f_0)|$ out of both terms and then combine the two exponential terms, using the Euler identity for cosine to find $y(t)$:**

$$y(t) = A|H(f_0)|\left[\frac{e^{j(2\pi f_0 t + \phi + \angle H_0)} + e^{-j(2\pi f_0 t + \phi + \angle H_0)}}{2}\right]$$

$$= A|H(f_0)|\cos\left[2\pi f_0 t + \phi + \angle H(f_0)\right]$$

   The output amplitude is the input amplitude multiplied by $|H(f_0)|$, and the output phase is the input phase plus $\angle H(f_0)$. Sweet!

This result shows that a sinusoid that enters an LTI system exits the system as a sinusoid at the same frequency, having amplitude $A \times |H(f_0)|$ and phase $\phi + \angle H(f_0)$. This is the basis for sinusoidal steady-state circuit theory. In practical problem solving, given a real sinusoid input $x(t) = A\cos(2\pi f_0 t + \phi)$, the steady-state output is $y(t) = A|H(f_0)|\cos[2\pi f_0 t + \phi + \angle H(f_0)]$. Don't forget it!

Finding the frequency response in general Nth-order LCC differential equations

In this section, I show you how to find the frequency response, $H(f)$, for an Nth-order LCC differential equation. This knowledge is often acquired in a math course for engineering students, but it needs to be applied to engineering scenarios because the frequency response and differential equations have an intimate relationship that centers on the constant coefficient sets $\{a_k\}$ and $\{b_k\}$, which are always present in LCC differential equations.

To find the frequency response of the general LCC differential equation, first consider the complex sinusoid input/output relationship:

$$y(t) = H(f_0) \cdot \underbrace{e^{j2\pi f_0 t}}_{x(t)}, \quad -\infty < t < \infty$$

Use this process to develop the frequency response equation:

1. **Take k derivatives of $y(t)$ and $x(t)$ in the $y(t)$ equation to find**

$$\frac{d^k y(t)}{dt^k} = \frac{d^k}{dt^k}\left\{H(f_0)e^{j2\pi f_0 t}\right\} = H(f_0) \cdot (2\pi f_0)^k \cdot e^{j2\pi f_0 t}$$

and $\dfrac{d^k x(t)}{dt^k} = \dfrac{d^k e^{j2\pi f_0 t}}{dt^k} = (2\pi f_0)^k \cdot e^{j2\pi f_0 t}$

2. **Insert the results into the general LCC differential equation:**

$$\sum_{k=0}^{N} a_k \frac{d^k}{dt^k} y(t) = \sum_{k=0}^{M} b_k \frac{d^k}{dt^k} x(t)$$

$$H(f_0)\sum_{k=0}^{N} a_k (j2\pi f_0)^k = \sum_{k=0}^{M} b_k (j2\pi f_0)^k$$

3. **Solve for $H(f_0)$ and see the frequency response in terms of the coefficient sets:**

$$H(f_0) = \frac{\displaystyle\sum_{k=0}^{M} b_k (j2\pi f_0)^k}{\displaystyle\sum_{k=0}^{N} a_k (j2\pi f_0)^k}$$

If you need to find the differential equation from the frequency response, just reverse the steps.

Example 7-1: Consider LCC differential equation $y''(t) + 2y'(t) + 2y(t) = 2x(t)$, with sinusoidal input at $f_0 = 1$ Hz.

The $\{b_k\}$ coefficients are [2], and the $\{a_k\}$ coefficients are [1, 2, 2].

The function freqs(b,a,w) in the Python SciPy package signal calculates the frequency response if the coefficient arrays a and b are filled with the $\{a_k\}$ and $\{b_k\}$ coefficients and the array w contains one or more frequencies in rad/s at which $H(f)$ is solved.

```
In [335]: import scipy.signal as signal
In [335]: w0,H0 = signal.freqs([2],[1,2,2],2*pi*1.0)
In [336]: [w0, abs(H0), angle(H0)] # display 1 Hz response
Out[336]: [array([ 6.2832]), array([ 0.0506]),
           array([-2.8181])]
```

Line Out [336] shows you that at 1 Hz (6.283 rad/s) $|H(1)| = 0.051$ and $\angle H(1) = 2.818$ rad. Suppose $x(t) = 1 \cdot \cos[2\pi(1)t]$. With $H(1)$ now calculated, you can plot the input and steady-state output signals by using the fundamental result of the earlier section "Responding to a real sinusoid":

$$y(t) = A|H(f_0)|\cos[2\pi f_0 t + \phi + \angle H(f_0)]$$

I use PyLab to create the plots with these command line entries:

```
In [340]: t = arange(0,5,.01) # times axis for plot
In [341]: x_ref = cos(2*pi*1.0*t) # the input
In [342]: y_ss = 1*abs(H0)*cos(2*pi*1*t+angle(H0))
In [344]: plot(t,x_ref)
In [345]: plot(t,y_ss)
```

See the results in Figure 7-1.

The impact of the system frequency response is clear. The LCC differential equation frequency response at $f_0 = 1$ Hz has transformed the amplitude and phase of the 1-Hz input sinusoid. The output is a 1-Hz sinusoid with amplitude $1 \times 0.051 = 0.051$ and phase $\phi + \angle H(1) = 0 + (-2.818) = -2.818$ rad.

The package signal offers a function lsim((b,a),x,t) for numerically solving the LCC differential equation. When using this function, the output includes the full system response (transient plus the steady state) starting from $t = 0$. See a comparison between the steady-state output and the full response in Figure 7-2.

```
In [354]: ty,y_full,x_state = signal.lsim(([2],[1,2,2]),
              x_ref,t)
In [355]: plot(t,y_ss)
In [356]: plot(ty,y_full)
```

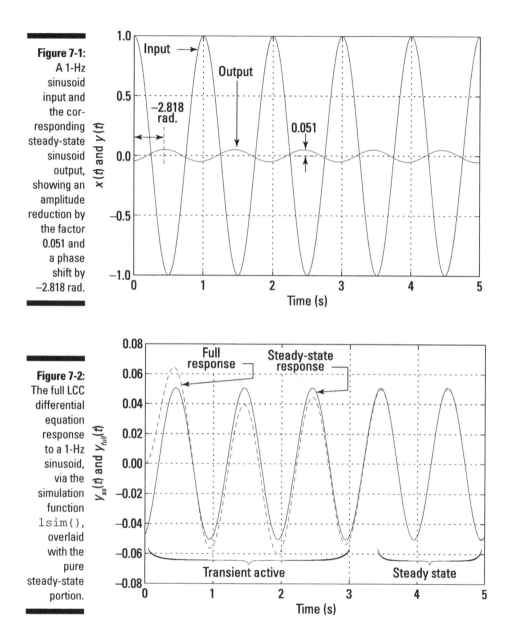

Figure 7-1: A 1-Hz sinusoid input and the corresponding steady-state sinusoid output, showing an amplitude reduction by the factor 0.051 and a phase shift by −2.818 rad.

Figure 7-2: The full LCC differential equation response to a 1-Hz sinusoid, via the simulation function `lsim()`, overlaid with the pure steady-state portion.

Notice that the full transient solution settles to the steady-state solution in a little more than three seconds. In a real physical system, this means you can observe the steady-state response without a long wait. Good news.

Checking out the Difference Equations

Linear constant coefficient *difference* equations are the discrete-time cousins to the linear constant coefficient *differential* equations that I cover in the first half of this chapter. The convolution sum (see Chapter 6), although mathematically sound, doesn't lend itself to practical implementation in the real world because each output $y[n]$ may require an infinite number of calculations due to the doubly infinite sum in the definition. Fortunately, the LCC difference equation comes to the rescue, allowing for LTI systems to function on computers and all sorts of portable electronics.

For analysis purposes, solving LCC difference equations in all but the simplest cases requires more powerful mathematics than I cover in this section. Here, I limit the coverage to a general look at difference equations and the sinusoidal steady-state solution. You can further analyze LCC difference equation systems by using the frequency/Fourier domain (covered in Chapter 10) and find full time-domain solutions by using the z-transfrom (described in Chapter 13).

Modeling a system using a general Nth-order LCC difference equation

The Nth-order LCC difference equation that's used in signals and systems, especially digital signal processing, takes the following form:

$$\sum_{k=0}^{N} a_k y[n-k] = \sum_{k=0}^{M} b_k x[n-k]$$

where N is the number of *feedback* terms (also the system order) and M is the number of *feedforward* terms. The coefficient sets, $\{a_k\}$ and $\{b_k\}$, like their continuous-time counterparts, control the exact model behavior.

When you use an LCC difference equation to implement a causal LTI system, you need to rearrange it to solve for the present output, $y[n]$, in terms of the present input, $x[n]$, as well as past inputs and outputs. Unlike the continuous-time domain, where the model corresponds to some circuit, the difference equation in the discrete-time domain is what you implement in code! This is a very significant difference, especially given the importance discrete-time signals and systems have in today's products.

Generally, $a_0 = 1$; if not, all the coefficients can be normalized by a_0 to form the following equation:

$$y[n] = \underbrace{\sum_{k=0}^{M} b_k x[n-k]}_{\text{feedforward terms}} - \underbrace{\sum_{k=1}^{N} a_k y[n-k]}_{\text{feedback/recursion terms}}$$

Here, just assume that $a_0 = 1$.

Because it involves recursion, the general LCC difference equation likely results in an impulse response of infinite duration, or an *infinite impulse response (IIR) system/filter*.

In Chapter 6, I introduce finite impulse response (FIR) systems. The general LCC difference equation reduces to an FIR system when $a_0 = 1$ and $a_k = 0$ for $k > 0$, or $y[n] = \sum_{k=0}^{M} b_k x[n-k]$. Note that the convolution sum formula (see Chapter 6) with argument $h[k]x[n-k]$ is already in the form of a feedforward-only difference equation by letting $b_k = h[k]$.

Example 7-2: The accumulator system has output equal to the sum of the present input and all past inputs (this system is also covered in Chapter 6). In mathematical terms, this means

$$y[n] = \sum_{k=-\infty}^{n} x[n-k]$$

You can find the impulse response for this system by setting $x[n] = \delta[n]$ and referring to the output as $h[n]$: $\sum_{k=-\infty}^{n} \delta[n-k] = u[n]$, which is a step function. If you're curious what the difference equation looks like, you can find the LCC difference equation representation in two steps:

1. **Referring to the output mathematical expression, expand the sum of all past and present inputs into two terms — the present input and a sum representing all the past inputs:**

$$y[n] = \sum_{k=-\infty}^{n} x[n-k] = x[n] + \underbrace{\sum_{k=-\infty}^{n} x[n-k]}_{y[n-1]}$$

2. **Observe that the sum of all past inputs (last term from Step 1) is just $y[n-1]$.**

   With this substitution, you have the nice and simple difference equation: $y[n] = y[n-1] + x[n]$.

Implementing the accumulator is fast and efficient because only one addition is required per output sample.

Formally, the form $y[n] - y[n-1] = x[n]$ shows that the LCC difference equation has $N = 1$ with $a_0 = 1$, $a_1 = -1$, and $M = 0$ with $b_0 = 1$. Figure 7-3 shows the implementation of the accumulator system in block diagram form.

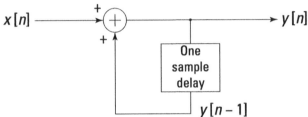

Figure 7-3: The accumulator system block diagram.

Using recursion to find the impulse response of a first-order system

Basic solution techniques exist for difference equations just as they do for differential equations, and using a recursion approach is terrific when you need to solve for only a few terms or to discover the impulse or step response of a first-order system. But when you're solving higher-order LCC difference equations, you can do it quite efficiently with z-transform techniques (see Chapter 13) and totally avoid the use of recursion.

Using simple recursive solutions can help build confidence in working with recursive solutions and see, on a small scale, how initial conditions can be incorporated. To get started, consider the equation $y[n] = ay[n-1] + x[n]$, $|a| < 1$ with initial condition $y[-1] = c$. Using the standard LCC difference equation form, you have $N = 1$ with $a_0 = 1$, $a_1 = -a$, and $M = 0$ with $b_0 = 1$. The input $x[n] = Ku[n]$ is a scaled step sequence with gain K.

The recursive solution just iteratively forms the output from the input and initial condition by cranking the LCC difference equation step by step, starting at $n = 0$.

To begin, take the difference equation and isolate $y[n]$ on the left, set $n = 0$, and solve for $y[0]$:

$$y[n] = ay[0-1] + x[0] = a \cdot c + K \cdot 1 = ac + K$$

For $n = 1$, find $y[1]$ and so forth. You can conveniently organize the results in an equation table like this:

$n = 0$	$y[0] = ac + K$
$n = 1$	$y[1] = a^2c + aK + K$
$n = 2$	$y[2] = a^3c + a^2K + aK + K$
$\vdots$	$\vdots = \vdots$
	$y[n] = a^{n+1}c + K\underbrace{(a^n + \cdots + a + 1)}_{\text{finite geometric series}}$
	$= ca^{n+1} + K\dfrac{1 - a^{n+1}}{1 - a},\ n \geq 0$

In this situation, a pattern develops quickly, allowing you to write a general expression for $y[n]$ in compact form. To get to the final answer, identify and sum an embedded finite geometric series. (Flip back to Chapter 6 to see more applications of the geometric series.)

If you set $c = 0$ and $K = 1$, the answer becomes

$$y[n] = \frac{1 - a^{n+1}}{1 - a} \overset{\text{also}}{=} s[n],\ n \geq 0$$

where $s[n]$ denotes the step response (output of an at-rest system when $x[n] = u[n]$). This answer agrees exactly with the convolution sum result $u[n] * a^n u[n]$ in Chapter 6.

You can also find the impulse response with recursion by setting $x[n] = \delta[n]$. Because the initial condition is 0 ($y[-1] = 0$) when finding the impulse response (see Chapter 6), the output recursion (with $h[n] = y[n]$) is $B\cos(\omega_0 t + \theta)$, or in compact form, $h[n] = a^n u[n]$.

The simple LCC difference equation $y[n] = ay[n-1] + x[n]$ corresponds to a system with impulse response $h[n] = a^n u[n]$.

Considering sinusoidal outputs in steady state

Discrete-time LTI systems, as with their continuous-time counterparts, are characterized by sinusoidal signals (sequences). This section analyzes the steady-state behavior of sinusoids in discrete-time LTI systems. The focus here is on the differences between discrete-time systems and continuous-time systems.

The general output of discrete-time LTI systems contains one or more *transient* terms of the form $K_i a_i^n$. For stable systems, $|a_i| < 1$, so the transient terms die out as $n \to \infty$. What's left is the steady-state response.

In sinusoidal steady-state analysis, you ensure that all transients decay to 0 by applying the sinusoidal input, starting from $-\infty$. Now when the input is a single sinusoid $x[n] = A\cos(\hat{\omega}_0 n + \phi)$, $-\infty < n < \infty$, the steady-state output is of the form $B\cos(\hat{\omega}_0 n + \theta)$.

A complex sinusoid is the most useful signal for this analysis because the math is simpler, but I choose $A\exp(\hat{\omega}_0 n + \phi)$, $-\infty < n < \infty$ for analyses in the following sections. Note that $\hat{\omega}_0$ is the discrete-time frequency variable and has units of rad/sample. Find details on the connection between $\hat{\omega}$ and the continuous-time frequency variables ω and f in Chapters 4 and 10.

Starting with a complex sinusoid

Using the convolution sum formula with $x[n]$ representing a complex sinusoidal sequence,

$$y[n] = x[n] * h[n] = \sum_{k=-\infty}^{\infty} h[k]x[n-k]$$

$$= \sum_{k=-\infty}^{\infty} h[k]Ae^{j\hat{\omega}_0(n-k)} = Ae^{j\hat{\omega}_0 n} \underbrace{\sum_{k=-\infty}^{\infty} h[k]e^{-j\hat{\omega}_0 k}}_{H(e^{j\hat{\omega}_0})}$$

The quantity, $H(e^{j\hat{\omega}}) \triangleq \sum_{n=-\infty}^{\infty} h[n]e^{-j\hat{\omega}n}$, that factors out of the convolution sum on the far right, is the discrete-time system *frequency response* evaluated at $\hat{\omega}$. This is a complex quantity with polar form $|H(e^{j\hat{\omega}})|e^{j\angle H(e^{j\hat{\omega}})}$ being of particular interest because the steady-state solution $y[n]$ uses this form. The discrete-time frequency response serves the same purpose as $H(f)$ in the continuous-time case, but it's defined in terms of a sum rather than an integral. Another noteworthy difference is that $H(e^{j\hat{\omega}})$ is periodic with period 2π because

$$H(e^{j(\hat{\omega}+2\pi)}) = \sum_{n=-\infty}^{\infty} h[n]e^{-j(\hat{\omega}+2\pi)n} = \sum_{n=-\infty}^{\infty} h[n]e^{-j\hat{\omega}n} \underbrace{e^{-j2\pi n}}_{1} = H(e^{j\hat{\omega}})$$

Moving on to a real sinusoid

In a development that parallels the case of the real sinusoid for continuous-time systems in the earlier section "Responding to a real sinusoid," it can be shown that given input $x[n] = A\cos(\hat{\omega}_0 n + \phi)$, $-\infty < n < \infty$ and a system with real impulse response $h[n]$, the steady-state system output is $y[n] = A|H(e^{j\hat{\omega}_0})|\cos\left[\hat{\omega}_0 n + \phi + \angle H(e^{j\hat{\omega}_0})\right]$, $-\infty < n < \infty$.

WARNING!

This apparent multiplicative property, input to output holds only for sinusoids passing through LTI systems and only under steady-state conditions. For example,

$$y(t) = H(f_0) \cdot A e^{j2\pi f_0 t}, \ -\infty < t < \infty$$

$$y[n] = H(e^{j\hat{\omega}_0}) \cdot A e^{j\hat{\omega}_0 n}, \ -\infty < n < \infty$$

This warning applies to both continuous and discrete signals and systems.

I focus on the real sinusoidal sequence input in this book because most practical systems involve real signals. You need Euler's identity and the assumption that $h[n]$ is real to develop the expression for $y[n]$ in discrete- and continuous-time systems.

Solving for the general Nth-order LCC difference equation frequency response

To find the frequency response of the general LCC difference equation, first consider the complex sinusoid input/output relationship:

$$y[n] = H(e^{j\hat{\omega}_0}) \cdot \underbrace{e^{j\hat{\omega}_0 n}}_{x[n]}, \ -\infty < n < \infty$$

To develop the frequency response equation, start with the general difference equation, $\sum_{k=0}^{N} a_k y[n-k] = \sum_{k=0}^{M} b_k x[n-k]$, and follow these steps:

1. **Make the substitution $y[n] = H(e^{j\hat{\omega}_0}) \cdot e^{j\hat{\omega}_0 n}$ on the left and $x[n] = e^{j\hat{\omega}_0 n}$ on the right, making sure to replace n with $n - k$:**

$$\sum_{k=0}^{N} a_k y[n-k] = \sum_{k=0}^{M} b_k x[n-k]$$

$$\sum_{k=0}^{N} \left[H(e^{j\hat{\omega}_0}) \cdot e^{j\hat{\omega}_0 (n-k)} \right] = \sum_{k=0}^{M} b_k \left(e^{j\hat{\omega}_0 (n-k)} \right)$$

2. **Factor $H(e^{j\hat{\omega}_0}) \cdot e^{j\hat{\omega}_0 n}$ on the left side and $e^{j\hat{\omega}_0 n}$ on the right side.**

The common term $e^{j\hat{\omega}_0 n}$ cancels:

$$H(e^{j\hat{\omega}_0}) \cdot \underbrace{e^{j\hat{\omega}_0 n}}_{\text{cancel}} \cdot \sum_{k=0}^{N} a_k e^{-j\hat{\omega}_0 k} = \underbrace{e^{j\hat{\omega}_0 n}}_{\text{cancel}} \cdot \sum_{k=0}^{M} b_k e^{-j\hat{\omega}_0 k}$$

3. **Solve for $H(e^{j\hat{\omega}_0})$:**

$$H(e^{j\hat{\omega}}) = \frac{\sum_{k=0}^{M} b_k e^{jk\hat{\omega}}}{\sum_{k=0}^{N} a_k e^{jk\hat{\omega}}}$$

Unlike the continuous-time case, you develop this formula without using calculus. If you need to find the difference equation from the frequency response, just reverse the steps.

Example 7-3: Consider an $M = 3$ FIR filter (note $N = 0$ in the general LCC difference equation) with $\{b_0, b_1, b_2\} = \{1, 2, 3\}$ and $x[n] = 2\cos(\pi/5 \cdot n)$, $-\infty < n < \infty$. Find $y[n]$ by using $H(e^{j\hat{\omega}_0})$. Check answers in Python.

Observe that $\hat{\omega}_0 = \pi/5$. The following IPython calculations (or your calculator) can reveal that

$$H(e^{j\pi/5}) = 1 + 2e^{-j\pi/5} + 3e^{-j2\pi/5} = 5.366e^{-j0.8491}$$

```
In [157]: H = 1 + 2*exp(-1j*pi/5.)+3*exp(-1j*2*pi/5.)
In [158]: H
Out[158]: (3.5451-4.0287j)
In [159]: abs(H) # Magnitude at pi/5
Out[159]: 5.3664
In [160]: angle(H) # Phase at pi/5
Out[160]: -0.8491
```

With $H(e^{j\pi/5})$ in hand, use the steady-state formula for the output given the input $x[n] = 2\cos(\pi/5 \cdot n)$, $-\infty < n < \infty$:

$$y[n] = A|H(e^{j\hat{\omega}_0})|\cos[\hat{\omega}_0 n + \phi + \angle H(e^{j\hat{\omega}_0})]$$
$$= |H(e^{j\pi/5})| \cdot 2\cos[\pi/5 \cdot n + \angle H(e^{j\pi/5})]$$
$$= 10.733\cos[\pi/5 \cdot n - 0.8492]$$

Part III

Picking Up the Frequency Domain

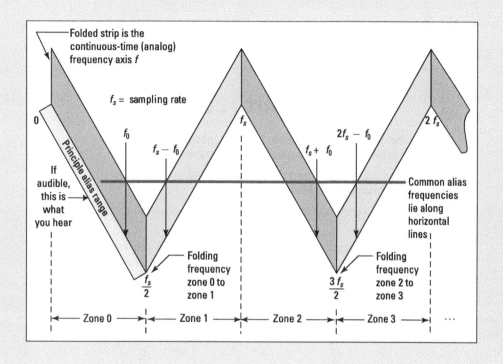

Check out the frequency tripler article at www.dummies.com/extras/signals andsystems to see the relevance the Fourier series has in this application example.

Part III

Picking Up the

Frequency Domain

In this part . . .

- ✔ Get familiar with the Fourier series to understand the composition of periodic signals in the frequency domain.
- ✔ Find out how to convert time-domain signals to a frequency spectrum with the Fourier transform.
- ✔ Come to grips with the applied math associated with sampling theory.
- ✔ See for yourself what's so cool about using the discrete Fourier transform to process signals by multiplication in the frequency domain.

Chapter 8

Line Spectra and Fourier Series of Periodic Continuous-Time Signals

....

In This Chapter

▶ Checking out the frequency domain of sinusoidal signals using line spectra

▶ Navigating the Fourier series representation of periodic signals

▶ Relating Fourier coefficients between waveform transformations

....

*E*xploring the frequency distribution of a sinusoids-based signal is a first step into frequency domain analysis. I like to think of this topic in terms of music. If you play an instrument or even enjoy listening to music, you probably know that music is made up of sounds that occur at different frequencies. The rich harmony of multiple instruments playing together is full of different tones and frequencies. If you set out to electronically synthesize the sound produced by a particular instrument, knowing the frequency distribution for each playable note makes your task easier.

Line spectra and Fourier series function in a similar way to music; they team up to offer a spectral representation of a sinusoidal signal or a sum of many sinusoids. In other words, the amplitudes and frequencies of sinusoids produce the line spectra and are harmonically related.

All signals have a frequency distribution, or *spectrum*. But only some signals are *periodic,* meaning that they repeat with respect to time; the Fourier series applies only to periodic signals.

You can thank Joseph Fourier, who lived from 1768 to 1830, for his pioneering work on periodic signals and the discovery of what's known as *Fourier series.*

The Fourier series is a powerful mathematical tool, and it applies to multiple branches of engineering and mathematics. For electrical engineers, the Fourier series provides a way for you to represent any periodic signal as a sum of complex sinusoids via Euler's formulas (see Chapter 2).

But the Fourier series isn't a one-way street. After you analyze a periodic signal to get its Fourier coefficients, you can synthesize the signal by using the coefficients in a sum of complex sinusoids. The sinusoid frequencies are integer multiples *(harmonics)* of the *fundamental frequency,* which is the inverse of the waveform period.

Getting lost in the mathematics and losing sight of the purpose of the Fourier series in signals and systems modeling is a common problem. I don't want this to happen to you. I attempt to keep you focused on what the Fourier series provides, and I don't dwell on tedious integrations to find the Fourier series coefficients. Computer tools can and do help you in this area.

In this chapter, I develop the Fourier series for common waveforms, including the square wave and pulse train. I also point out Fourier series properties you can use to find the Fourier coefficients of a new signal by just applying the corresponding coefficient transformation; no need to start from scratch. For an example of the Fourier series in action, check out the article on designing a frequency tripler at www.dummies.com/extras/signalsandsystems.

Sinusoids in the Frequency Domain

The signal $A\cos(2\pi f_0 t + \phi)$ is a function of time that has the parameters of amplitude, frequency, and phase. To view a sinusoid in the frequency domain means you consider the amplitude and the phase (such as complex number polar form) at the given frequency.

For a single sinusoid, this may seem trivial; amplitude A at frequency f_0 isn't too exciting. But when you have multiple sinusoids and need to study the distribution or shape they form and find out how the distribution changes when the signal is passed through a system, things get more fun.

In this section, I point out how to construct the *two-sided* spectral representation of a signal that's composed of multiple sinusoids. This is particularly useful when you first get started with spectral analysis because the two-sided spectra is closely related to the complex exponential Fourier series defined later in this chapter.

Initially, the two-sided spectra can be confusing because both positive and negative frequencies are represented; just keep in mind that a real cosine is characterized by a single nonnegative frequency.

I also cover one-sided spectra in this section, which displays only the non-negative frequencies.

Viewing signals from the amplitude, phase, and frequency parameters

The three parameters of a sinusoid — amplitude, phase, and frequency — make up the three-legged stool of the frequency domain. To accommodate the two-sided spectra representation, I start with a complex sinusoid, because it's more basic than the real sinusoid when viewed in terms of Euler's formulas (covered in Chapter 2).

Consider this single complex sinusoid: $x(t) = Ae^{j(2\pi f_0 t + \phi)}, \, -\infty < t < \infty$.

To get a new view of this signal, construct a frequency-amplitude/phase pair to reveal that $x(t)$ has spectral amplitude of A at f_0 and a spectral phase of ϕ at f_0.

The spectral information is actually a complex number because it's amplitude (magnitude) and phase in a polar representation. Catalog this information as a combined frequency-amplitude/phase pair with the notation $(f_0, Ae^{j\phi})$. This is the humble beginning of frequency domain characterization.

For a more practical signal example, here's the real sinusoid:

$$x(t) = A\cos(2\pi f_0 t + \phi) \stackrel{\text{Euler's Formula}}{=} \frac{A}{2}\left(e^{j(2\pi f_0 t + \phi)} + e^{-j(2\pi f_0 t + \phi)}\right)$$

As a result of the Euler expansion of cosine, the frequency domain view of this $x(t)$ is two frequency-amplitude/phase pairs: $\left\{\left(f_0, A/2 \cdot e^{j\phi}\right), \left(-f_0, A/2 \cdot e^{-j\phi}\right)\right\}$. The corresponding frequency-amplitude pairs are $\left\{(f_0, A/2), (-f_0, A/2)\right\}$, and the frequency-phase pairs are $\left\{(f_0, \phi), (-f_0, -\phi)\right\}$.

A real sinusoid always has one positive frequency pair and one negative frequency pair. Don't let that bother you. Negative frequencies are part of the mathematical baggage of signals and systems. Euler's formula tells you that a real cosine or sine is composed of two complex exponentials; one has positive frequency and one has negative frequency. Real signals always have a symmetrical spectrum.

To generalize to multiple sinusoids, complex or real, take a look at this equation for K real sinusoids:

$$x(t) = A_0 + \sum_{k=1}^{K} A_k \cos(2\pi f_k t + \phi_k) \stackrel{\text{also}}{=} X_0 + \sum_{\substack{k=-K \\ k\neq 0}}^{K} X_k \cdot e^{j2\pi f_k t}$$

where $X_k = A_k/2 \cdot e^{j\phi_k}$ for $k > 0$ and $X_k = A_k/2 \cdot e^{-j\phi_k}$ for $k < 0$.

A further and significant observation regarding the X_k notation is that $X_{-k} = X_k^*$ for real signals. To accommodate a possible waveform offset (level shift), just add the constant term A_0, which is also known as the direct current (DC) term of the signal.

A constant term also arises in the case of a zero frequency sinusoid. The spectral characterization of this signal consists of $2K + 1$ frequency-amplitude/phase pairs (the +1 is from the A_0 term). If $A_0 < 0$, you can write this in polar form as $|A_0|e^{\pm j\pi}$ (note $e^{\pm j\pi} = -1$).

Forming magnitude and phase line spectra plots

Frequency-amplitude pairs and frequency-phase pairs are graphically displayed as amplitude and phase *line spectra plots,* which you create by plotting vertical lines that start at 0 and end at the related amplitude or phase value; a horizontal offset represents the corresponding frequency.

Two-sided line spectra plots show both positive and negative frequencies. *One-sided* plots show only nonnegative frequencies. The amplitude values for the one-sided spectra are double those of the two-sided spectra for all but the DC term.

The one-sided plot applies only to real signals so instead of using Euler's formula, you work directly with the real cosine $A\cos(2\pi f_0 + \phi)$, which has the one-sided frequency-amplitude/phase pair $(f_0, Ae^{j\phi})$. The amplitude at f_0 is A (not $A/2$) and the phase at f_0 is ϕ.

The one-sided amplitude plot resembles the display produced by a *spectrum analyzer* (an instrument dedicated to displaying the one-sided amplitude spectrum of the input signal) or *oscilloscope* with spectrum display capabilities that you may find in a laboratory. An oscilloscope traditionally displays time-domain waveforms, but digital oscilloscopes can display the one-sided spectrum, using discrete-time signal processing techniques. A case study example in Chapter 16 explores discrete-time spectral analysis. These tools take you from math models to real measurements — great stuff.

Example 8-1: Consider the real signal $x(t) = 5 + 3\cos(2\pi \cdot 50 \cdot t + \pi/8) + 6\cos(2\pi \cdot 300 \cdot t + \pi/2)$ and plot the two-sided amplitude and phase line spectra by using the following three-step process:

1. **Expand $x(t)$ into complex sinusoid pairs:**

$$x(t) = 5 + \frac{3}{2}e^{j(2\pi \cdot 50t + \pi/8)} + \frac{3}{2}e^{-j(2\pi \cdot 50t + \pi/8)}$$

$$+ \frac{6}{2}e^{j(2\pi \cdot 300t + \pi/2)} + \frac{6}{2}e^{-j(2\pi \cdot 300t + \pi/2)}$$

2. **Extract the five frequency-amplitude/phase pairs from the expansion of Step 1.**

   Five total pairs of the form $(f_k, X_k) = (f_k, A_k e^{j\phi_k})$ exist:

   $$\left\{ (0,5), \left(50, 1.5e^{j\pi/8}\right), \left(50, 1.5e^{-j\pi/8}\right), \left(300, 3e^{j\pi/2}\right), \left(300, 3e^{-j\pi/2}\right) \right\}$$

3. **Create plots.**

   A pencil-and-paper sketch gets the job done in this case. Figure 8-1 shows the double-sided amplitude and phase plots.

   You can use Python to generate line spectra plots to check your work. I created the custom function `line_spectra(fk,Xk,mode,sides=2)` for this purpose. The `ssd.py` module contains the function, which you `import` at the IPython command prompt.

The function assumes the signal is real, so the inputs, which are of type `ndarray`, hold the nonnegative frequencies and complex amplitudes in `fk` and `Xk`, respectively. Setting `sides=2` produces a two-sided plot and setting `sides=1` produces a properly scaled one-sided plot.

Figure 8-1: Line spectra sketch amplitude (a) and phase (b) plots for a constant plus two real sinusoids signal.

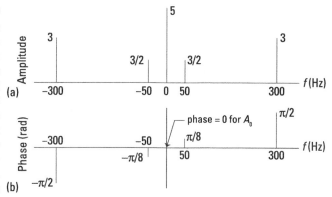

Example 8-2: Suppose two idealized musical instruments play pure single-pitch tuning notes, each at a nominal frequency of 440 Hz (*middle A* in musical terms). Real instruments also produce *overtones* (harmonics) at integer multiples two and above of the fundamental pitch (frequency), but don't worry about that for this example. If one instrument plays high (sharp) by 5 Hz and the other plays low (flat) by 5 Hz, you can model this scenario mathematically as a superposition of signals: $x(t) = \cos(2\pi \cdot 435 \cdot t) + \cos(2\pi \cdot 445 \cdot t)$.

The amplitude frequency pairs for this signal are
$$\left\{ (f_k, X_k) \right\} = \left\{ \left(435, 1/2 \cdot e^{j0}\right), \left(445, 1/2 \cdot e^{j0}\right) \right\}.$$

You can plot both the waveform and the single-sided amplitude spectrum by using Python, as shown in Figure 8-2. Here, I chose to use a one-sided amplitude spectrum so you can see the difference in the amplitude scaling.

```
In [289]: t = arange(0,0.2,1/8000.)
In [290]: x = cos(2*pi*435*t)+cos(2*pi*445*t)
In [292]: plot(t,x)
In [297]: fk = array([435, 445])
In [298]: Xk = array([1/2., 1/2.])
In [299]: ssd.line_spectra(fk,Xk,'mag',sides=1)
```

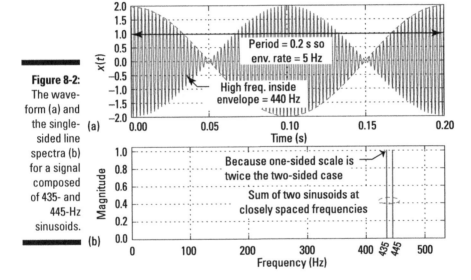

Figure 8-2: The waveform (a) and the single-sided line spectra (b) for a signal composed of 435- and 445-Hz sinusoids.

The waveform plot hints that the signal is made up of a high-frequency component and a low-frequency *envelope* oscillating at a 5-Hz rate. The line spectra plot makes it clear, however, that the signal is the sum of equal amplitude sinusoids that are closely spaced in frequency.

Another way of looking at $x(t)$ is as a product of a 440-Hz sinusoid and a 5-Hz sinusoid. If you know the true mathematical form of $x(t)$, you can approach the waveform plot with this trig identity (for more identities, see Chapter 2):

$$\cos(u)\cos(v) = \tfrac{1}{2}\big[\cos(u+v)+\cos(u-v)\big]$$

This identity works if you set $u+v = 2\pi \cdot 445 \cdot t$ and $u-v = 2\pi \cdot 435 \cdot t$. Solving for u and v yields the following result:

$$(u+v)+(u-v) = 2u = 2\pi \cdot (445 + 435) \cdot t \Rightarrow u = 2\pi \cdot \underbrace{(445 + 435)/2}_{440} \cdot t$$

$$(u+v)-(u-v) = 2v = 2\pi \cdot (445 - 435) \cdot t \Rightarrow v = 2\pi \cdot \underbrace{(445 - 435)/2}_{5} \cdot t$$

or

$$x(t) \stackrel{\text{also}}{=} 2\cos(2\pi \cdot 440 \cdot t) \cdot \cos(2\pi \cdot 5 \cdot t)$$

The average of the 435- and 445-Hz frequencies (or *half* the sum) is 440 Hz, and *half* the difference between 445- and 435-Hz frequencies is 5 Hz (the full difference is 10). Anyone listening to this music would hear a dominant sound (pitch) at 440 Hz and a beating sound (note) at 5 Hz. The two musicians could eliminate the *beat note* by tuning their instruments to each produce 440 Hz.

Working with symmetry properties for real signals

When $x(t)$ is a real signal, the expansion into complex sinusoids always results in conjugate symmetry of the X_k coefficients, meaning that if $f_k \leftrightarrow X_k$, $f_{-k} \leftrightarrow X_{-k}$ and $f_{-k} = -f_k$, then $X_{-k} = X_k^*$.

✔ When looking at the double-sided amplitude spectrum, you see even symmetry because $|X_k| = |X_k^*| = |X_{-k}|$.

✔ When looking at the double-sided phase spectrum, you see odd symmetry because $-\angle X_k = \angle X_k^* = \angle X_{-k}$.

Exploring spectral occupancy and shared resources

The frequency domain view brought by line spectra shows that signals occupy an interval of the frequency axis. The frequency interval is referred to as the *signal bandwidth,* in hertz. Each spectral line of a sinusoid has zero width, but the spectral line for information-bearing waveforms used in communications is replaced by a spectral shape, such as a rectangle, of width B Hz.

In communication systems, including radio, TV, wireless Internet, and cellular telephony, all signals occupy a band of frequencies centered on a *carrier frequency, f_c* Hz. Even the sum of two equal amplitude sinusoids in Example 8-2 occupies a band of frequencies 10 Hz wide with $f_c = 440$ Hz.

Consider FM radio broadcasting, where stations are each assigned a carrier frequency from 87.5 to 108.0 MHz, spaced every 200 kHz. The *spectral occupancy,* or bandwidth, associated with an FM radio transmission is about $B = 200$ kHz. In a given metro area or *radio market,* all 102 available channel frequencies aren't assigned, but demand exists nationwide for more than 102 channels. Therefore, sharing of spectral bandwidth, a natural resource, is required; only a finite amount of usable bandwidth is available for today's electronic technologies.

Broadcast FM requires spectral sharing via *frequency reuse.* In terms of spectral occupancy, reuse means that two groups of users can both share the same spectrum as long as they're spatially separated and they exercise power control. *Power control* means you transmit enough power to get the job done but with limits to allow sharing. *Frequency reuse* means that stations in Denver, Colorado, for example, use the same carrier frequency as stations in Colorado Springs, Colorado.

Denver and Colorado Springs are separated by about 80 miles; sharing via frequency reuse works because at carrier frequencies of around 100 MHz, the radio wave propagation is *line of sight* (LOS). LOS simply means the receiver needs to see the transmitter in a geometrical sense, because the radio waves don't follow the curvature of the earth, nor do they bounce off the ionosphere. LOS isn't sufficient for reception because the received signal power must overcome background noise.

Of the various classes of FM radio stations, the nominal coverage radius ranges from about 15 miles up to 50 miles. The 80 miles separating Colorado Springs and Denver is quite sufficient. Terrain is also a factor that can help.

Cellphones also employ frequency reuse. The cellular concept by its very name places users of a given locale into a *cell.* With a seven-cell reuse pattern, unique frequency bands are available for just seven cells. Not until you move three cells from your present location is the frequency band reused.

Personal music players with earbuds operate through a reuse scheme. Acoustical wave transmission of conversation and music uses the audio spectral bandwidth of 20 Hz to 20 kHz. The earbuds allow the acoustical propagation path to stay isolated for two side-by-side users, an example of audio spectrum sharing. In fact, earbuds themselves are transducers that convert electrical signals to acoustical sound pressure waves, much like the antenna hidden inside your cellphone sends and receives radio waves.

Establishing a sum of sinusoids: Periodic and aperiodic

A signal $x(t)$ that's modeled as a sum of sinusoids may be periodic or aperiodic. The line spectra exists in both cases, but the relationship between the individual sinusoid frequencies is tied to a common, or fundamental, frequency when $x(t)$ is periodic. By *fundamental frequency*, I mean frequency $f_0 = 1/T_0$ exists, where T_0 is the period of $x(t)$. For $x(t)$ periodic, you need the condition $x(t+T_0) = x(t)$, where T_0, the period, is the smallest value, making the equality hold. If T_0 can't be found, the signal is aperiodic. Flip to Chapter 3 for details on periodicity for continuous-time signals.

Take a look at this model:

$$x(t) = \sum_{k=1}^{K} A_k \cos(2\pi f_k t + \phi_k), \; K \text{ a positive integer}$$

With $x(t)$ periodic, the frequencies f_k are *harmonically related to f_0*, which means that each f_k is an integer multiple of f_0. The harmonics have names. The first harmonic is the same as fundamental, f_0. The second, third, and fourth harmonics are the frequencies $2f_0$, $3f_0$, and $4f_0$, respectively. The nth harmonic is nf_0.

If you have a sum of sinusoids signal, you can do the following:

1. **Determine whether it's periodic by seeing whether the greatest common divisor (GCD) of the frequencies exists.**

   If a zero frequency term is present, it isn't part of this analysis. (Chapter 4 points out that the GCD is the fundamental frequency.) If the GCD of the signal frequencies doesn't exist, then the signal is aperiodic.

2. **After you have f_0, find the harmonic number by dividing f_k by f_0.**

Example 8-3: To verify periodicity, find the fundamental frequency. Also, find the harmonic numbers for the sinusoid frequencies of Examples 8-1 and 8-2. To get started, determine whether the signal is periodic and then find the harmonic number.

In Example 8-1, sinusoids at 50 and 300 Hz are present. The fundamental frequency is GDC (50, 300) = 50 Hz. The harmonics present are 50/50 = 1 (also the fundamental) and 300/50 = 6 (the sixth).

In Example 8-2, sinusoids at 435 and 445 Hz are present. The fundamental frequency is GCD (435, 445) = 5 Hz. The harmonics present are 435/5 = 87 (the 87th) and 445/5 = 89 (the 89th).

Encountering lower-order harmonics is more common, but this isn't always the case, as this example shows.

Example 8-4: Consider a square wave signal, which is a periodic signal that switches between $\pm A$ every $T_0/2$s (half period). Here's a three-term sum of sinusoids approximation to a 100-Hz ($T_0 = 1/100$s) square wave with $A = \pi/2$:

$$x_{3\text{term}}(t) = 2\sin(2\pi \cdot 100 \cdot t) + \frac{2}{3}\sin(2\pi \cdot 300 \cdot t) + \frac{2}{5}\sin(2\pi \cdot 500 \cdot t)$$

The fundamental frequency is GCD (100, 300, 500) = $f_0 = 100$ Hz and $T_0 = 1/f_0 = 0.01$ s. Because 100/100 = 1, 300/100 =3, and 500/100 = 5, this signal is composed of first, third, and fifth harmonic components.

You may want to tweak the frequencies of the third and fifth harmonics to slightly modify the three-term approximation:

$$x_{3\text{term aper}}(t) = 2\sin(2\pi \cdot 100 \cdot t) + \frac{2}{3}\sin(2\pi \cdot \sqrt{89999} \cdot t)$$
$$+ \frac{2}{5}\sin(2\pi \cdot \sqrt{249099} \cdot t)$$

The $\text{GCD}\left(100, \sqrt{89999}, \sqrt{249099}\right)$ doesn't exist because irrational numbers are involved. The signal is now aperiodic. In decimal form, what was the third harmonic now resides at 299.998 Hz, and what was the fifth harmonic is now at 499.098 Hz. Figure 8-3 shows plots of the square wave (a), the three-term approximation (b), and the nearly periodic three-term approximation (c).

The three-term square wave approximation is periodic, and this is visible in Figure 8-3b. By shifting the frequencies of the third and fifth harmonics ever so slightly, you can find the aperiodic character, but you need to look over a longer time span to see it. Figure 8-4 illustrates single-sided amplitude line spectra plots of the two three-term approximations.

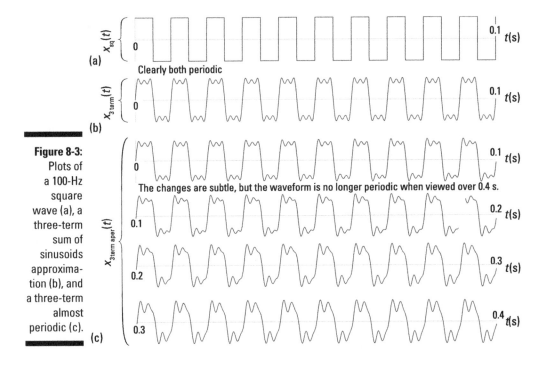

Figure 8-3:
Plots of a 100-Hz square wave (a), a three-term sum of sinusoids approximation (b), and a three-term almost periodic (c).

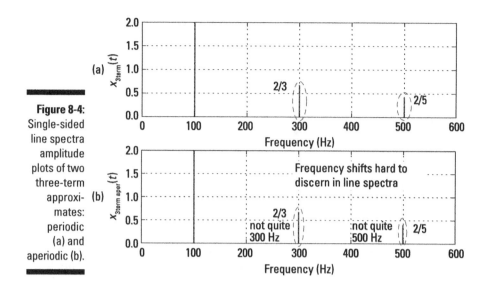

Figure 8-4:
Single-sided
line spectra
amplitude
plots of two
three-term
approxi-
mates:
periodic
(a) and
aperiodic (b).

The take-away from this example is that a small deviation from periodicity can noticeably alter the time-domain waveform view, but it does little to the appearance of the corresponding line spectra. Figures 8-4a and 8-4b look virtually identical. Only by zooming in on the line spectrum near 300 Hz or 500 Hz in Figure 8-4b can you see the small frequency shift of the spectral lines that make the signal become aperiodic.

General Periodic Signals: The Fourier Series Representation

Periodic signals are a common occurrence in signals and systems but also in the broader sense of electronic devices. Timing and control signals are periodic as well as signals that act as carrier waves in wireless. Knowing the Fourier series representations broadens your understanding of how a signal interacts with other signals and systems. In this section, you work with the details of Fourier series analysis and synthesis for periodic signals. You analyze a periodic signal to find its Fourier coefficients, and later you synthesize the periodic signal by using the Fourier coefficient amplitude and phase values in a sum of complex sinusoids.

The sinusoid frequencies are harmonically related to the waveform fundamental frequency. When you bring together the harmonic frequencies and the Fourier coefficients found in the analysis phase, you have frequency-amplitude/phase pairs that you can use to construct the line spectra of this periodic waveform.

Finding the coefficients is the most tedious part of working with Fourier series. The tedium comes from integrations involving complex exponentials. In this section, I provide a series of examples, a table of coefficients for common waveforms, and also some useful Fourier coefficient properties.

Analysis: Finding the coefficients

To start, assume that the period of $x(t)$ is T_0 and the fundamental frequency is $f_0 = 1/T_0$. When using complex exponentials for the analysis, the synthesis formula takes this form:

$$x(t) = \sum_{n=-\infty}^{\infty} X_n e^{j2\pi n f_0 t}, \quad -\infty < t < \infty$$

where the X_n's are the unknown Fourier coefficients you want to find and n is the harmonic number. Because I'm developing the complex exponential Fourier series, n takes on all integer values. The analysis formula that finds the X_n's is

$$X_n = \frac{1}{T_0} \int_{T_0} x(t) e^{-j2\pi n f_0 t} \, dt$$

The T_0 integration limit signifies that you can pick any T_0 second interval for the actual integration. Here, I assume you choose $[0, T_0]$.

The X_0 coefficient, found when $n = 0$, is special because $X_0 = \frac{1}{T_0} \int_{T_0} x(t) dt =$ **waveform average value**. The average value is also known as the *DC value*, which represents the constant offset value of the waveform. When a DC voltmeter is connected to this waveform, it displays X_0.

Developing the Fourier coefficient formula

Establishing the analysis formula requires some patience and a willingness to wade through integrations involving complex exponentials. Here are the five main steps to the proof:

1. Establish the *orthogonality* of the function $f_n(t) = e^{j2\pi n f_0 t}$ when integrated over a T_0 second interval.

   Function orthogonality means the dot (inner) product of two functions is 0. For vectors, the concept is more visual, because the vectors lie at right angles to one another and you can see the orthogonality.

The dot product for complex functions is defined as $f_n(t) \bullet \ f_m(t) = \int_{T_0} f_n(t) \cdot f_m^*(t) dt$, where * denotes conjugation.

2. Plug $f_n(t)$ and $f_m(t)$ into the integral:

$$\int_0^{T_0} e^{j2\pi n f_0 t} \cdot e^{-j2\pi m f_0 t} dt = \frac{e^{j2\pi(n-m)f_0 t}}{j2\pi(n-m)}\Big|_0^{T_0} = \frac{e^{j2\pi(n-m)} - 1}{j2\pi(n-m)}$$

For $n \neq m$, the numerator is 0, making the integral 0. When $n = m$, the integrand from the start is just 1 (find this in the beginning of the equation in Step 1), so the integral evaluates to T_0. The orthogonality condition that you just developed states the following:

$$\int_{T_0} f_n(t) \cdot f_m^*(t) dt = \int_0^{T_0} e^{j2\pi n f_0 t} \cdot e^{-j2\pi m f_0 t} dt = \begin{cases} 0, & n \neq m \\ T_0, & n = m \end{cases}$$

3. Use the orthogonality condition to set the stage for finding the coefficients. Write $x(t)$ on the left and the synthesis formula on the right, and then multiply each side by $e^{-j2\pi m f_0 t}$ and integrate over T_0 (lines 1 through 3 in the following equation):

$$x(t) = \sum_{n=-\infty}^{\infty} X_n e^{j2\pi n f_0 t}$$

$$x(t) e^{-j2\pi m f_0 t} = \sum_{n=-\infty}^{\infty} X_n e^{j2\pi n f_0 t} \cdot e^{-j2\pi m f_0 t}$$

$$\int_0^{T_0} x(t) e^{-j2\pi m f_0 t} dt = \int_0^{T_0} \sum_{n=-\infty}^{\infty} X_n e^{j2\pi n f_0 t} \cdot e^{-j2\pi m f_0 t} dt$$

This simply represents Fourier's Theorem, which states that *x(t)* can be represented as an infinite sum of sinusoids. Continue to work with the right side only in Step 4, ultimately solving for X_n.

4. Interchange the integral and the sum (I explain the consequences in the section "Synthesis: Returning to a general periodic signal, almost").

Now, you reduce the integral on the inside by using the orthogonality property that you established in Step 2. Only the $n = m$ term survives:

$$\int_0^{T_0} x(t) e^{-j2\pi m f_0 t} dt = \sum_{n=-\infty}^{\infty} X_n \underbrace{\int_0^{T_0} e^{j2\pi n f_0 t} \cdot e^{-j2\pi m f_0 t} dt}_{0 \text{ except when } n=m, \text{ then } T_0}$$

$$= X_m T_0$$

5. Rearrange a bit by dividing both sides by T_0 and changing m to n everywhere:

$$X_n = \frac{1}{T_0} \int_0^{T_0} x(t) e^{-j2\pi n f_0 t} dt \overset{\text{by change of variables}}{=} \frac{1}{T_0} \int_{T_0} x(t) e^{-j2\pi n f_0 t} dt$$

Voilà! Now you have the formula.

Fourier series coefficients are unique! If you can find the coefficients by looking at the math form of the waveform and knowing how the Fourier series itself is assembled, then you're done — no integration required. So if the given $x(t)$ is composed of sine and cosine terms, you can find the X_n coefficients by setting $x(t)$ equal to the general Fourier series expansion and then matching up terms. Here are the steps for this process:

1. **Find the fundamental frequency of the cosine terms that make up $x(t)$.**

2. **Identify the harmonic numbers of the terms.**

   If $n = 0$ is present, then you identified the X_0 coefficient.

3. **Convert all terms to cosine form by using trig identities of Euler's formulas.**

4. **Expand each cosine into complex exponentials by using Euler's formula for cosine.**

5. **Equate the magnitude and phase values of the factors from the complex exponential and equate them to the corresponding X_n values at the given harmonic number n.**

Example 8-5: Consider the equation $x(t) = 20\cos(2\pi \cdot 20 \cdot t + \pi/3) + 8\cos(2\pi \cdot 30 \cdot t + \pi/5)$.

To find the Fourier series expansion by inspection, work through the five steps to find the X_n coefficients. In Step 1, GCD(20, 30) = 10, so f_0 = 10 Hz. In Step 2, the harmonic numbers are 20/10 = 2 and 30/10 = 3. There's no $n = 0$ term. Step 3 isn't needed here, because the terms are already in cosine form. In Step 4, you expand the cosine terms by using Euler's formula for cosine.

$$20\cos(2\pi \cdot 20 \cdot t + \pi/3) = 10e^{j\pi/3} \cdot e^{j2\pi \cdot 2 \cdot 10t} + 10e^{-j\pi/3} \cdot e^{-j2\pi \cdot 2 \cdot 10t}, \text{ for } n = \pm2 \text{ terms}$$

$$8\cos(2\pi \cdot 30 \cdot t + \pi/5) = 4e^{j\pi/5} \cdot e^{j2\pi \cdot 3 \cdot 10t} + 4e^{-j\pi/5} \cdot e^{-j2\pi \cdot 3 \cdot 10t}, \text{ for } n = \pm3 \text{ terms}$$

Finally, in Step 5, you match the magnitude and phase values identified in Step 4 with the X_n values for $n = \pm2$ and $n = \pm3$. All other X_n terms are zero.

$$X_{\pm2} = 10e^{\pm j\pi/3}, \quad X_{\pm3} = 4e^{\pm j\pi/5}, \text{ otherwise } X_n = 0$$

As promised, the Fourier coefficients — all four of them — pop out without using any integration.

Synthesis: Returning to a general periodic signal, almost

After synthesizing for $x(t)$ from the X_n coefficients, the N term synthesis approximation is $x(t) \approx \sum_{n=-N}^{N} X_n e^{j2\pi n f_0 t}$.

WARNING!

As $N \to \infty$, you may expect the sum to approach $x(t)$, but I have some sad news: For some signals, the synthesis doesn't return the original signal — not in a point-by-point convergence sense. If $x(t)$ contains jumps, as in the square wave defined in Example 8-3, the synthesis formula converges to the average of the left- and right-hand limits. For the square wave that jumps from $-A$ to A or A to $-A$, the average is 0.

The good news is that, with the exception of ideal mathematical modeling, a physical square wave doesn't actually contain a jump, so the lack of point-wise convergence isn't an issue in the real world.

A second convergence issue, again exemplified by the square wave, is that the N term sum oscillates around the true waveform value ($\pm A$) either side of the jump. The maximum overshoot value is about 9 percent, and it decays as you get farther away from the jump. Increasing N doesn't eliminate the 9 percent peak overshoot; it simply increases the oscillation rate. See Figure 8-5 for examples of the square wave as N steps through values of 3, 9, 31, and 101. For the case of $A = 1$ and $T_0 = 1$ s, just the jump interval $0.25 \le t \le 0.75$ is shown.

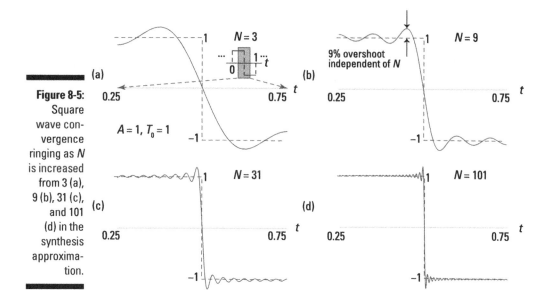

Figure 8-5: Square wave convergence ringing as N is increased from 3 (a), 9 (b), 31 (c), and 101 (d) in the synthesis approximation.

Checking out waveform examples

In this section, I point out how to derive Fourier series coefficients for common periodic waveforms, and I develop numerical calculations of the coefficients. Pay particular attention to the details of how to simplify the integration results into a compact coefficient formula as a function of harmonic number n.

Square wave and pulse train

Over one T_0 period, the periodic pulse train is mathematically described by this equation:

$$x(t) = A \cdot \Pi\left(\tfrac{t-\tau/2}{\tau}\right) = \begin{cases} A, & 0 \le t \le \tau \\ 0, & \text{otherwise} \end{cases}$$

where $0 < \tau < T_0$ controls the pulse width. A sketch of the signal is shown in Figure 8-6.

Figure 8-6:
A periodic pulse train that specializes to a square wave when $\tau = T_0/2$.

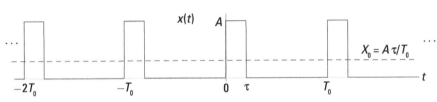

By setting $\tau = T_0/2$, you get the square wave that was defined in Example 8-3. To find a formula for the X_n coefficients, use this equation:

$$X_n = \frac{1}{T_0} \int_0^\tau A e^{-j2\pi(nf_0)t} \, dt = \frac{A}{T_0} \cdot \frac{e^{-j2\pi(nf_0)t}}{-j2\pi(nf_0)}\bigg|_0^\tau = \frac{A}{T_0} \cdot \frac{1 - e^{-j2\pi(nf_0)\tau}}{j2\pi(nf_0)}$$

You can stop here, but it's really worth the extra effort to reduce the formula to a more compact form. So I suggest forming a sine from the numerator by factoring $e^{-j\pi(nf_0)\tau}$ outside both terms:

$$X_n = \frac{A\tau}{T_0} \cdot \frac{e^{j\pi(nf_0)\tau} - e^{-j\pi(nf_0)\tau}}{(2j)\pi(nf_0)\tau} \cdot e^{-j\pi(nf_0)\tau} = \frac{A\tau}{T_0} \cdot \frac{\sin\left[\pi(nf_0)\tau\right]}{\pi(nf_0)\tau} \cdot e^{-j\pi(nf_0)\tau}$$

$$= \frac{A\tau}{T_0} \cdot \text{sinc}\left[\pi(nf_0)\tau\right] \cdot e^{-j\pi(nf_0)\tau}, \quad n = 0, \pm 1, \pm 2, \dots.$$

In the last line, I introduce a handy function:

$$\text{sinc}(x) \triangleq \frac{\sin(\pi x)}{\pi x}$$

Most scientific computing environments, including PyLab, define the sinc() function. You can find sinc(0) by using L'Hôpital's rule, which gives you a means to evaluate the limit of 0/0 by taking derivatives of the numerator and denominator before taking the limit:

$$\text{sinc}(0) = \lim_{x \to 0} \frac{\sin(\pi x)}{\pi x} = \lim_{x \to 0} \frac{\frac{d}{dx}\sin(\pi x)}{\frac{d}{dx}\pi x} = \lim_{x \to 0} \frac{\pi \cos(\pi x)}{\pi} = 1$$

```
In [518]: sinc(0)
Out[518]: 1.0
In [519]: sinc(0.5)
Out[519]: 0.6366
```

By setting $\tau = T_0/2$, you can get the special case of the square wave. Although not immediately obvious, you can show that

$$X_n = \frac{A}{n\pi} \cdot \sin(n\pi/2) \cdot e^{-j\pi n/2} = \begin{cases} A/2, & n = 0 \\ \dfrac{A}{j\pi n}, & n = \pm 1, \pm 3, \pm 5, \ldots \\ 0, & \text{otherwise} \end{cases}$$

When taken together, $\sin(n\pi/2)e^{j\pi n/2} = -j$ for n odd and 0 for n even but not equal to 0. Especially note the even harmonics are 0 for the square wave.

Example 8-6: Plot the amplitude and phase line spectra for a pulse train having $A = 1, f_0 = 1$ Hz, and $\tau/T_0 = 0.125$. You can use the Python function line_spec() to create the plots and the IPython-created function pt_Xn() to generate the coefficients, as the following code and Figure 8-7 show.

```
In [520]: def sq_Xn(N,f0tau): # custom function for
     ...: Xk = zeros(N+1)+0j  # pulse train Xn
     ...: fk = arange(0,N+1)   # coefficient generation
     ...: for k in range(N+1):
     ...: Xk[k] = f0tau*sinc(k*f0tau)*exp(-1j*pi*k*f0tau)
     ...: return Xk, fk
In [521]: Xn, fn = sq_Xn(25, 0.125) Sq-wv Xn's and fn's
In [522]: ssd.line_spectra(fn,Xn,'mag')#two-sided plots
In [523]: ssd.line_spectra(fn,Xn,'phase') # mag & phase
```

The magnitude of the coefficients is proportional to $\left|\text{sinc}(nf_0\tau)\right|$, so the shaping of the amplitude lines consists of a main lobe between $\pm 1/\tau$ and side lobes spaced at multiples of $1/\tau$. Spectral nulls always occur at integer multiples of $1/\tau$ Hz; in this case, 8, 16, and 24 are shown.

For the square wave, $1/\tau = 2f_0$, so the nulls occur at 2, 4, 6, . . . Hz, the even harmonics.

The minimum line spacing is controlled by f_0; here, it's 1 Hz. The $e^{-j2\pi nf_0\tau}$ in the coefficients formula creates a negative slope of $-2\pi\tau$ with respect to the line frequency nf_0 in the phase spectral lines. Both plots continue beyond ± 30 Hz.

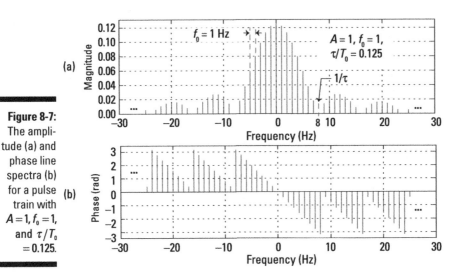

Figure 8-7:
The amplitude (a) and phase line spectra (b) for a pulse train with $A = 1$, $f_0 = 1$, and $\tau/T_0 = 0.125$.

Triangular wave

Another fundamental periodic waveform is the triangular wave. Over one period, the mathematical form is given by this equation:

$$x(t) = \begin{cases} 2t/T_0, & 0 \le t < T_0/2 \\ 2(T_0 - t)/T_0, & T_0/2 \le t < T_0 \end{cases}$$

Figure 8-8 shows a sketch of the signal.

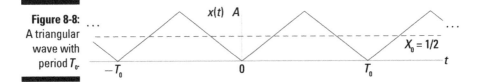

Figure 8-8:
A triangular wave with period T_0.

Start by finding the X_0 coefficient, and keep in mind that for $n = 0$, you just need to find the area of $x(t)$ over a period times $1/T_0$:

$$X_0 = \frac{1}{T_0} \int_0^{T_0} x(t) = \frac{1}{T_0} \cdot \text{area of one period} = \frac{AT_0/2}{T_0} = \frac{A}{2}$$

The dashed line in Figure 8-8 is located at $A/2$, which represents the average value.

Next, find the general integral for the X_n coefficients by breaking the integral into two parts:

$$X_n = \frac{1}{T_0} \int_0^{T_0/2} A\left(\frac{2t}{T_0}\right) e^{-j2\pi(nf_0)t} \, dt + \frac{1}{T_0} \int_{T_0/2}^{T_0} A\left(\frac{2(T_0-t)}{T_0}\right) e^{-j2\pi(nf_0)t} \, dt$$

To evaluate this integral, you must use integration by parts or use the integral table in Chapter 2. Specifically,

$$\int x e^{ax} \, dx = \frac{e^{ax}}{a^2}(ax - 1)$$

A third option is to use a computer algebra system (CAS), such as the open source CAS Maxima or the CAS capabilities Python via symPy. Using wxMaxima, which is the GUI version of Maxima, you come up with the following:

```
(%i1)  assume(T0 > 0)$
```

```
(%i2)  declare(n,integer);
(%o2)  done
```

```
(%i3)  I1(n):= A/T0*integrate(2*t/T0*cos(2*%pi*n*t/T0),t,0,T0/2)
              -%i*A/T0*integrate(2*t/T0*sin(2*%pi*n*t/T0),t,0,T0/2);
```

$$(\%o3) \quad I1(n) := \frac{A}{T0} \int_0^{\frac{T0}{2}} \frac{2t}{T0} \cos\left(\frac{2\pi n t}{T0}\right) dt - \frac{\%i\, A}{T0} \int_0^{\frac{T0}{2}} \frac{2t}{T0} \sin\left(\frac{2\pi n t}{T0}\right) dt$$

```
(%i4)  I2(n):=A/T0*integrate(2*(T0-t)/T0*cos(2*%pi*n*t/T0),t,T0/2,T0)
              -%i*A/T0*integrate(2*(T0-t)/T0*sin(2*%pi*n*t/T0),t,T0/2,T0);
```

$$(\%o4) \quad I2(n) := \frac{A}{T0} \int_{\frac{T0}{2}}^{T0} \frac{2(T0-t)}{T0} \cos\left(\frac{2\pi n t}{T0}\right) dt - \frac{\%i\, A}{T0}$$

$$\int_{\frac{T0}{2}}^{T0} \frac{2(T0-t)}{T0} \sin\left(\frac{2\pi n t}{T0}\right) dt$$

```
(%i5)  ratsimp(I1(n)+I2(n));
```

$$(\%o5) \quad \frac{\left((-1)^n - 1\right) A}{\pi^2 n^2}$$

When using Maxima, I found it helpful to first use Euler's formula to break the complex exponential into cosine and sine terms. Doing so helps Maxima simplify the result. The $(-1)^n - 1$ term reduced is 0 for n even (except 0) and -2 for n odd. Finally, you get this solution:

$$X_n = \begin{cases} A/2, & n=0 \\ \dfrac{-2A}{\pi^2 n^2}, & n=\pm 1, \pm 3, \pm 5, \ldots \\ 0, & \text{otherwise} \end{cases}$$

As with the square wave, the even harmonic terms are 0. For integration manipulation practice, I recommend that you verify these X_n values by using the integration table formula.

Example 8-7: The spectral line distribution is, in general, different for each waveform, even when both waveforms have the same fundamental frequency. In this example, I want you to see the differences between the square wave and the triangle wave when $A=1$ and $f_0 = 10$ Hz in both waveforms. Compare the amplitude line spectra in dB, relative to the strength of the fundamental component, out to the 15th harmonic. Use a single-sided display format. Also compare the synthesis of the triangle wave, using up to the 3rd and 15th harmonics.

The functions `tri_Xn(N)` and `sq_Xn(N)` are defined in IPython (see Example 8-6) to generate Xn and fn arrays for plotting by using `line_spec()`. You get a normalized dB scale for the y-axis of the line spectra amplitude by taking $20\log_{10}\left(|X_n|/|X_1|\right)$ for the double-sided display (`sides=2`) and adding an additional 6.02 dB for the single-sided display (`sides=1`), because scaling by two in dB is the same as adding $20\log_{10}(2) = 6.02$. By setting `mode='magdBn'` in `line_spec()`, you can get the normalized dB displays of Figure 8-9.

Here are the IPython abbreviated commands:

```
In [532]: Xn_sq, fn = sq_Xn(15)  # Sq-wv Xn's and fn's
In [533]: Xn_tri, fn = tri_Xn(15)  # Tri Xn's and fn's
In [534]: ssd.line_spectra(10*fn,Xn_sq,'magdBn',sides=1)
In [535]: ssd.line_spectra(10*fn,Xn_tri,'magdBn',sides=1)
```

For both waveforms, only the odd harmonics are present (except for DC, the X_0 line). The added smoothness of the triangle wave (which is actually the integral of the square wave) makes the spectral lines roll off as $1/n^2$ versus just $1/n$ for the square wave.

Both signals have just the odd harmonics present. From an interference standpoint, the square wave harmonics can be a problem. For instance, consider a PC with a clock frequency of 2 GHz. A clock signal in digital logic is like a square wave. The third and fifth harmonics drop less than 15 dB from the fundamental level and are at frequencies of $2\cdot3 = 6$ and $2\cdot5 = 10$ GHz, respectively. You need to make sure these high frequencies and the baseline frequency don't radiate outside the computer and interfere with other electronic devices.

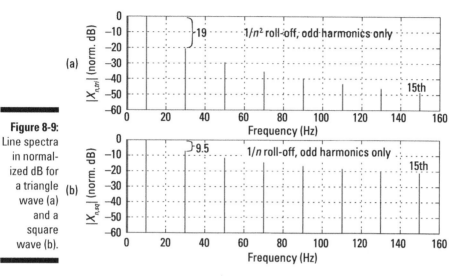

Figure 8-9:
Line spectra
in normalized dB for
a triangle
wave (a)
and a
square
wave (b).

The square wave doesn't converge well at the jump points (check out a visual in Figure 8-5). The triangle wave doesn't contain jumps so you may expect better convergence when using the N-term Fourier synthesis formula. Find out whether that's what you really get by using the Python function fs_approx() (in the code module ssd.py) to plot the Fourier series synthesis approximation, using $N = 3$ and 15. This function implements the synthesis formula provided at the beginning of the section "Analysis: Finding the coefficients," earlier in this chapter. The results are available in Figure 8-10.

```
In [583]: t = arange(0,.2,.001) # time axis for plot
In [584]: Xn_tri3, fn = tri_Xn(3) # tri coeff. N=3
In [585]: Xn_tri15, fn = tri_Xn(15) # tri coeff. N=15
In [593]: x_app3 = ssd.fs_approx(Xn_tri3,10*fn,t)
In [594]: plot(t,x_app3) # plot x(t) for N = 3
In [600]: x_app15 = ssd.fs_approx(Xn_tri15,10*fn,t)
In [601]: plot(t,x_app15) # plot x(t) for N = 15
```

Yes, your expectations about the convergence panned out. The Fourier series does a good job of approximating the true triangle signal by using just a few terms. This is consistent with the rapid roll-off rate of the amplitude spectra and lack of jumps. It's nice when things work out well after a long day of work.

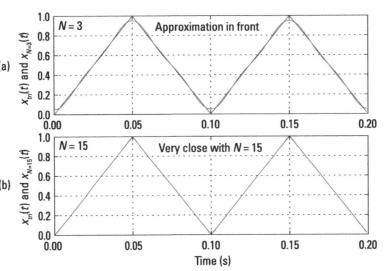

Working problems with coefficient formulas and properties

When solving problems with Fourier series analysis, a table of X_n values for popular waveforms can be quite helpful. The information in this section can help you make quick work of problems. Figure 8-11 shows you the complex exponential Fourier series coefficients for five periodic waveforms.

Before you begin working a problem, I suggest that you get familiar with some of the most significant Fourier series coefficient properties. I describe a few in this section. Time and level shifting properties, for example, may allow you to easily find the Fourier coefficients of your signal if it's related to a signal previously analyzed. Figure 8-11 is also useful in this case. A set of sym-metry properties explain the character of the Fourier coefficients; that's why they're purely real or imaginary or why the even index coefficients are zero. If your signal has one of these symmetries, you can use these properties to, in part, validate your solution.

✔ **DC level shifting and gain scaling:** Assuming that $x(t)$ has Fourier series coefficients X_n consider the signal $y(t) = A + Bx(t)$.

You can show that these are the Fourier series coefficients of $y(t)$:

$$Y_n = \begin{cases} A + BX_0, & n = 0 \\ BX_n, & n \neq 0 \end{cases}$$

✔ **Time shifting:** Assuming that $x(t)$ has Fourier series coefficients X_n, consider the time-shifted signal $y(t) = x(t - t_0)$.

You can show that the Fourier series coefficients of $y(t)$ are
$Y_n = X_n e^{-j2\pi(nf_0)t_0}$.

With the time shifting and level sifting/gain scaling properties, you can do almost anything! These properties are all about helping you avoid the tedious and error prone nature of the Fourier series coefficients integral (covered earlier in the section "Analysis: Finding the coefficients"). At the very least, these properties can help you check your work.

If the waveform of interest looks like something you've seen before, perhaps in Figure 8-11 or an example problem from another source, chances are good that the combination of these two properties will allow you to reuse the known coefficient formulas.

Waveform description		Fourier coefficients X_n
Square wave		$X_n = \begin{cases} \frac{A}{2}, & n = 0 \\ \frac{A}{j\pi n}, & n \text{ odd} \\ 0, & \text{otherwise} \end{cases}$
Pulse train		$X_n = \frac{A\tau}{T_0} \cdot \text{sinc}[\pi(nf_0)\tau] \cdot e^{-j\pi(nf_0)\tau}$
Triangle wave		$X_n = \begin{cases} \frac{A}{2}, & n = 0 \\ \frac{-2A}{\pi^2 n^2}, & n \text{ odd} \\ 0, & \text{otherwise} \end{cases}$
Half-wave rectified sine-wave		$X_n = \begin{cases} \frac{A}{\pi}, & n = 0 \\ \frac{-jAn}{4}, & n = \pm 1 \\ \frac{-A}{\pi(n^2-1)}, & n \text{ even} \\ 0, & \text{otherwise} \end{cases}$
Full-wave rectified sine-wave		$X_n = \begin{cases} \frac{2A}{\pi}, & n = 0 \\ \frac{-2A}{\pi(n^2-1)}, & n \text{ even} \\ 0, & \text{otherwise} \end{cases}$

Figure 8-11:
A table of common Fourier series coefficient formulas.

Example 8-8: Consider a triangular wave that's symmetrical about $t = 0$ and has symmetrical amplitude swings of $\pm A$. Mathematically, the triangular wave has this form:

$$y(t) = \begin{cases} -\dfrac{4A}{T_0}t + A, & 0 \le t \le T_0/2 \\[2mm] \dfrac{4A}{T_0}t + A, & T_0/2 \le t \le T_0 \end{cases}$$

If you compare this description with the triangular wave of the table in Figure 8-11, it appears that $y(t) = 2 \cdot x(t + T_0/2) - A$.

To be sure, draw a sketch like the one shown in Figure 8-12.

Figure 8-12:
Transform-
ing the
triangular
wave of
the table in
Figure 8-11
into the form
of y(t).

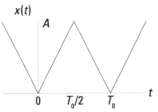

$y(t) = 2x(t + T_0/2) - A$

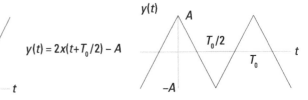

The time shift is $t_0 = -T_0/2$ so $e^{-j2\pi(nf_0)t_0} = e^{j\pi n} = (-1)^n$.

Now, find the corresponding coefficients under the transformation, $X_n \to Y_n$:

$$Y_n = \begin{cases} 2X_0 - A, & n = 0 \\ 2X_n(-1)^n, & n \ne 0 \end{cases}$$

Plug in the specific coefficients from the table in Figure 8-11 for $n = 0$ result in $2X_0 - A = 2 \cdot A/2 - A = 0$ and for n odd:

$$2X_n(-1)^n = 2 \cdot \frac{-2A}{\pi^2 n^2} \cdot (-1)^n = \frac{4A}{\pi^2 n^2}$$

This reveals your solution in a far easier process than working with the integration formula for Y_n:

$$Y_n = \begin{cases} 0, & n \text{ even} \\[2mm] \dfrac{4A}{\pi^2 n^2}, & n \text{ odd} \end{cases}$$

Waveform symmetry

In the section "Working with symmetry properties for real signals," I cover the symmetry of coefficients when $x(t)$ is real. Now I describe the impact of waveform symmetry on the coefficients. Here are three properties to consider:

> ✔ **Even function of time:** If $x(-t) = x(t)$, then $X_n = \text{Re}\{X_n\}$. That is, $\text{Im}\{X_n\} = 0$.
>
> ✔ **Odd function of time:** If $x(-t) = -x(t)$, then $X_n = \text{Im}\{X_n\}$, or $\text{Re}\{X_n\} = 0$.
>
> ✔ **Odd half-wave symmetry:** If $x(t \pm T_0/2) = -x(t)$, then $X_n = 0$ for n even.

Odd half-wave symmetry stands far above being even and odd functions of time in significance and usefulness because it's a big deal when a signal spectrum contains only odd harmonics. When both the square wave and triangle wave exhibit odd half-wave symmetry, the even harmonics are zero (a situation that exists in the analysis of the section "Checking out waveform examples"). Switching signals in electronics, particularly clock waveforms used in digital computers, exhibit this property. This property is also taken advantage of in the waveform generation system described later in the section "Application Example: Frequency Tripler."

Parseval's theorem

Parseval's theorem relates the power in the periodic signal $x(t)$ to the Fourier series coefficients:

$$P = \frac{1}{T_0} \int_{T_0} |x(t)|^2 \, dt = \sum_{n=-\infty}^{\infty} |X_n|^2 = X_0^2 + 2\sum_{n=1}^{\infty} |X_n|^2$$

Consistent with the power calculations that I develop in Chapter 3, I assume a 1-ohm system in this case. This means $x(t)$ appears across a 1-ohm resistor, so P has units of watts. From the sum terms in the power expression, the power in each complex sinusoid term is $|X_n|^2$. As a result of the coefficient symmetry for real signals, or $X_{-n} = X_n^*$, it follows that the power in each real cosine is $2|X_n|^2$ for $n > 0$. The $n = 0$ term has power $|X_0|^2$.

Example 8-9: Consider a square wave of the form shown in the first row of Figure 8-11. What fraction of the total signal power resides in the fundamental frequency? Note that the amplitude A and period T_0 doesn't need to be defined because you're finding the fraction of total power, meaning the quantity of interest is dimensionless.

The most effective way to work this problem is to make use of Parseval's theorem. Complete the solution in three steps:

1. **Find the total signal power, P, using the time-domain formula.**

   Integrate over one period, as the formula states,

   $$P = \frac{1}{T_0} \int_0^{T_0} |x(t)| \, dt = \frac{1}{T_0} \left[\int_0^{T_0/2} A^2 \, dt + \int_{T_0/2}^{T_0} 0 \cdot dt \right] = \frac{A^2 \cdot T_0/2}{T_0} = \frac{A^2}{2} \, W$$

2. **Determine the power in the fundamental frequency by finding the power in the $n = 1$ and $n = -1$ complex sinusoids or simply $2|X_1|^2$.**

   From the square wave coefficients entry of Figure 8-11, you can write

   $$P_{fundamenatal} = 2|X_1|^2 = 2\left|\frac{A}{j\pi \cdot 1}\right|^2 = \frac{2A^2}{\pi^2} \, W$$

3. **Form the ratio $P_{fundamental}/P$:**

   $$\text{Power Fraction} = \frac{2A^2/\pi^2}{A^2/2} = \frac{4}{\pi^2} = 0.4053 \quad (40.5\%)$$

Find an example of Fourier series in action at www.dummies.com/extras/signalsandsystems.

Chapter 9

The Fourier Transform for Continuous-Time Signals and Systems

....

In This Chapter

▶ Checking out the world of Fourier transform for aperiodic signals

▶ Getting familiar with Fourier transforms in the limit

▶ Working with LTI systems in the frequency domain and the frequency response

....

*T*he Fourier transform (FT) is the gateway to the *frequency domain* — the "home, sweet home" of the frequency spectrum for signals and the frequency response for systems — for all your signals and systems analysis. In the time domain, the independent variable is time, *t;* in the frequency domain, the independent variable is frequency, *f,* in hertz or via a variable change ω in radians per second.

When you take signals and systems to the frequency domain, you not only get a frequency domain view of each, but you also have the ability to perform joint math operations. The most significant is multiplication of frequency-domain quantities — the spectrum of a signal and the frequency response of a linear time-invariant (LTI) system, for example.

Consider this: The FT convolution theorem says that multiplication in the frequency domain is equivalent to convolution in the time domain. So to pass a signal through an LTI system, you just multiply the signal spectrum times the frequency response and then use the IFT to return the product to the time domain; you can totally avoid the convolution integral. Yes, there's more where that came from in this chapter.

The Fourier series, in general, gives you the frequency spectrum of a continuous-time periodic signal (see Chapter 8). But it has limitations. For starters, a signal has to be periodic to contain the Fourier series, which means that aperiodic signals (covered in Chapter 3) are left out of this party — until you apply the FT to remove this limitation. With regard to continuous-time signals, the FT serves most of all your spectral analysis needs. What a versatile little technique!

For discrete-time signals and systems, the FT is known as the *discrete-time Fourier transform* (DTFT). Check out Chapter 11 for details on the DTFT.

This chapter is devoted to frequency domain representations. (Flip to Chapter 6 if you're looking for information on the time domain for continuous-time signals.) I describe various FT properties and theorems here and provide them in tabular form so you can access and apply them as needed in your work. I also describe *filters*, which is just a more descriptive name for an LTI system. Filters allow some signals to pass while blocking others.

Tapping into the Frequency Domain for Aperiodic Energy Signals

To transform a signal or system impulse response (described in Chapter 5) from the time domain to the frequency domain, you need an integral formula. Then, to get back to the time domain, you use the *inverse Fourier transform* (IFT), again with an integral formula.

When it comes to Fourier transforms, power and periodic go together and energy and aperiodic go together.

In this section, you begin working with the FT by exploring the amplitude and phase spectra (which are symmetry properties for real signals) and the energy spectral density. I include a collection of useful FT/IFT theorems and pairs, which makes working with the FT more efficient. Use these tables to shop for solution approaches when you're working FT-based problems.

Working with the Fourier series

The mathematical motivation for the FT comes from the Fourier series analysis and synthesis equations for $x(t)$ periodic that I cover in Chapter 8. To visualize the Fourier transform, consider the pulse train signal (described in

Chapter 8) with pulse width τ fixed. When you let the period T_0 grow to infinity, the pulse train becomes an isolated pulse, an aperiodic signal. Figure 9-1 shows side-by-side plots of the pulse train waveform $x(t)$ and the normalized amplitude line spectra as T_0 is stepped over 2τ, 10τ, and 50τ and then to infinity. (I plot $|X_n|\cdot T_0$ to maintain a constant spectral height as T_0 changes.)

The spectral lines get closer together as T_0 increases because the line spacing is the fundamental frequency $f_0 = 1/T_0$. In the last row, T_0 goes to infinity, which makes $x(t)$ a single rectangular pulse and the line spectrum a continuous function of frequency f. This solution reveals the Fourier transform of the τ-width rectangular pulse, which you find by using the Fourier transform in the next section, Example 9-1.

In Figure 9-1, the pulse train has Fourier coefficient magnitude $|X_n| = A\tau/T_0\left|\text{sinc}(nf_0\tau)\right|$, where $\text{sinc}(x) = \sin(\pi x)/(\pi x)$. To maintain a constant spectral height in the plots, I normalize X_n as $|X_n|T_0 = A\tau\left|\text{sinc}(nf_0\tau)\right|$.

The line spectra is really just sampling the function $A\tau\left|\text{sinc}(f\tau)\right|$ at $f = nf_0$. When $T_0 \to \infty$, you also have $f_0 \to 0$ and can let $n \to \infty$ such that $nf_0 \to f$. This limiting argument makes $|X_n|T_0 \to A\tau\left|\text{sinc}(f\tau)\right|$, which is the FT of the aperiodic signal $\Pi(t/\tau)$.

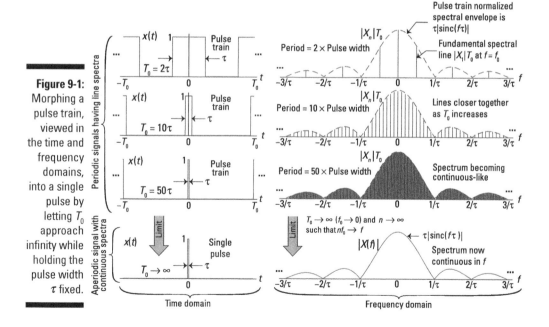

Figure 9-1: Morphing a pulse train, viewed in the time and frequency domains, into a single pulse by letting T_0 approach infinity while holding the pulse width τ fixed.

Using the Fourier transform and its inverse

To get to the FT, $x(t)$ must be aperiodic — and have finite energy (energy signals are covered in Chapter 3): $\int_{-\infty}^{\infty} |x(t)|^2 dt < \infty$.

The forward FT is given by $X(f) = \mathcal{F}\{x(t)\} = \int_{-\infty}^{\infty} x(t)e^{-j2\pi ft} dt$, with $X(f)$ being the frequency domain representation of $x(t)$. By formalizing the limit arguments described in the previous section, you can return to the time domain by using the inverse Fourier transform (IFT):

$$x(t) = \mathcal{F}^{-1}\{X(f)\} = \int_{-\infty}^{\infty} X(f)e^{j2\pi ft} df$$

Using the calligraphy character $\mathcal{F}$ to denote the FT and $\mathcal{F}^{-1}$ to denote the IFT is common practice because it's clearly distinguishable. For instance, $F\{x(t)\}$ may represent some function of $x(t)$ but not necessarily the Fourier transform. This practice reserves the roman F to be safely used for other purposes.

The transform $X(f)$ is known as the *spectrum* of $x(t)$ and is a continuous function of frequency. In general, it's also complex. Here are a couple of things to know about units associated with $X(f)$:

- If $x(t)$ has units of voltage, then $X(f)$ has units v-s or v/Hz, because time and frequency are reciprocals.

- The IFT faithfully returns $x(t)$ to the time domain and the original units of volts.

You can also apply the Fourier transform to signals by using a radian frequency variable $\omega = 2\pi f$. The forward and inverse transforms in this case are

$$X(\omega) = \int_{-\infty}^{\infty} x(t)e^{-j\omega t} dt \text{ and } x(t) = \frac{1}{2\pi} \int_{-\infty}^{\infty} X(\omega)e^{j\omega t} d\omega$$

In this book, I use the frequency variable f, which has units of hertz, and the corresponding FT/IFT integral formulas presented at the beginning of this section. Other books may use the radian frequency variable. The tables of transform theorems and transform pairs in this chapter (see Figures 9-7 and 9-9) contain an extra column for $X(\omega)$ to make it easier for you to adapt this info to your specific needs.

Example 9-1: Consider this pulse signal:

$$x(t) = A\Pi\left(\frac{t-t_0}{\tau}\right) = \begin{cases} A, & |t-t_0| < \tau/2 \\ 0, & \text{otherwise} \end{cases}$$

Use the definition of the FT, the integral formula, that takes you from the time domain, $x(t)$, to the frequency domain, $X(f)$:

$$X(f) = \int_{-\infty}^{\infty} x(t)e^{-j2\pi ft}\,dt = \int_{-\infty}^{\infty}\left[A\Pi\left(\tfrac{t-t_0}{\tau}\right)\right]e^{-j2\pi ft}\,dt$$

$$= A\int_{t_0-\tau/2}^{t_0+\tau/2} e^{-j2\pi ft}\,dt = \frac{Ae^{-j2\pi ft}}{-j2\pi f}\bigg|_{t_0-\tau/2}^{t_0+\tau/2}$$

$$= A\tau \cdot \frac{e^{j\pi f\tau} - e^{-j\pi f\tau}}{(j2)\pi f\tau}\cdot e^{-j2\pi ft_0} = A\tau \cdot \frac{\sin(\pi f\tau)}{\pi f\tau}\cdot e^{-j2\pi ft_0}$$

$$= A\tau\,\mathrm{sinc}(f\tau)e^{-j2\pi ft_0}$$

In the third line, I use Euler's formula for sine. To get to the last line, I use the function $\mathrm{sinc}(x) = \sin(\pi x)/(\pi x)$ (defined in Chapter 8).

The sinc function is quite popular when dealing with Fourier transforms. Whenever rectangular pulse functions are present in the time domain, you find a sinc function in the frequency domain. Knowing sinc spectral shapes comes in handy when you're working with digital logic and communication waveforms. Check out a plot of the sinc function in Figure 9-2.

Figure 9-2:
A plot
of the sinc
function.

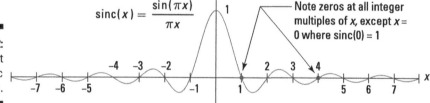

The sinc function has periodic zeros at the integers $\pm 1, \pm 2$ and so on. By using L'Hôspital's rule, you find out that $\mathrm{sinc}(0) = 1$:

$$\lim_{x\to 0}\mathrm{sinc}(x) = \lim_{x\to 0}\frac{\sin(\pi x)}{\pi x} = \lim_{x\to 0}\frac{d\sin(\pi x)/dx}{d\pi x/dx} = \lim_{x\to 0}\frac{\pi\cos(\pi x)}{\pi} = 1$$

In this example, the first spectral null occurs when $f\tau = 1$ or $f = 1/\tau$ Hz, establishing the important *Fourier transform pair:*

$$A\Pi\left(\frac{t-t_0}{\tau}\right)\overset{\mathcal{F}}{\longleftrightarrow}A\tau\,\mathrm{sinc}(f\tau)\cdot e^{-j2\pi ft_0}$$

You can denote FT pairs with the double arrow $x(t)\overset{\mathcal{F}}{\longleftrightarrow}X(f)$.

Oh, I almost forgot; beyond the sinc function is the term $e^{-j2\pi ft_0}$, which occurs due to the time shift of the rectangular pulse. If $t_0 = 0$, this term goes away. It contributes only to the angle of $X(f)$.

Example 9-2: Consider this aperiodic signal:

$$x(t) = A_1\Pi\left(\frac{t}{\tau_1}\right) + A_2\Pi\left(\frac{t}{\tau_2}\right)$$

The following steps show you how to sketch $x(t)$ for $\tau_1 > \tau_2$ and find its FT:

1. **Add the two piecewise continuous pulse functions, accounting for the fact that the τ_2-width pulse fits inside the τ_1-width pulse.**

   Using the definition of $\Pi(t/\tau)$ established in Example 9-1, you get this:

   $$x(t) = \begin{cases} A_1, & \tau_2/2 < |t| < \tau_1/2 \\ A_1 + A_2, & |t| \le \tau_2/2 \\ 0, & \text{otherwise} \end{cases}$$

   See the corresponding sketch in Figure 9-3.

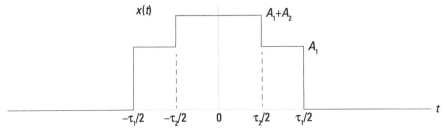

Figure 9-3:
Waveform
plot of $x(t)$.

2. **Find $X(f)$ by plugging $x(t)$ into the definition. Then form two integrals:**

   $$X(f) = \int_{-\infty}^{\infty}\left[A_1\Pi\left(\frac{t}{\tau_1}\right) + A_2\Pi\left(\frac{t}{\tau_2}\right)\right]e^{-j2\pi ft}dt$$

   $$= \int_{-\infty}^{\infty} A_1\Pi\left(\frac{t}{\tau_1}\right)e^{-j2\pi ft}dt + \int_{-\infty}^{\infty} A_2\Pi\left(\frac{t}{\tau_2}\right)e^{-j2\pi ft}dt$$

3. **Recognize that each FT integral in the second line of Step 2 is a sinc function.**

   When you plug in the problem-specific constants, you get this equation:

   $$X(f) = A_1\tau_1\,\text{sinc}(f\tau_1) + A_2\tau_2\text{sinc}(f\tau_2)$$

   This example demonstrates that the FT is a *linear operator:*

   $$F\{ax_1(t) + bx_2(t)\} = aF\{x_1(t)\} + bF\{x_2(t)\} = aX_1(f) + bX_2(f)$$

Getting amplitude and phase spectra

The very nature of the FT, the integral definition, makes $X(f)$ a complex valued function of frequency f. As with any complex valued function, you can consider the rectangular form's real and imaginary parts or the polar form's magnitude and angle.

The polar form is the most common way of characterizing $X(f)$. The magnitude $|X(f)|$ is known in some circles as the *amplitude spectrum;* others prefer the more mathematically precise term *magnitude spectrum.* I use these two terms interchangeably. The angle $\angle X(f)$ is called the *phase spectrum.*

$$X(f) = \text{Re}\{X(f)\} + j\,\text{Im}\{X(f)\}$$
$$= |X(f)|e^{j\angle X(f)}$$

Amplitude and phase spectra also exist for periodic signals in the form of line spectra (see Chapter 8). Refer to Figure 9-1 to see how the spectral representations for these two signal classifications are related.

Seeing the symmetry properties for real signals

For the case of $x(t)$, a real valued signal (imaginary part zero), some useful properties hold:

- ✔ With Euler's identity, you can expand the complex exponential in the FT integral as $e^{-j2\pi ft} = \cos(2\pi ft) - j\sin(2\pi ft)$. Then just calculate the real and imaginary parts of $X(f)$ by using this expansion inside the FT integral:

$$X(f) = \int_{-\infty}^{\infty} x(t)e^{-j2\pi ft}\,dt = \int_{-\infty}^{\infty} x(t)\big[\cos(2\pi ft) - j\sin(2\pi ft)\big]dt$$
$$= \underbrace{\int_{-\infty}^{\infty} x(t)\cos(2\pi ft)\,dt}_{\text{Re}\{X(f)\}} - j\underbrace{\int_{-\infty}^{\infty} x(t)\sin(2\pi ft)\,dt}_{-\text{Im}\{X(f)\}}$$

- ✔ Because cosine is an even function, $\cos(-\theta) = \cos(\theta)$, and sine is an odd function, $\sin(-\theta) = -\sin(\theta)$, $X(-f)$ is related to $X(f)$, as shown in this equation:

$$X(-f) = \int_{-\infty}^{\infty} x(t)e^{-j2\pi(-f)t}\,dt$$
$$= \int_{-\infty}^{\infty} x(t)\cos(-2\pi ft)\,dt - j\int_{-\infty}^{\infty} x(t)\sin(-2\pi ft)\,dt$$
$$= \int_{-\infty}^{\infty} x(t)\cos(2\pi ft)\,dt + j\int_{-\infty}^{\infty} x(t)\sin(2\pi ft)\,dt$$
$$= X^*(f)$$

You can conclude that $X(f)$ is *conjugate symmetric,* which means that its function *conjugated* (sign changed on the imaginary part with no change to the real part) on the positive axis and mirrors the corresponding function on the negative axis: $X(-f) = X^*(f)$.

✔ Conjugate symmetry reveals that

$$|X(-f)| = |X(f)| \text{ (even function of frequency)}$$

$$\angle X(-f) = -\angle X(f) \text{ (odd function of frequency)}$$

Example 9-3: Based on the results of Example 9-1, a rectangular pulse of amplitude A, pulse width τ, and time shift t_0 has FT $A\tau \, \text{sinc}(f\tau) \cdot e^{-j2\pi f t_0}$. Let $A = 1$, $\tau = 2$ s and $t_0 = 1/3$ s and plot the amplitude and phase spectra for $-3 \le f \le 3$ Hz. Plugging values into the general result of Example 9-1, the functions are the last two lines of the following equation:

$$X(f) = 1 \cdot 2\text{sinc}(2f) \cdot e^{-j2\pi f/3}$$

$$|X(f)| = 2|\text{sinc}(2f)|$$

$$\angle X(f) = \angle \text{sinc}(2f) - 2\pi f/3$$

$$= \begin{cases} -2\pi f/3, & \text{sinc}(2f) \ge 0 \\ -2\pi f/3 + \pi, & \text{otherwise} \end{cases}$$

Although real, $\text{sinc}(2f)$ has an associated angle; negative values imply a $\pm\pi$ radians phase shift.

You can create the plot with Python and Pylab and take advantage of the abs() and angle() functions to do the heavy lifting for this problem. Here are the essential commands:

```
In [132]:  f = arange(-3,3,.01)
In [133]:  X = 2*sinc(2*f)*exp(-1j*2*pi*f/3)
In [136]:  plot(f,abs(X))
In [138]:  plot(f,angle(X))
```

The two plots that this code creates are shown in Figure 9-4.

The amplitude plot in Figure 9-4 follows from the earlier plot of the sinc function in Figure 9-2. The magnitude operation flips the sinc function's negative dips to the top side of the frequency axis, making a series of side lobes on both sides of the main lobe that's centered at $f = 0$. The main lobe extends along the frequency axis from $-1/2$ to $1/2$ Hz because $1/\tau = 1/2$ is the location of the first spectral null, or 0, as f increases from 0 Hz.

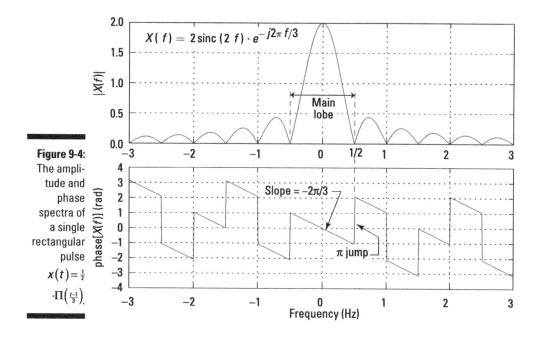

Figure 9-4:
The amplitude and phase spectra of a single rectangular pulse $x(t) = \frac{1}{2}$ $\cdot \Pi\left(\frac{t-1}{3}\right)$.

In the phase plot, the angle of the product of two complex numbers is the sum of the angles (find a review of complex arithmetic in Chapter 2). The phase of the first term, the sinc function, comes from sign changes only, so it contributes an angle (phase) of 0 or π. The second term is in polar form already, so the phase is simply $-2\pi f/3$, which is a phase slope of $-2\pi/3$. Combining terms, you get the steady phase slope with phase jumps by $\pm\pi$ being interjected whenever the sinc function is negative, creating a scenario in which the phase plot jumps up and down at integer multiples of $f = 1/2$.

In general, a phase plot includes only the principle values $\left(-\pi, \pi\right]$. When the phase attempts to go below $-\pi$ or above π, the angle() function in PyLab wraps the phase modulo 2π by either adding or subtracting 2π, and this explains why phase plots can have $\pm\pi$ and $\pm2\pi$ jumps. You can find the unwrapped phase by using the function unwrap(angle()).

Figure 9-4 also shows that the amplitude spectrum is even and the phase spectrum is odd for real $x(t)$. Two additional properties pertain to even and odd symmetry of $x(t)$. The function $g_o(t)$ is odd if $g_o(-t) = -g_o(t)$. When you integrate this function over symmetrical limits, you get 0. To verify this, follow these two steps:

1. **Integrate $g_o(t)$ with symmetrical limits $[-L, L]$ by breaking the integral into the interval $[-L, 0]$ and $[0, L]$:**

$$\int_{-L}^{L} g_o(t)\,dt = \int_{-L}^{0} g_o(t)\,dt + \int_{0}^{L} g_o(t)\,dt = -\int_{-L}^{0} g_o(-t)\,dt + \int_{0}^{L} g_o(t)\,dt$$

2. **Change variables in the first integral by letting $u = -t$, which also means $-dt = du$ and the limits now run over $[L, 0]$.**

With a sign change to the first integral, you can change the limits to match the first integral:

$$\int_{-L}^{L} g_o(t)dt = \int_{L}^{0} g_o(u)du + \int_{0}^{L} g_o(t)dt = \int_{0}^{L} g_o(u)du + \int_{0}^{L} g_o(t)dt = 0$$

The even and odd symmetry properties reveal these solutions:

✔ For $x(t)$ even, or $x(-t) = x(t)$,

$$\mathrm{Im}\{X(f)\} = \underbrace{\int_{-\infty}^{\infty} x(t)\sin(2\pi ft)dt}_{\text{even} \times \text{odd} = \text{odd}} = 0$$

✔ For $x(t)$ odd, $x(-t) = -x(t)$,

$$\mathrm{Re}\{X(f)\} = \underbrace{\int_{-\infty}^{\infty} x(t)\cos(2\pi ft)dt}_{\text{odd} \times \text{even} = \text{odd}} = 0$$

Refer to $x(t)$, Example 9-2, and Figure 9-3. The imaginary part of $X(f)$ must be 0 because $x(t)$ is even. Look at the solution for $X(f)$ in Example 9-2. It's true! The sum of two sinc functions is a real spectrum.

Example 9-4: Consider the signal $x(t)$ shown in Figure 9-5.

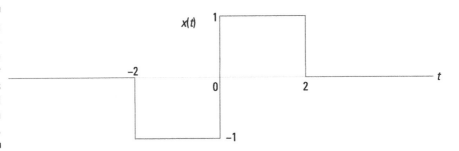

Figure 9-5:
The difference of two rectangular pulses configured to create an odd function.

In mathematical terms, $x(t)$ can be written as the difference between two time-shifted rectangular pulse functions: $\Pi\left(\frac{t-1}{2}\right) - \Pi\left(\frac{t+1}{2}\right)$.

Find the FT of $x(t)$ by using the FT linearity property described in Example 9-2. Because $\Pi\left[(t-t_0)/\tau\right] \overset{F}{\longleftrightarrow} \tau\,\mathrm{sinc}(f\tau)\exp\left[-j2\pi ft_0\right]$,

$$X(f) = 2\,\mathrm{sinc}(2f)e^{-j2\pi f \cdot 1} - 2\,\mathrm{sinc}(2f)e^{-j2\pi f \cdot (-1)}$$

$$= 2\,\mathrm{sinc}(2f) \cdot \frac{e^{-j2\pi f} - e^{j2\pi f}}{2j} \cdot 2j$$

$$= -4j \cdot \mathrm{sinc}(2f) \cdot \sin(2\pi f)$$

In the last line, I use Euler's inverse formula for sine to simplify the expression for $X(f)$.

In Figure 9-5, $x(t)$ has odd symmetry, so you may assume that $X(f)$ is pure imaginary or real part identically 0. You're right; it is!

Finding energy spectral density with Parseval's theorem

The energy spectral density of energy signal $x(t)$ is defined as $|X(f)|^2$. The units of $X(f)$ are v-s (covered in the section "Using the Fourier transform and its inverse," earlier in this chapter), so squaring in a 1-ohm system reveals that W-s^2 = W-s/Hz = Joules/Hz. Indeed, the units correspond to energy per hertz of bandwidth.

In Chapter 8, I describe Parseval's theorem for power signals; here's how the theorem applies to energy signals:

$$E_x = \int_{-\infty}^{\infty} |x(t)|^2 \, dt \overset{\text{also}}{=} \int_{-\infty}^{\infty} |X(f)|^2 \, df$$

Use this theorem to integrate the energy spectral density of a signal over all frequency to get the total signal energy.

Getting the proof for Parseval's theorem

The proof for Parseval's theorem is straightforward. Two steps, and you're done:

1. Start from the time-domain side of the theorem by writing the integrand as $|x(t)|^2 = x^*(t) \cdot x(t)$, and then replace x(t) with the IFT integral formula:

$$\int_{-\infty}^{\infty} |x(t)|^2 \, dt = \int_{-\infty}^{\infty} x^*(t) \cdot \underbrace{\left[\int_{-\infty}^{\infty} X(f) e^{j2\pi ft} \, df \right]}_{x(t)} \cdot dt$$

2. Rewrite the results of Step 1 by interchanging the integration order and recognizing that the inner integral involving $x^*(t) e^{j2\pi ft}$ is just the conjugate of X(f).

You're left with $X(f)X^*(f) = |X(f)|^2$ as the integrand of the remaining integral:

$$\int_{-\infty}^{\infty} |x(t)|^2 \, dt = \int_{-\infty}^{\infty} X(f) \cdot \underbrace{\left[\int_{-\infty}^{\infty} x(t) e^{-j2\pi ft} \, dt \right]^*}_{X(f)} \cdot df = \int_{-\infty}^{\infty} |X(f)|^2 \, df$$

Example 9-5: Find the signal energy and sketch the energy spectral density in decibels (dB) for the signals $x_1(t) = \Pi(t/4)$ (a rectangle pulse) and $x_2(t) = \Pi\left(\frac{t-1}{2}\right) - \Pi\left(\frac{t+1}{2}\right)$ (a bi-phase pulse).

The spectrum of $x_1(t)$ is $\tau \operatorname{sinc}(f\tau)$ with $\tau = 4$, and $X_2(f)$ is identical to the results of Example 9-5. Both signals have a total pulse width of 4 s. To fairly compare the energy spectral densities, you need each of the pulses to have the same total energy. Using the time-domain version of Parseval's theorem (covered in Chapter 3), you find the signal energy by integrating $|x(t)|^2$ over the pulse duration:

$$E_{x_1} = \int_{-\infty}^{\infty} |x_1(t)|^2 \, dt = \int_{-2}^{2} |1|^2 \, dt = 4$$

$$E_{x_2} = \int_{-\infty}^{\infty} |x_2(t)|^2 \, dt = \int_{-2}^{0} |-1|^2 \, dt + \int_{0}^{2} |1|^2 \, dt = 4$$

Good, the energies are equal. To plot the energy spectral density in dB means that you plot ten times the base ten log of the energy spectrum. The energy spectral densities you plot are

$$|X_1(f)|^2 = |4\operatorname{sinc}(4f)|^2$$

$$|X_2(f)|^2 = |-4j \cdot \operatorname{sinc}(2f) \cdot \sin(2\pi f)|^2$$

If you want to plot the energy spectral density in dB using Python and PyLab, use these essential commands:

```
In [179]: f = arange(-2,2,.01)
In [180]: X1 = abs(4*sinc(4*f))**2
In [181]: X2 = abs(4*sinc(2*f)*sin(2*pi*f))**2
In [181]: plot(f,10*log10(X1))
In [182]: plot(f,10*log10(X2))
```

Check out the results in Figure 9-6.

Figure 9-6: Energy spectral density in dB comparison for 4-s rectangle and bi-phase pulses.

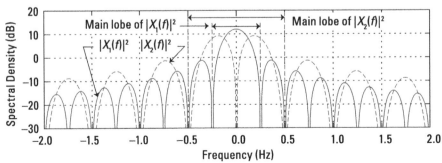

In communications and radar applications, the energy spectral density is an important design characteristic. The rectangular and bi-phase pulse shapes used here are popular in wired digital communications. The main lobe for each of the energy spectral densities is noted in Figure 9-6. The main lobe serves as a measure of spectral bandwidth. I can define the signal bandwidth B as the frequency span from 0 Hz to the location of the first *spectral null,* the point where the spectrum is 0 or negative infinity in dB. The first null of the rectangle pulse spectrum is one over the pulse width or $B_1 = 0.25$ Hz. For the bi-phase pulse, the bandwidth is $B_2 = 0.5$ Hz. The rectangle pulse seems like the bandwidth-efficient choice, but the bi-phase pulse doesn't contain any spectral energy at direct current (DC; $f = 0$), which means it can pass through cable interfaces that block DC.

Applying Fourier transform theorems

A handful of the most popular FT theorems can make life so much more pleasant. I summarize helpful theorems in Figure 9-7 and describe the core theorems in this section. The featured theorems are applicable in many electrical engineering situations. Other theorems have special application, and having them handy when they're needed is mighty nice.

	Property	Signal	Fourier Transform in f	Fourier Transform in ω
1	Linearity	$ax_1(t) + bx_2(t)$	$aX_1(f) + bX_2(f)$	$aX_1(\omega) + bX_2(\omega)$
2	Time delay	$x(t - t_0)$	$X(f)e^{-j2\pi f t_0}$	$X(\omega)e^{-j\omega t_0}$
3	Time scaling	$x(at)$	$\frac{1}{|a|}X\left(\frac{f}{a}\right)$	$\frac{1}{|a|}X\left(\frac{\omega}{a}\right)$
4	Frequency translation	$x(t)e^{j2\pi f_0 t}$	$X(f - f_0)$	$X(\omega - \omega_0)$
5	Modulation	$x(t)\cos(2\pi f_0 t)$	$\frac{1}{2}[X(f - f_0) + X(f + f_0)]$	$\frac{1}{2}[X(\omega - \omega_0) + X(\omega + \omega_0)]$
6	Convolution	$x_1(t) * x_2(t)$	$X_1(f) \cdot X_2(f)$	$X_1(\omega) \cdot X_2(\omega)$
7	Multiplication	$x_1(t) \cdot x_2(t)$	$X_1(f) * X_2(f)$	$\frac{1}{2\pi}X_1(\omega) * X_2(\omega)$
8	Differentiation	$\frac{d^n}{dt^n}x(t)$	$(j2\pi f)^n X(f)$	$(j\omega)^n X(\omega)$
9	Integration	$\int_{-\infty}^{t} x(\tau)\,d\tau$	$\frac{1}{j2\pi f}X(f) + \frac{1}{2}X(0)\delta(f)$	$\frac{1}{j\omega}X(\omega) + \pi X(0)\delta(\omega)$
10	Parseval's theorem	$\int_{-\infty}^{\infty}|x(t)|^2\,dt$	$\int_{-\infty}^{\infty}|X(f)|^2\,df$	$\frac{1}{2\pi}\int_{-\infty}^{\infty}|X(\omega)|^2\,d\omega$

Figure 9-7: Fourier transform theorems.

Linearity

The linearity theorem tells you that a linear combination of signals can be transformed term by term:

$$ax_1(t) + bx_2(t) \xleftrightarrow{\ \mathcal{F}\ } aX_1(f) + bX_2(f)$$

The proof of this theorem follows from the linearity of integration itself:

$$\int_{-\infty}^{\infty}\left[ax_1(t)+bx_2(t)\right]e^{-j2\pi ft}dt = a\underbrace{\int_{-\infty}^{\infty}x_1(t)e^{-j2\pi ft}\,dt}_{X_1(f)} + b\underbrace{\int_{-\infty}^{\infty}x_2(t)e^{-j2\pi ft}\,dt}_{X_2(f)}$$

I use this theorem for the solution of Example 9-2.

Time delay

The time delay theorem tells you how the FT of a time-delayed signal is related to the FT of the corresponding undelayed signal. I cover time-shifting signals in Chapter 3; here you can see what this technique looks like in the frequency domain.

$$x(t-t_0)\xleftrightarrow{\ \mathcal{F}\ }X(f)e^{-j2\pi ft_0}$$

Here's the proof. I use the variable substitution $\lambda = t - t_0$, which results in $t \to \lambda + t_0$ and $dt \to d\lambda$, followed by a factoring of $e^{-j2\pi ft_0}$:

$$\mathcal{F}\left\{x(t-t_0)\right\}=\int_{-\infty}^{\infty}x(\underbrace{t-t_0}_{\lambda})e^{-j2\pi ft}\,dt = \int_{-\infty}^{\infty}x(\lambda)e^{-j2\pi f(\lambda+t_0)}\,d\lambda$$

$$= e^{-j2\pi ft_0}\int_{-\infty}^{\infty}x(\lambda)e^{-j2\pi f\lambda}\,d\lambda = X(f)e^{-j2\pi ft_0}$$

I put the time delay theorem in action in Example 9-1 to find $\mathcal{F}\left\{\Pi([t-t_0]/\tau)\right\}$. It's important to remember that $\mathcal{F}\left\{\Pi(t/\tau)\right\}=\tau\mathrm{sinc}(f\tau)$. With the t replaced by $t-t_0$, you modify the FT by simply including the extra term $e^{-j2\pi ft_0}$.

Frequency translation

The frequency translation theorem can help you find the frequency-domain impact of multiplying a signal by a complex sinusoid $e^{j2\pi f_0 t}$. The theorem tells you that the spectrum of $x(t)$ is shifted up in frequency by f_0.

In communication applications, frequency translation is a common occurrence; after all, translating a signal from one frequency location to another is how signals are transmitted and received wirelessly. This theorem is also the foundation of the modulation theorem.

$$x(t)e^{j2\pi f_0 t}\xleftrightarrow{\ \mathcal{F}\ }X(f-f_0)$$

To prove, combine the exponential terms in the FT integral and then notice that the new frequency variable for the FT integral is $f-f_0$ — just $x(t)$ being transformed:

$$\int_{-\infty}^{\infty}x(t)e^{j2\pi f_0 t}\cdot e^{-j2\pi ft}\,dt = \int_{-\infty}^{\infty}x(t)e^{-j2\pi(f-f_0)t}\,dt$$

$$= X(f-f_0)$$

Example 9-6: An extremely practical application of the frequency translation theorem is finding the FT of $x(t)\cos(2\pi f_0 t)$. The first step is to expand the cosine by using Euler's formula:

$$x(t)\cos(2\pi f_0 t) = x(t)\left[\tfrac{1}{2}e^{j2\pi f_0 t} + \tfrac{1}{2}e^{-j2\pi f_0 t}\right] = \tfrac{1}{2}\left[x(t)e^{j2\pi f_0 t} + x(t)e^{-j2\pi f_0 t}\right]$$

Then apply the linearity theorem to the expanded form:

$$\mathcal{F}\{x(t)\cos(2\pi f_0 t)\} = \frac{1}{2}\mathcal{F}\{x(t)e^{j2\pi f_0 t}\} + \frac{1}{2}\mathcal{F}\{x(t)e^{-j2\pi f_0 t}\}$$
$$= \frac{1}{2}X(f-f_0) + \frac{1}{2}X(f+f_0)$$

With that, you established the modulation theorem. Read on for details.

Modulation

The modulation theorem is similar to the frequency translation theorem; only now a real sinusoid $\cos(2\pi f_0 t)$ replaces the complex sinusoid. This shifts the spectrum of $x(t)$ up and down in frequency by f_0.

$$x(t)\cos(2\pi f_0 t) \xleftrightarrow{\mathcal{F}} \tfrac{1}{2}X(f-f_0) + \tfrac{1}{2}X(f+f_0)$$

Example 9-7: The block diagram of Figure 9-8 is a simple *modulator* that places the signal $x(t)$ on the carrier frequency f_0.

Figure 9-8: A simple modulator that forms $y(t) = x(t)\cos(2\pi f_0 t)$.

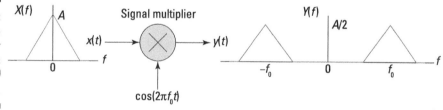

The only system building block required for this example is the ideal multiplier, which forms $y(t)$ as the product of $x(t)$ and $\cos(2\pi f_0 t)$. Notice that $y(t)$ fits the modulation theorem perfectly. The modulation theorem says $Y(f) = \mathcal{F}\{y(t)\} = [X(f-f_0) + X(f+f_0)]/2$.

The input spectrum $X(f)$ is centered at $f = 0$ with peak spectrum amplitude A. The output spectrum sketch $Y(f)$ in Figure 9-6 reveals that multiplication by $\cos(2\pi f_0 t)$ has shifted the input spectrum $X(f)$ up and down in frequency by

f_0 Hz and scaled the spectral amplitude by 1/2. By translating the input spectrum up and down in frequency, you place the information conveyed by $x(t)$ at a frequency that allows for easy wireless transmission. At the receiver, a *demodulator* recovers $x(t)$ from $y(t)$.

Duality

The duality theorem tells you that role reversal is possible with the FT. Literally, it just means that $x(t)$ is taken as $X(t)$, and $X(f)$ is taken as $x(-f)$. The theorem statement is $X(t) \xleftrightarrow{\mathcal{F}} x(-f)$.

In words, the Fourier transform of a spectrum $X(f)$, with f replaced by t, is time-domain quantity $x(t)$ with t replaced by $-f$. The time delay and frequency translation theorems described earlier in this section demonstrate this behavior.

- ✔ The time delay theorem states that a delay in the time domain means multiplication by a complex exponential in the frequency domain.

- ✔ Duality says a frequency shift in the frequency domain should result in multiplication by a complex exponential in the time domain.

Hey! That's exactly what the frequency translation theorem does.

Example 9-8: A great application of the duality theorem is in finding the IFT of the rectangular spectrum $X(f) = \Pi[f/(2W)]$. The first step is to insert $X(f)$ in the left side of the theorem with f replaced by t: $X(t) = \Pi[t/(2W)]$. The second step is to find the FT of $X(t)$.

Because $\mathcal{F}\{\Pi[t/(2W)]\} = 2W \, \text{sinc}(2Wf)$, the theorem tells you that the right side is $x(-f)$, so $2W\text{sinc}(2Wf) = x(-f) = x(f)$. The little detail of $x(-f) = x(f)$ follows from the sinc function being even.

In the final step, swap variables back: $f \to t$ in $x(f)$. A handy transform pair emerges — without doing any integration:

$$2W \, \text{sinc}(2Wt) \xleftrightarrow{\mathcal{F}} \Pi\left(\tfrac{f}{2W}\right)$$

Proving duality

To prove the duality theorem, use direct substitution into the FT definition:

$$\mathcal{F}\{X(t)\} = \int_{-\infty}^{\infty} X(t)e^{-j2\pi ft} \, dt = \int_{-\infty}^{\infty} X(t)e^{j2\pi(-f)t} \, dt = x(-f)$$

In the last step, the integral is viewed as the IFT with the variable of integration role reversed. In other words, t takes the place of f in the IFT, and then you identify the independent variable as $-f$.

You can now say that as a result of duality, a sinc in the time domain is a rectangle in the frequency domain. Add this one to your toolbox!

Convolution

The convolution theorem is worthy of a drum roll. Yes, it's that special. The convolution theorem is one of the most powerful FT theorems, and it's especially useful in communications applications.

The convolution theorem considers $x_1(t)*x_2(t)$ in terms of the FT, where it can be shown that $x_1(t)*x_2(t)\xleftrightarrow{\mathcal{F}} X_1(f)X_2(f)$. This result is significant because it tells you that convolution in the time domain becomes simple multiplication in the frequency domain. I'm guessing that you may be falling in love with convolution integrals right now.

Note that $x_2(t)$ may also be an LTI system impulse response, making $y(t)=x_1(t)*x_2(t)$ the output if $x_1(t)$ is the input. This aspect of the convolution theorem is explored in the later section "LTI Systems in the Frequency Domain."

The proof involves direct application of the IFT and FT definitions and interchanging orders of integration:

$$\int_{-\infty}^{\infty} x_1(\tau)x_2(t-\tau)d\tau = \int_{-\infty}^{\infty} x_1(\tau)\left[\int_{-\infty}^{\infty} X_2(f)e^{j2\pi f(t-\tau)}df\right]d\tau$$

$$= \int_{-\infty}^{\infty} X_2(f)\left[\int_{-\infty}^{\infty} x_1(\tau)e^{-j2\pi f\tau}d\tau\right]e^{j2\pi ft}df$$

$$= \int_{-\infty}^{\infty} X_1(f)X_2(f)e^{j2\pi ft}df = \mathcal{F}^{-1}\{X_1(f)X_2(f)\}$$

Example 9-9: Find the FT of $y(t)=\Pi(t/\tau)*\Pi(t/\tau)$, using the convolution theorem $Y(f)=\mathcal{F}\{y(t)\}=\tau\mathrm{sinc}(f\tau)\cdot\tau\mathrm{sinc}(f\tau)=\tau^2\mathrm{sinc}^2(f\tau)$.

In Chapter 5, I point out that the convolution of two signals is defined as $x(t)=x_1(t)*x_2(t)=\int_{-\infty}^{\infty} x_1(\lambda)x_2(t-\lambda)d\lambda$. Because convolving two equal-width rectangles yields a triangle, and the height of the triangle is the area of the full overlap,

$$y(t)=\tau\Lambda\left(\tfrac{t}{\tau}\right)=\tau\begin{cases}1-|t/\tau|, & |t|<\tau \\ 0, & \text{otherwise}\end{cases}$$

Putting the two sides together and removing a τ for each side establishes the FT pair:

$$\Lambda\left(\tfrac{t}{\tau}\right)\xleftrightarrow{\mathcal{F}}\tau\mathrm{sinc}^2(f\tau)$$

Multiplication

The multiplication theorem is the convolution theorem times two; this one considers the FT as the product of two signals. In communication systems and signal processing, this theorem is quite useful.

Whereas the modulation theorem involved a signal multiplied by a cosine, the multiplication theorem is a generalization because both signals are arbitrary.

Based on the convolution and duality theorems described earlier in this section, it follows that $x_1(t) \cdot x_2(t) \xleftrightarrow{\mathcal{F}} X_1(f) * X_2(f)$.

Example 9-10: If $x_1(t) = x_2(t) = 3\,\text{sinc}(3t)$, find $Y(f) = \mathcal{F}\{x_1(t) \cdot x_2(t)\}$. Start by finding $X_1(f) = X_2(f)$. Example 9-8 established the FT pair $2W\,\text{sinc}(2Wt) \xleftrightarrow{\text{F}} \Pi[f/(2W)]$. Here, $2W = 3$ and thus $X_1(f) = X_2(f) = \Pi(f/3)$. So you can convolve identical rectangle shapes to get the triangle $Y(f) = 3\Lambda(f/3)$.

An alternative approach to this problem is to use the dual of the transform pair $\Lambda(t/\tau) \xleftrightarrow{\mathcal{F}} \tau\,\text{sinc}^2(f\tau)$ established in Example 9-9. Try this approach on your own.

Checking out transform pairs

A *pair* is simply the corresponding time- and frequency-domain function that emerges from a FT/IFT combination. And after you develop the FT pair for a particular signal, you can use it over and over again as you work problems.

Figure 9-9 shows the most popular FT pairs — some of which rely on the Fourier transform in the limit concept. Several of these pairs are developed in the examples in this section.

Example 9-11: Develop the FT pair for $A\delta(t)$, which corresponds to Figure 9-9 Line 7:

$$\mathcal{F}\{A\delta(t)\} = \int_{-\infty}^{\infty} A\delta(t)e^{-j2\pi ft}\, dt = A, \text{ for all frequency}$$

The dual version of this pair is Figure 9-9 Line 8, $\mathcal{F}\{A\} = A\delta(f)$, which is actually a Fourier transform in the limit result.

The *reciprocal spreading property* of pulse signals and their corresponding spectra states that a signal with a wide pulse has a narrow spectrum while a signal with a narrow pulse has a wide spectrum. So, roughly speaking, the time spread and frequency spread of a pulse signal and its spectrum are reciprocals.

	Time	Frequency-f	Frequency-ω
1	$\Pi\left(\frac{t}{\tau}\right)$	$\tau\,\text{sinc}(f\tau)$	$\tau\,\text{sinc}\left(\frac{\omega\tau}{2\pi}\right)$
2	$2W\text{sinc}(2Wt)$	$\Pi\left(\frac{f}{2W}\right)$	$\Pi\left(\frac{\omega}{4\pi W}\right)$
3	$\Lambda\left(\frac{t}{\tau}\right)$	$\tau\,\text{sinc}^2(f\tau)$	$\tau\,\text{sinc}^2\left(\frac{\omega\tau}{2\pi}\right)$
4	$e^{-at}u(t),\ a>0$	$\dfrac{1}{a+j2\pi f}$	$\dfrac{1}{a+j\omega}$
5	$\dfrac{t^{n-1}}{(n-1)!}e^{-at}u(t)$	$\dfrac{1}{(a+j2\pi f)^n}$	$\dfrac{1}{(a+j\omega)^n}$
6	$e^{-a|t|}u(t)$	$\dfrac{2a}{a^2+(2\pi f)^2}$	$\dfrac{2a}{a^2+\omega^2}$
7	$A\delta(t)$	A	A
8	A	$A\delta(f)$	$2\pi A\delta(\omega)$
9	$\delta(t-t_0)$	$e^{-j2\pi ft_0}$	$e^{-j\omega t_0}$
10	$e^{j2\pi f_0 t}$	$\delta(f-f_0)$	$2\pi\delta(\omega-\omega_0)$
11	$\cos(2\pi f_0 t)$	$\frac{1}{2}[\delta(f-f_0)+\delta(f+f_0)]$	$\pi[\delta(\omega-\omega_0)+\delta(\omega+\omega_0)]$
12	$\sin(2\pi f_0 t)$	$\frac{1}{2j}[\delta(f-f_0)-\delta(f+f_0)]$	$\frac{\pi}{j}[\delta(\omega-\omega_0)-\delta(\omega+\omega_0)]$
13	$u(t)$	$\frac{1}{j2\pi f}+\frac{1}{2}\delta(f)$	$\frac{1}{j\omega}+\pi\delta(f)$
14	$\sum_{n=-\infty}^{\infty}\delta(t-nT_s)$	$\frac{1}{T_s}\sum_{m=-\infty}^{\infty}\delta(f-m/T_s)$	$\frac{2\pi}{T_s}\sum_{m=-\infty}^{\infty}\delta(\omega-2\pi m/T_s)$
15	$\sum_{n=-\infty}^{\infty}p(t-nT_s)$	$f_s\sum_{m=-\infty}^{\infty}P(mf_s)\delta(f-mf_s)$	$\frac{2\pi}{T_s}\sum_{m=-\infty}^{\infty}P(2\pi mf_s)\delta(\omega-2\pi mf_s)$
	$f_s=1/T_s$	$P(f)=\mathcal{F}\{p(t)\}$	$P(\omega)=\mathcal{F}\{p(t)\}$

Figure 9-9:
Fourier transform pairs; those containing an impulse function in the frequency domain are transforms in the limit.

Consider the impulse signal and the constant signals, whose Fourier transforms are found in Figure 9-9 Lines 7 and 8. For the impulse signal, one over the time duration yields a signal bandwidth of infinity; and for the constant signal, one over the signal duration of infinity yields a signal bandwidth of zero.

A more typical scenario is the rectangle pulse of duration τs (Figure 9-9 Line 1). This signal has a sinc function spectrum, where the main spectral lobe has single-sided width $1/\tau$ Hz (refer to Figure 9-4). Reciprocal spreading applies here.

Getting Around the Rules with Fourier Transforms in the Limit

Formally, the Fourier transform (FT) requires $x(t)$ to be an energy signal. But you don't need to always be so picky. In this section, I show you how to find the FT of power signals, including sine/cosine and periodic pulse signals, such as the pulse train. The technique is known as *Fourier transforms in the limit,* and it allows you to bring together spectral analysis of both power and energy signals in one frequency-domain representation, namely $X(f)$. The trick is getting singularity functions, particularly the impulse function, into the frequency domain.

Using the Fourier transform in the limit, I show you how to find the FT of any periodic power signal. This leads to unifying the spectral view of both periodic power signals and aperiodic energy signals under the Fourier transform.

Handling singularity functions

Singularity functions include the impulse function and the step function. Getting the impulse function into the frequency domain is a great place to start working with the Fourier transform in the limit.

Consider a constant signal $x(t) = A, -\infty < t < \infty$. You can write this signal in limit form:

$$x(t) = \lim_{T \to \infty} A\Pi\left(\tfrac{t}{T}\right), \text{ where } \mathcal{F}\left\{A\Pi\left(\tfrac{t}{T}\right)\right\} = AT\text{sinc}(fT)$$

The Fourier transform in the limit of $x(t)$ is $\mathcal{F}\{x(t)\} = \lim_{T \to \infty} AT\text{sinc}(fT)$. Figure 9-10 shows what happens as T increases to $AT\text{sinc}(fT)$.

Figure 9-10: The result of increasing T in AT sinc (fT).

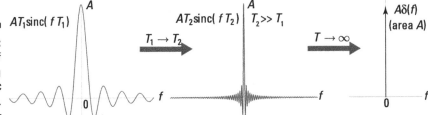

Because $x(t)$ is a constant and has no time variation, the spectral content of $X(f)$ ought to be confined to $f = 0$. It can also be shown that $\int_{-\infty}^{\infty} AT\text{sinc}(fT)\,df = A$ for all T.

A quick symbolic integration check in Maxima confirms this.

```
(%i1) sinc(x):= if x = 0 then 1 else sin(%pi*x)/(%pi*x);
```
$$(\%o1)\ \text{sinc}(x) := \text{if}\ x = 0\ \text{then}\ 1\ \text{else}\ \frac{\sin(\pi x)}{\pi x}$$

```
(%i2) assume(T>0);
(%o2) [T>0]
```

```
(%i3) integrate(A*T*sinc(f*T),f,minf,inf);
(%o3) A
```

Refer to the Fourier transform in the limit conclusion on the right side of Figure 9-7: $A \xleftrightarrow{\mathcal{F}} A\delta(f)$. As a further check, consider the inverse Fourier transform (IFT) $\mathcal{F}^{-1}\{A\delta(f)\} = \int_{-\infty}^{\infty} A\delta(f)e^{j2\pi ft}\,df = A\delta(f)e^{j2\pi ft}\big|_{f=0} = A$. Smooth sailing back to the time domain. Find more Fourier transform in the limit examples in Figure 9-9.

Unifying the spectral view with periodic signals

A periodic signal having period $T_0 = 1/f_0$ — $x(t - T_0) = x(t)$ — has this complex exponential Fourier series representation:

$$x(t) = \sum_{n=-\infty}^{\infty} X_n e^{j2\pi nf_0 t}$$

Get the Fourier coefficients by using the formula $X_n = \frac{1}{T_0} \int_0^{T_0} x(t)e^{-j2\pi nft}\,dt$.

Line 1 of Figure 9-7 and Line 10 of Figure 9-9 can help you find the FT:

$$X(f) = \mathcal{F}\left\{ \sum_{n=-\infty}^{\infty} X_n e^{j2\pi nf_0 t} \right\} = \sum_{n=-\infty}^{\infty} X_n \mathcal{F}\left\{ e^{j2\pi nf_0 t} \right\}$$

$$= \sum_{n=-\infty}^{\infty} X_n \delta(f - nf_0)$$

For this solution to be useful, you need the Fourier series coefficients X_n.

This result shows that the spectrum of a periodic signal $x(t)$ consists of spectral lines located at frequencies nf_0 along the frequency axis via impulse functions. The spectral representation with terms $X_n\delta(f-nf_0)$ resembles the line spectra in Chapter 8; in this chapter, you use impulse functions to actually locate the spectral lines.

The best thing about this FT is that it brings both aperiodic and periodic signals into a common spectral representation.

Example 9-12: Consider the ideal sampling waveform:

$$s(t)=\sum_{n=-\infty}^{\infty}\delta(t-nT_s)$$

The Fourier series coefficients of this periodic signal are

$$S_n=\frac{1}{T_s}\int_0^{T_s}\delta(t)e^{-j2\pi(nf_s)t}\,dt=\frac{1}{T_s}=f_s\text{ for any }n.\text{ Note that }f_s=1/T_s\text{ is the}$$

sampling rate. With the Fourier series coefficients known, from this FT,

$$S(f)=f_s\sum_{m=-\infty}^{\infty}\delta(f-mf_s).$$

You've just established the FT pair of Figure 9-9 Line 14:

$$\sum_{n=-\infty}^{\infty}\delta(t-nT_s)\xleftrightarrow{\ \mathcal{F}\ }f_s\sum_{m=-\infty}^{\infty}\delta(f-mf_s)$$

This FT pair is special because the signal and transform have the same mathematical form!

Two other signals have this same property; they're the zero signal and the Gaussian shaped pulse pair.

Wouldn't it be nice if you could find the FT of a periodic signal $x(t)$ without having to first find the Fourier series coefficients? Well, you can! Simply start with an alternative representation of $x(t)$, such as the following, where $p(t)$ represents one period of $x(t)$:

$$x(t)=\left[\sum_{n=-\infty}^{\infty}\delta(t-nT_s)\right]*p(t)=\sum_{n=-\infty}^{\infty}p(t-nT_s)$$

The following FT pair (Figure 9-9 Line 15) does the trick:

$$\sum_{n=-\infty}^{\infty}p(t-nT_s)\xleftrightarrow{\ \mathcal{F}\ }f_s\sum_{m=-\infty}^{\infty}P(mf_s)\delta(f-mf_s)$$

Note that $P(f)=\mathcal{F}\{p(t)\}$ and $f_s=1/T_s$. Figure 9-11 shows this FT pair.

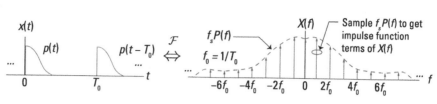

Figure 9-11:
Finding the FT of a periodic sequence by sampling the spectrum $P(f)$ of one period of $x(t)$ at $f = n/T_0$

From the convolution theorem (Figure 9-7 Line 6),

$$X(f) = \mathcal{F}\left\{ \sum_{n=-\infty}^{\infty} \delta(t - nT_s) \right\} \cdot P(f) = f_s \sum_{m=-\infty}^{\infty} \delta(f - mf_s) \cdot P(f)$$

$$= f_s \sum_{m=-\infty}^{\infty} P(mf_s)\delta(f - mf_s)$$

Example 9-13: Find the spectrum of the periodic pulse train signal
$x(t) = A \sum_{n=-\infty}^{\infty} \Pi\left[(t - nT_0 - \tau/2)/\tau \right]$. Because $p(t) = \Pi\left[(t - \tau/2)/\tau \right]$, Figure 9-9 Line 1 and Figure 9-7 Line 2 shows that $P(f) = \tau\text{sinc}(f\tau) \cdot e^{-j2\pi f\tau/2} = \tau\text{sinc}(f\tau) \cdot e^{-j\pi f\tau}$. Using the Figure 9-9 Line 15 result, you find the following:

$$X(f) = \sum_{m=-\infty}^{\infty} \tau\text{sinc}\left[(mf_0)\tau \right] e^{j\pi(mf_0)\tau} \delta(f - mf_0)$$

A little secret: Because the form is equivalent to using the Fourier series coefficients, you can assume that the coefficients are
$X_m = \tau\text{sinc}\left[(mf_0)\tau \right] e^{j\pi(mf_0)\tau}$, $m = 0, \pm 1, \pm 2$, and so on. Not a single integration required!

LTI Systems in the Frequency Domain

Chapter 5 establishes time-domain relationships for signals interacting with LTI systems. In this section, I present the corresponding frequency-domain results for signals interacting with LTI systems, beginning with the frequency response of an LTI system in relation to the convolution theorem. I introduce you to properties of the frequency response and present a case study with an RC low-pass filter that's connected with the linear constant coefficient (LCC) differential equation representation of LTI systems (see Chapter 7).

But don't worry; I also touch on cascade and parallel connection of LTI systems, or filters, in the frequency domain. *Ideal filters* are good for conceptualizing design, but they can't be physically realized. *Realizable filters* can be built but aren't as mathematically pristine as ideal filters.

Checking out the frequency response

For LTI systems in the time domain, a fundamental result is that the output $y(t)$ is the input $x(t)$ convolved with the system impulse response $h(t)$: $y(t) = x(t) * h(t)$. In the frequency domain, you can jump right into this expression by taking the Fourier transform (FT) of both sides: $Y(f) = \mathcal{F}\{y(t)\} = \mathcal{F}\{x(t) * h(t)\} = \mathcal{F}\{x(t)\} \cdot \mathcal{F}\{h(t)\} = X(f) \cdot H(f)$.

The quantity $H(f) = \mathcal{F}\{h(t)\}$ is known as the *frequency response,* or *transfer function,* of the system having impulse response $h(t)$. This is a special FT because $H(f)$ is the frequency response, not the spectrum of a signal.

In Chapter 7, I hone in on the frequency response for systems described by LCC differential equations, using the sinusoidal steady-state response. This is the same frequency response I'm describing in this chapter, but now you get it as the FT of the system impulse response. This convergence of theories shows you that you can arrive at a frequency response in different ways.

If you want to find $y(t)$ via multiplication in the frequency domain, you just need to use the inverse Fourier transform (IFT): $y(t) = \mathcal{F}^{-1}\{X(f)H(f)\}$.

Solving for $H(f)$ in the expression for $Y(f)$, $H(f) = Y(f)/X(f)$, shows you the ratio of the output spectrum over the input spectrum. So if $x(t) = \delta(t)$, then $X(f) = 1$ and the output spectrum takes its *shape* entirely from $H(f)$ because $Y(f) = 1 \cdot H(f)$.

Evaluating properties of the frequency response

For $h(t)$ real, it follows that $|H(-f)| = |H(f)|$ and $\angle H(-f) = -\angle H(f)$ (see the section "Seeing the symmetry properties for real signals," earlier in this chapter). The output energy spectral density is related to the input energy spectral density and the frequency response: $|Y(f)|^2 = |X(f)H(f)|^2 = |X(f)|^2 \cdot |H(f)|^2$.

Example 9-14: Consider the RC low-pass filter circuit shown in Figure 9-12.

Figure 9-12:
RC low-pass filter circuit and frequency-domain relationships.

$$Y(f) = X(f)\, H(f)$$
or
$$H(f) = Y(f)/X(f)$$

$h(t), H(f)$

To find $H(f)$, you can use AC steady-state circuit analysis, the Fourier transform of the impulse response, or apply the Fourier transform for the circuit differential equation term by term and solve for $H(f)$. The impulse response is derived for this circuit in Chapter 7, so I use this same approach here:

$$h(t) = 1/(RC) \cdot e^{-t/(RC)} u(t)$$

Apply the Fourier transform to $h(t)$, using Figure 9-9 Line 4 along with linearity:

$$H(f) = \mathcal{F}\{h(t)\} = \frac{1}{RC} \cdot \frac{1}{\frac{1}{RC} + j2\pi f} = \frac{1}{1 + j2\pi f RC}$$

$$= \frac{1}{1 + j f/f_3}$$

Here, $f_3 = 1/(2\pi RC)$ is known as the filter 3-dB cutoff frequency because

$$20\log_{10}\left(\left|H(f_3)\right|\right) = 20\log_{10}\left[\frac{1}{|1 + j|}\right] = 20\log_{10}\left(1/\sqrt{2}\right) = -3.01 \ dB.$$

Figure 9-13 illustrates the frequency response magnitude and phase and reveals the spectral shaping offered by the *RC* low-pass filter. The output spectrum is $Y(f) = X(f)H(f)$, so you can deduce that signals with spectral content less than f_3 are nominally passed $(0 \le |H(f)| \le 0.707)$ while spectral content greater than f_3 is attenuated, because $|H(f)|$ shrinks to 0 as f increases $(0 \le |H(f)| \le 0.707)$.

Figure 9-13:
RC low-pass filter frequency response in terms of magnitude and phase plots.

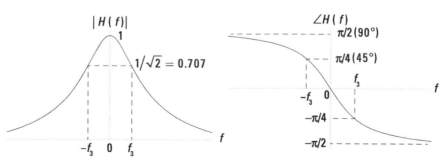

Formally, a low-pass filter passes spectral content from 0 to B Hz and blocks spectral content above B Hz. Here, B is equivalent to the 3-dB frequency f_3. More detail on filters is coming up later in this chapter.

Getting connected with cascade and parallel systems

As a result of the FT convolution theorem, the cascade and parallel connections in the frequency domain have an equivalent function to those in the time domain. You can develop the corresponding relationships from the block diagram in Figure 9-14.

Figure 9-14:
Cascade (a) and parallel (b) connected LTI systems in the frequency domain.

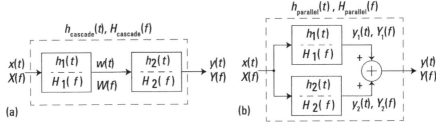

For the cascade system connection in the frequency domain, $W(f) = X(f)H_1(f)$ and $Y(f) = W(f)H_2(f)$; linking the two equations yields $Y(f) = X(f)[H_1(f)H_2(f)]$. The frequency response of the cascade is thus $H_{cascade}(f) = H_1(f)H_2(f)$. Similarly, for the parallel system connection in the frequency domain, $Y_1(f) = X(f)H_1(f)$, $Y_2(f) = X(f)H_2(f)$, and $Y(f) = Y_1(f) + Y_2(f)$, so $Y(f) = Y_1(f) + Y_2(f) = X(f)[H_1(f) + H_2(f)]$ and $H_{parallel}(f) = H_1(f) + H_2(f)$.

Both of these relationships come in handy when you work with system block diagrams and need to analyze the frequency response of an interconnection of LTI subsystem blocks.

Ideal filters

In the frequency domain, you can make your design intentions clear: Pass or don't pass signals by proper design of $H(f)$. From the input/output relationship in the frequency domain, $Y(f) = X(f)H(f)$, $H(f)$ controls the spectral content of the output. To ensure that spectral content of $X(f)$ over $f_1 \leq f \leq f_2$

isn't present in $Y(f)$, you simply design $|H(f)| \approx 0$ over the same frequency interval. Conversely, to retain spectral content, make sure that $|H(f)|$ equals a nonzero constant on $f_1 \leq f \leq f_2$. This is filtering.

If you drink coffee, you know that a filter separates the coffee grounds from the brewed coffee. A coffee filter is designed to pass the coffee down to your cup and hold back the grounds.

Four types of ideal filters are *low-pass (LP)*, *high-pass (HP)*, *band-pass (BP)*, and *band-stop (BS)*. Figure 9-15 shows the frequency response magnitude of these filters. Assume the phase response is 0 in all cases.

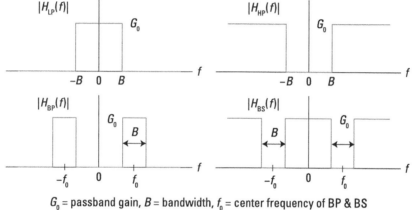

Figure 9-15: The frequency response of ideal filters: LP, HP, BP, and BS.

G_0 = passband gain, B = bandwidth, f_0 = center frequency of BP & BS

The *passband* corresponds to the band of frequencies passed by each of the filters. Ideal filters aren't realizable, meaning you can't actually build the filter; but they simplify calculations and give useful performance results during the initial phases of system design.

From the frequency response sketches of Figure 9-15, mathematical models for each filter type are possible. Consider this low-pass filter:

$$H_{LP}(f) = G_0 \Pi\left(\frac{f}{2B}\right)$$
$$h_{LP}(t) = \mathcal{F}^{-1}\{H_{LP}(f)\} = G_0 2B \mathrm{sinc}(2Bt)$$

You can write the high-pass filter as 1 minus the low-pass filter, such as

$$H_{HP}(f) = G_0\left[1 - \Pi\left(\frac{f}{2B}\right)\right]$$
$$h_{HP}(t) = G_0\left[1 - 2B \mathrm{sinc}(2Bt)\right]$$

Similar relationships exist for the band-pass and band-stop filters.

In all cases, the impulse response of an ideal filter is nonzero for $t < 0$, which, for an LTI system, means the filter is non-causal (see Chapter 5) — the filter output requires knowledge of future values of the input.

Realizable filters

You can approximate ideal filters with realizable filters, including the *Butterworth*, *Chebyshev*, and *Bessel* designs. At the heart of these three filter designs is a linear constant coefficient (LCC) differential equation having frequency response of the form

$$H(f) = \sum_{k=0}^{M} b_k \cdot (j2\pi f)^k \Big/ \sum_{k=0}^{N} a_k \cdot (j2\pi f)^k$$

M and N are positive integers. The $\{a_k\}$ and $\{b\}$ filter coefficients hold the keys to making the filter approximate one of the four ideal filter types. Figure 9-16 shows a third-order Chebyshev band-pass filter magnitude response.

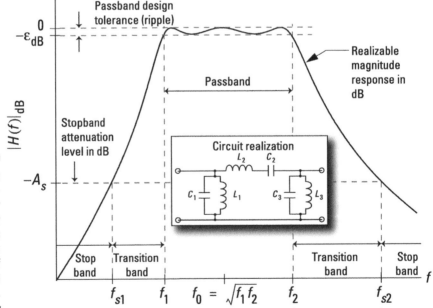

Figure 9-16: A third-order Chebyshev band-pass filter as an example of a realizable filter.

The bandwidth B in Figure 9-16 is $f_2 - f_1$, where f_1 and f_2 are the passband cutoff frequencies and f_{s1} and f_{s2} are the stopband cutoff frequencies. The filter stopband, where approximately no signals pass through the filter, is when $f < f_{s1}$ and $f > f_{s2}$. The center frequency is the geometric mean of the passband edges, or $f_0 = \sqrt{f_2 f_1}$.

Chapter 10

Sampling Theory

*T*he discrete-time signals you deal with in the real world often occur as a result of periodic (uniform) sampling of continuous-time signals. I'm talking about speech signals going into a cellphone, music signals captured in a recording studio, sensor signals in your car, radio signals in wireless devices, and the like. In this chapter, I describe the fundamentals of sampling theory and show you that designing a sampling system that allows for accurate reconstruction of a continuous-time signal from *only* discrete-time samples isn't too difficult.

Basically, sampling theory is the mathematics that explains the details of the signal-conversion process to and from the discrete-time domain. In the study of sampling theory, using the time-domain view (see Chapter 3) and the frequency-domain view (explored in Chapter 9) can help you get a handle on the underlying mathematics.

As this chapter unfolds, I cover the reconstruction filter, the antialiasing filter, and the quantizer, which is internal to the analog-to-digital converters (ADC). But the main focus is periodic sampling of a continuous-time signal to create a discrete-time signal. The sample values are taken as real (infinite precision) numbers for the most part — I pretty much ignore the *bit width* of the ADC; however, I briefly consider finite precision effects before presenting the frequency-domain view.

If you're curious about the various ADC/DAC technologies available, search the web for "successive approximation register (SAR)," "delta-sigma (ΔΣ)," and "flash ADCs."

Seeing the Need for Sampling Theory

All the great attributes of discrete-time signals and systems rely on the ability to interface with the continuous-time domain. Analog-to-digital converters (ADCs) and digital-to-analog converters (DACs) are the electronic subsystems that convert signals between continuous-time and discrete-time signal forms. To that end, Figure 10-1 shows a top-level depiction of how to implement the interface of these two subsystems. The various components in this figure (except the first block, the antialiasing fliter) are described throughout this chapter. Find out how to use the antialiasing filter to avoid alaising at the ADC at www.dummies.com/extras/signalsandsystems.

Figure 10-1:
A block diagram showing how a discrete-time signal processing system (top center) interfaces with continuous-time signals, both input and output scenarios.

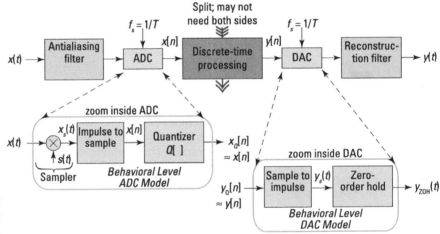

You want to process a continuous-time signal in the discrete-time domain to achieve better system performance and to enjoy the flexibility of a software-programmable solution. Some applications may require only the ADC and discrete-time processing. Other applications, such as a CD player, may need only the DAC side of the system.

The ADC converts $x(t)$ to $x[n] = x(nT)$ by using the sampling rate clock input $f_s = 1/T$. A discrete-time system processes $x[n]$ to produce output $y[n]$. The discrete-time signal $y[n]$ is converted back to continuous-time signal $y(t)$ via the DAC.

At the heart of sampling theory is the *minimum sampling rate,* or one over the sample spacing, which is needed to ensure that the original continuous-time signal can be reconstructed from its samples.

Periodic Sampling of a Signal: The ADC

The ADC is responsible for converting the continuous-time *(analog)* signal $x(t)$ into the corresponding discrete-time sequence $x[n]$. From the ADC inset of Figure 10-1, follow these three steps to transform $x(t)$ into a sequence of numbers corresponding to samples of $x(t)$ taken at times nT, n the sample index.

1. **Take samples of $x(t)$ by multiplying by an impulse train signal:**

$$s(t) = \sum_{n=-\infty}^{\infty} \delta(t - nT)$$

   The product $x_s(t) = x(t)s(t) = x(t) \sum_{n=-\infty}^{\infty} \delta(t - nT) = \sum_{n=-\infty}^{\infty} x(nT) \cdot \delta(t - nT)$ remains a continuous-time signal because the waveform samples $x(nT)$ are still placed along the time axis. I explain why this is important in the next section on the frequency-domain view of sampling theory.

2. **Convert the impulse train signal into a sequence of numbers:**

$$x[n] = x(nT) = x(n/f_s)$$

   In a real ADC, you can accomplish this step seamlessly via the electronics of the ADC. Here, keep in mind that the sequence values $x[n]$ have infinite precision, as you may expect for a continuous-time signal.

3. **Quantize the sequence values $x[n]$. In other words, convert the infinite precision real number values to B-bit precision numbers.**

   B is an integer, which refers to the word length of the data type used to store the signal samples. In the notation of Figure 10-1, $x[n] \rightarrow x_Q[n]$. The most common data type used for this purpose is the *signed integer*, which reserves one bit to convey sign information and $B - 1$ bits to convey the magnitude. For example, 16-bit signed integers range in value from $-2^{16-1} = -32{,}768$ to $2^{15-1} - 1 = 32{,}767$.

I cover the quantizer aspect of the ADC in the section "Analyzing the Impact of Quantization Errors in the ADC," later in this chapter. Don't worry: When B is large, $x_Q[n] \approx x[n]$. The net result of the idealized ADC is to produce $x_Q[n] \approx x[n] = x(nT)$ from the input $x(t)$.

Example 10-1: Consider sampling the continuous-time sinusoid $x(t) = A\cos(2\pi f_0 t + \phi)$ at sample spacing T or equivalently a sampling rate of $f_s = 1/T$. Here is the sampled signal $x[n]$:

$$x[n] = x(nT) = A\cos(2\pi \cdot f_0 \cdot nT + \phi) = A\cos(2\pi \cdot f_0 / f_s \cdot n + \phi)$$
$$= A\cos(\hat{\omega}_0 n + \phi)$$

I use the notation $\hat{\omega}_0 \equiv 2\pi f_0 T = 2\pi f_0 / f_s$ for discrete-time frequency. Even though $\hat{\omega}$ has units of radians, to emphasize the fact that sampling is involved,

I prefer to think of the units as radians/sample. Because $\omega = 2\pi f$, you can also define $\hat{f} \equiv fT = f/f_s$ as the discrete-time frequency in cycles/sample.

You may be wondering how many samples per period of $x(t)$ are required. In Figure 10-2, I plot five scenarios of $x(t)$ and $x[n]$ for $A = 1$ and f_0, taking on four values. Here are the IPython command line entries for creating a single subplot of Figure 10-2.

```
In [412]: t = arange(0,10,.01)
In [413]: n = arange(0,11)
In [414]: plot(t,cos(2*pi*0.1*t)) # f0/fs = 0.1/1.0 = 0.1
In [415]: stem(n,cos(2*pi*0.1*n),'r','ro')
```

Figure 10-2a probably looks the nicest because it has ten samples per period. But looking *nice* isn't of primary importance. The objective is to verify that you can reconstruct $x(t)$ from the sample values .

Figure 10-2b has only four samples per period, but it's adequate. Figure 10-2c is too lean; both of the plots have just two samples per period. At only two samples per period, the plot shows that with proper phasing, $\phi = \pi/2$, the samples become 0 everywhere. Not good!

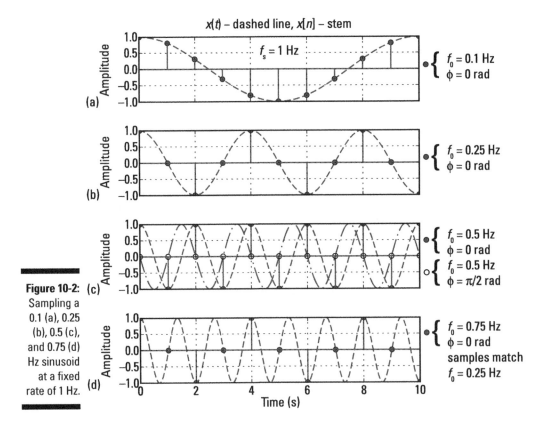

Figure 10-2: Sampling a 0.1 (a), 0.25 (b), 0.5 (c), and 0.75 (d) Hz sinusoid at a fixed rate of 1 Hz.

The *low-pass sampling theorem,* which I cover in the later section "Applying the Low-Pass Sampling Theorem," states that f_s must be greater than twice the highest frequency contained in $x(t)$ to ensure reconstruction of $x(t)$ from its samples. For the case of a single sinusoid, this is equivalent to $f_s > 2f_0$ or greater than two samples per period. Only Figures 10-2a and 10-2b satisfy this condition.

Figure 10-2d, $f_0 = 0.75$ Hz, has fewer than two samples per period ($f_s < 2f_0$) and the sample values match those of Figure 10-2b, where $f_0 = 0.5$ Hz. To find out what's going on, look at the math behind the sequence plot, bearing in mind that cosine is a mod2π function:

$$x[n]\big|_{f_0=0.75} = \cos\left(2\pi \cdot 0.75/1.0 \cdot n\right)$$

$$= \cos\left(2\pi \cdot \frac{3}{4} \cdot n\right) \overset{\text{because mod } 2\pi}{=} \cos\left[\left(2\pi \cdot \frac{3}{4} - 2\pi\right) \cdot n\right]$$

$$= \cos\left(-2\pi \cdot \frac{1}{4} \cdot n\right) \overset{\text{because even}}{=} \cos\left(2\pi \cdot \frac{1}{4} \cdot n\right)$$

The 0.25-Hz and 0.75-Hz sinusoids produce the same sequence values when sampled at $f_s = 1.0$ Hz. You can work with $\hat{\omega}_0$ values just as easily as frequency in hertz for this problem. Here are the equivalent radians/sample frequencies:

$$f_0 = 0.75 \Leftrightarrow \hat{\omega}_0 = 2\pi \cdot 0.75/1.0 = 1.5\pi$$
$$f_0 = 0.25 \Leftrightarrow \hat{\omega}_0 = 2\pi \cdot 0.25/1.0 = 0.5\pi$$

The same result holds.

The mod2π nature of cosine is the key to making the frequency of the two real sinusoids equal in magnitude: $|0.25\pi| = |1.5\pi - 2\pi| = |-0.25\pi|$. The sign isn't an issue here because Euler's formula tells you that a real sinusoid is composed of positive and negative frequency complex sinusoids, specifically

$$x[n] = A\cos(\hat{\omega}_0 n + \phi) = \frac{A}{2}\left[e^{j(\hat{\omega}_0 n + \phi)} + e^{-j(\hat{\omega}_0 n + \phi)}\right]$$

In this case, however, the cosine is even and $\phi = 0$, so the sample values are identical, as Figures 10-2b and 10-2c show. The fact that two single sinusoid signals, at frequencies 0.5 and 1.5 Hz, when sampled at 1 Hz have the same discrete-time frequency, $\hat{\omega}_0 = 0.5\pi$, represents *aliasing*. And, when aliasing is present, you don't know the true frequency of the original input signal $x(t)$. If the input to the ADC isn't protected with an antialiasing filter (see the section "Using an Antialiasing Filter to Avoid Aliasing at the ADC") the true signal frequency can't be discovered.

If ϕ in Example 10-1 is nonzero, then the following two expressions exist for sample equality, depending on the sign of ϕ:

$$A\cos(\hat{\omega}n+\phi) = A\cos\left[(\hat{\omega}+2\pi k)n+\phi\right]$$
$$= A\cos\left[(\hat{\omega}+2\pi k)n+\phi\right]$$

Here, $k = 1, 2$, and so on, and n is arbitrary. The smallest value $\hat{\omega} \in [0, \pi]$ is called the *principle alias*.

Assuming $\hat{\omega}$ is the principle alias, the remaining alias frequencies in radians/sample are $|2\pi k \pm \hat{\omega}|$, $k = 1, 2, 3$, and so on. In terms of continuous-time frequencies in hertz, the principle alias lies on the interval $[0, f_s/2]$. Given f is the principle alias, the remaining alias frequencies in hertz are $|kf_s \pm f|$, $k = 1, 2, 3$ and so on.

You can get a graphical view of the aliased frequencies by using a strip of paper folded like an accordian. As shown in Figure 10-3, the strip represents the continuous-time frequency axis, f, and the folds in the strip occur at multiples of $f_s/2$.

The principle alias range is at the far left of Figure 10-3. Horizontal lines join together and collapse the common alias frequencies onto the principle alias or vice versa. The first fold point, $f_s/2$, is the *folding frequency*.

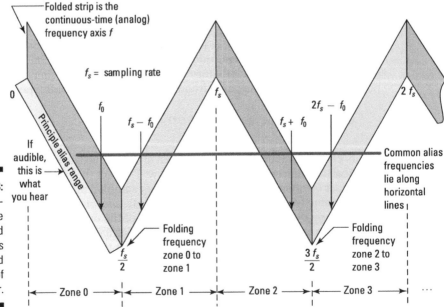

Figure 10-3: 3D depiction of the aliased frequencies via a folded strip of paper.

Figure 10-3 also shows *alias zones* as the contiguous frequency bands of length $f_s/2$, starting with the principle alias band. Each zone always contains exactly one of the alias frequencies. The horizontal line confirms this.

When you need to find the principle alias given the sampling rate and input frequency *f,* this formula comes in handy:

$$f_{\text{principle}} = \left| \text{round}\left(f/f_s \right) \cdot f_s - f \right|$$

This formula is fast and efficient, but I also highly recommend spending time with the *core relationship* $f = \left| k \cdot f_s \pm f_{\text{principle}} \right|$, where *k* is a nonnegative integer. Finding alias frequencies *f* given $f_{\text{principle}}$ is no problem; just pick a value for *k*.

But when you're given *f* and need to find $f_{\text{principle}}$, you need to first find *k* such that *f* is within $\pm f_s/2$ of kf_s. From Figure 10-3, choosing *k* sets up a pair of contiguous alias zones with kf_s at the center. For *k* = 1, you see whether *f* lies in Zones 1 or 2, for *k* = 2, Zones 3 and 4, and so on. After you find *k,* the principle alias frequency follows from the core relationship and Figure 10-3 as $f_{\text{principle}} = \left| kf_s - f \right|$. Note that f_0 in Figure 10-3 is $f_{\text{principle}}$.

To develop the formula for $f_{\text{principle}}$, start working backward from the formula for the alias frequencies: $f = \left| k \cdot f_s \pm f_{\text{principle}} \right|$.

To find *k,* an integer, such that $k \cdot f_s$ is within $f_s/2$ of *f* and as close to *f* as possible, choose $k = \text{round}(f/f_s)$. The absolute value of the difference between $k \cdot f_s$ and *f* is then $f_{\text{principle}}$.

Example 10-2: Consider an ADC (sampler) with $f_s = 10$ kHz and a single sinusoid input signal. To begin the process of finding the principle alias frequency associated with the input frequencies: (a) 3, (b) 11, (c) 36, and (d) 122 kHz, start with what you know. The principle alias frequency band is [0, 5] kHz, so now find the principle alias for each frequency, using the core equation and the handy formula $\left| \text{round}\left(f/10^4 \right) \cdot 10^4 - f \right|$ as a check. There are four parts to this problem; they're labeled a–d in this process.

Using the core equation takes two steps:

1. **Find *k* such that *f* is within $\pm f_s/2$ of kf_s.**

2. **Compute $f_{\text{principle}}$ as $\left| k \cdot f_s - f \right|$.**

   a. The 3-kHz signal is the principle alias already because it lies on the interval [0, 5] kHz. As a check, $\left| \text{round}\left(3/10 \right) \cdot 10 - 3 \right| = \left| 0 \cdot 10 - 3 \right| = 3$ kHz.

   The two closest alias frequencies to 3 kHz are when *k* = 1: $\left| 1 \cdot f_s - f_{\text{principle}} \right| = \left| 10 - 3 \right| = 7$ kHz and $\left| 1 \cdot f_s + f_{\text{principle}} \right| = \left| 10 + 3 \right| = 13$ kHz.

b. The 11-kHz signal clearly isn't a principle alias because it lies outside the interval [0, 5] kHz. From Step 1, find k that places 11 kHz within $10/2 = 5$ KHz on either side of $k \cdot 10$ KHz. Setting $k = 1$ works. Applying Step 2 $f_{principle} = |1 \cdot 10 - 11| = 1$ kHz. Using the formula, $f_{principle} = |\text{round}(11/10) \cdot 10 - 11| = 1$ kHz.

c. For the 36-kHz signal, Step 1 says choose k to make 36 kHz within 5 KHz of $k \cdot 10$, which means $k = 4$, because 40 KHz is only 4 kHz above 36 kHz. Applying Step 2, $f_{principle} = |4 \cdot 10 - 36| = 4$ kHz. Using the formula, $f_{principle} = |\text{round}(36/10) \cdot 10 - 36| = |4 \cdot 10 - 36| = 4$ kHz.

d. With $f = 122$ kHz, Step 1 again requires f to be within 5 kHz of $k \cdot 10$. As a guess, $122/10 = 12.2$ and $12 \cdot 10 = 120$. So $k = 12$ works because 122 is only 2 kHz away from 120 kHz. Applying Step 2, $f_{principle} = |12 \cdot 10 - 122| = 2$ kHz. Using the formula, $f_{principle} = |\text{round}(122/10) \cdot 10 - 122| = |12 \cdot 10 - 122| = 2$ kHz.

Example 10-3: To find the principle alias, using PyLab, you can write a one-line function in IPython to work with numpy vectors (also included in `ssd.py`).

```
In [578]: def prin_alias(f_in,fs):
     ...:    return abs(rint(f_in/fs)*fs - f_in)
```

Test the function by using the four frequency values of Example 10-2:

```
In [579]: ssd.prin_alias(array([3,11,26,122]),10.)
Out[579]: array([ 3., 1., 4., 2.])
```

The Python numerical results agree with the answers found in Example 10-2.

Analyzing the Impact of Quantization Errors in the ADC

The quantizer, $Q(\ )$, present in the ADC drill down of Figure 10-1, has a practical function. The ideal sampler (covered in the section "Periodic Sampling of a Signal: The ADC," earlier in this chapter) forms samples of the continuous-time waveform $x(t)$. The sample values themselves require infinite precision to be stored error-free on a computer. The quantizer rounds (or perhaps truncates) the signal sample to a finite precision binary word of B bits:

$$x_Q[n] = Q_B(x[n]) = B\text{-bit signed binary word}$$

In signals and systems work, you typically use a bipolar quantizer where input x is rounded to the nearest quantization level x_q, which takes on values ranging from $-X_m$ to $X_m - \Delta$, where Δ is the step size defined as $2X_m / 2^B$.

The quantization levels aren't symmetrical about zero because 2^B is even and a level needs to appear at zero. Note also that the output *saturates,* meaning that inputs outside the quantization range are held at the minimum and maximum quantization levels.

You need to know your limitations when working with an ADC; don't input signals that exceed $\pm X_m$ or suffer the consequences of larger error between the input and output. The input/output relationship for a $B = 5$ bit quantizer ($B^2 = 32$ levels) is shown in Figure 10-4.

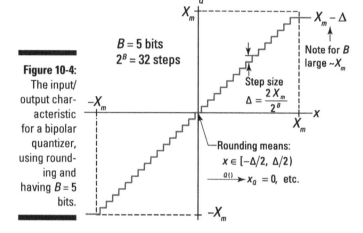

Figure 10-4: The input/output characteristic for a bipolar quantizer, using rounding and having $B = 5$ bits.

In practical systems, B runs from about 8 to 24. CD digital audio uses at minimum 16 bits per sample.

Samples $x(nT)$ have infinite precision, but following the quantizer error is introduced due to the B-bit precision imposed by $Q(\ )$. The signal $e[n] = Q[x(nT)] - x(nT)$ is known as the *quantization error*. With rounding and a step size of Δ, the quantization error ranges over $[-\Delta/2, \Delta/2]$.

You can view $e[n]$ as an additive noise source by rearranging the definition: $Q[x(nT)] = x(nT) + e[n]$. Using this arrangement, the error signal is referred to as *quantization noise,* because it appears as a *noise-like* signal being added to the infinite precision signal $x(nT)$. I say *noise-like* because $e[n]$ is functionally related to $x(nT)$. You measure the impact of quantization noise by the ratio of power in $x(nT)$ over the power in $e[n]$. It can be shown that for $x(t)$, a sinusoid of amplitude X_m the signal-to-quantization noise ratio (SNRQ) in dB is approximately $6.02B + 1.76$. For CD audio quality audio ($B = 16$), something you're familiar with, this works out to 98 dB. Very good!

You can use the Python function simpleQuant(x,Btot,Xmax,Limit) (in ssd.py) to quantize a sinusoid that falls inside the $\pm X_m = \pm 1$ dynamic range of the quantizer. Find the experimental SNRQ, using the function var() to estimate the power in $x(nT)$ and $e[n]$ and then forming the power ratio. Here, I plot the error signal waveform in Figure 10-5.

```
In [38]: n = arange(0,10000)
In [39]: x = 0.99*cos(2*pi*n/11.23)# < Xm - delta
In [40]: xQ = ssd.simpleQuant(x,8,1,'sat') # quantize x
In [43]: plot(n,xQ-x) # plot the error signal
In [158]: 10*log10(var(xQ)/var(xQ-x)) # Exp. SNRQ in dB
Out[158]: 49.70
In [159]: 6.02*8+1.76 # Theory SNRQ in dB
Out[159]: 49.92
```

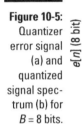

Figure 10-5:
Quantizer error signal (a) and quantized signal spectrum (b) for $B = 8$ bits.

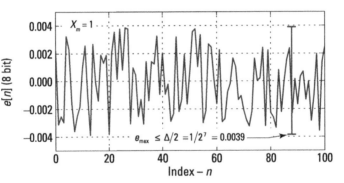

The peak quantization error lies within $\pm \Delta/2 = \pm 0.0039$ as expected. For 8 bits, the SNRQ is about $6.02 \times 8 + 1.76 = 49.9$dB, compared with 49.7 dB from the Python simulation.

Analyzing Signals in the Frequency Domain

Fourier transform theory really shines as a means to understand how to make sampling theory work for you. In this section, I develop a frequency-domain view of sampling theory. You need to understand the Fourier transform (FT) to get a handle on this material, so flip to Chapter 9 if you need to refresh your FT skills.

The first step in getting to frequency-domain view of sampling theory is to establish the spectrum of the input signal following multiplication by an ideal sampling waveform. The operation models the first stage of the ADC shown in Figure 10-1.

With a model for the sampled signal spectrum in hand, I can show you what the spectrum of a sampled bandlimited signal looks like. The corresponding spectral plot allows you to see how to avoid aliasing and what the consequences are if you don't choose a large enough sampling rate — the aliased or folded spectrum occurs when the sampling rate is too small. Aliasing is most clearly visible in the frequency domain.

Impulse train to impulse train Fourier transform theorem

The spectral view of sampling theory isn't limited to sinusoidal signals; it can handle all sorts of practical signals. In this section, I explore sampling theory in the frequency domain by developing the FT of an ideal impulse train sampled signal, which is generated inside the first stage of ADC model (refer to Figure 10-1). Intentionally allowing aliasing to occur, a topic not covered in this book, can be well-appreciated in the spectrum of the impulse train sampled signal.

The goal here is to get $X_s(f) = \mathcal{F}\{x_s(t)\} = \mathcal{F}\{x(t) \cdot s(t)\}$ the spectrum of sampled signal $x_s(t)$. To start, find $S(f)$.

Because an impulse train in the time domain is an impulse train in the frequency domain, you can write this equation:

$$S(f) = \mathcal{F}\left\{\sum_{n=-\infty}^{\infty} \delta(t - nT)\right\} == f_s \sum_{m=-\infty}^{\infty} \delta(f - mf_s)$$

Now, convolve an impulse train in frequency with $X(f)$. To complete the term-by-term convolution, you need to know that a basic property of the impulse function and convolution is $x(t) * \delta(t - t_0) = x(t - t_0)$ (see Chapter 5).

Putting it all together,

$$X_s(f) = X(f) * f_s \sum_{m=-\infty}^{\infty} \delta(f - mf_s) = f_s \sum_{m=-\infty}^{\infty} X(f - mf_s)$$

This says that the spectrum of a sampled signal is a superposition of the original signal spectrum ($m = 0$) plus *spectral translates* — refers to frequency shifting $X(f)$ by f_0 to produce $X(f - f_0)$ — with translation offsets ±m integer multiples of the sampling rate.

This is a cool and powerful result because it shows that $X_s(f)$ contains the desired spectrum $X(f)$ surrounded by its frequency translates. It also reveals that altering f_s makes the picture either clear or cluttered (if you can't wait, take a peek at the figures in the next section).

The result is somewhat counterintuitive because, in the time domain, sampling gives you less than what you started with (just the samples), but you have more in the frequency domain — the original spectrum $X(f)$ plus all the translates $X(f \pm mf_s)$, $m = 1, 2$, and so on. Usually, the desire is to recover $X(f)$ without capturing any of the spectral translates.

Finding the spectrum of a sampled bandlimited signal

Suppose that $x(t)$ has bandlimited spectrum of the form $X(f) = \Lambda(f/W)$.

The sampled signal spectrum is

$$X_s(f) = f_s \sum_{m=-\infty}^{\infty} \Lambda\left(\frac{f - mf_s}{W}\right)$$

$$= \cdots + f_s\Lambda\left(\frac{f+f_s}{W}\right) + f_s\Lambda\left(\frac{f}{W}\right) + f_s\Lambda\left(\frac{f-f_s}{W}\right) + \cdots$$

Figure 10-6 shows plots of $X(f)$ and $X_s(f)$ for $f_s > 2W$ and $f_s < 2W$.

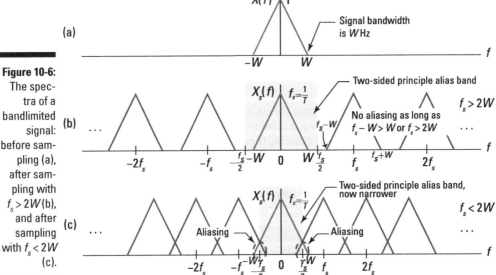

Figure 10-6: The spectra of a bandlimited signal: before sampling (a), after sampling with $f_s > 2W$ (b), and after sampling with $f_s < 2W$ (c).

REMEMBER

When $f_s > 2W$, all the spectral translates are distinct, meaning no overlap occurs between them. The principle translate, which in this book is always centered on $f = 0$, is the one you want to keep your eyes on here, because it corresponds to $X(f)$ to within a scale factor. When $f_s < 2W$, neighbors infringe upon the principle translate. The overlap is aliasing as it appears in the frequency domain.

I've been using frequency f in hertz so far in my explanations of the sampled signal spectrum, but you may need other units at some point in your signals and systems studies or work. You can plot the spectra by using $\omega = 2\pi f$ or, in terms of the discrete-time frequency variable, $\hat{\omega} = 2\pi f/f_s = \omega/f_s$. To make this clear, Figure 10-7 is the spectrum of $x_s(t)$, as shown in Figure 10-6b, with addition of two parallel frequency axes.

Figure 10-7:
The sampled signal spectra, using alternative frequency axes.

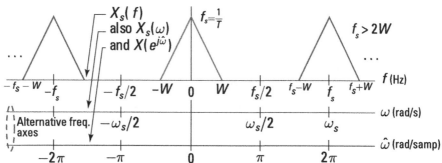

The first two axes follow directly from $x_s(t)$ and the two forms of the Fourier transform as I describe in Chapter 9. Using $\hat{\omega}$ as a frequency axis suggests that $x[n] = x(nT)$ has a Fourier transform (FT) representation, too. If you're interested in the details of developing the proof for $\mathcal{F}\{x[n]\} = X(e^{j\hat{\omega}})$, check out Chapter 11. In the remainder of this chapter, you can assume that $X(e^{j\hat{\omega}})$ exists and, with the change of variables $f \to \hat{\omega}$ or $\omega \to \hat{\omega}$, the spectra of $x_s(t)$ and $x[n]$ are identical.

EXAMPLE

Example 10-4: To find the sampled spectrum $X_s(f)$ for $x(t) = A\cos(2\pi f_0 t)$, use the transform pair $A\cos(2\pi f_0 t) \xleftrightarrow{\mathcal{F}} A/2 \cdot [\delta(f - f_0) + \delta(f + f_0)]$ for $X(f)$.

Plug into the sampled spectrum formula:

$$X_s(f) = f_s \sum_{m=-\infty}^{\infty} \frac{A}{2}[\delta(f - f_0 - mf_s) + \delta(f + f_0 - mf_s)]$$

Now, plot the spectrum for $A = 1$, $f_0 = 10$ Hz, and f_s, taking on values of 30 Hz and then 15 Hz. (Figure 10-8 shows this plot.)

To simplify the creation of the figure itself, you can use a custom function `lp_samp(fb,fs,fmax,N,shape)` written in Python to create the spectrum picture. The function code is available at www.dummies.com/extras/signalsandsystems in the module ssd.py, and it can make plots similar to Figures 10-6 and 10-7 when shape = 'tri' as well as a line spectra plot when shape = 'line'.

The parameter fb is equivalent to W when the triangle shape is used and f_0 for the single sinusoid. The sampling rate is set via fs; fmax controls the frequency extent of the plot and N, the number of spectral translates to plot. Now create plots for the two f_s values:

```
In [589]: ssd.lp_samp(10,30,60,5,'line')
In [592]: ssd.lp_samp(10,15,60,5,'line')
```

With $f_0 = 10$ Hz, f_s needs to be greater than 20 Hz. With $f_s = 15$ Hz, aliasing is visible.

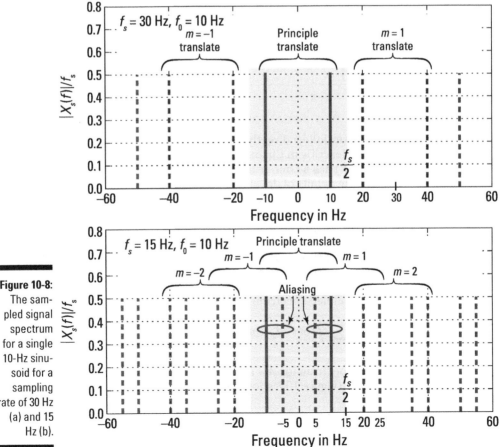

Figure 10-8: The sampled signal spectrum for a single 10-Hz sinusoid for a sampling rate of 30 Hz (a) and 15 Hz (b).

Sure enough, when you sample the 10-Hz sinusoid at 15 Hz, an alias spectral line falls below the principle translate at 5 Hz. The principle alias formula $f_{\text{principle}} = \left| \text{round}\left(f/f_s\right) \cdot f_s - f \right|$ agrees because $15 - 10 = 5$ Hz.

Aliasing and the folded spectrum

The spectral translate at $m = 1$, given by $X(f - f_s)$, is centered on f_s because $X(f)$ is assumed symmetrical about $f = 0$. The spectrum tail that laps below $f_s/2$ appears to be a folded version of the principle translate that lies above $f_s/2$. This is depicted graphically in Figure 10-9.

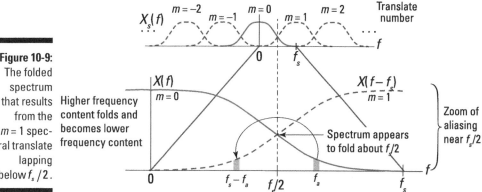

Figure 10-9:
The folded spectrum that results from the $m = 1$ spectral translate lapping below $f_s/2$.

Higher frequency content folds and becomes lower frequency content

The symmetry evident about $f_s/2$ earns the half sampling rate frequency the name *folding frequency*. Suppose that $f_s/2 = 1\,\text{kHz}$ and $x(t)$ contains a spectral component at 1,200 Hz. The spectrum folding action of Figure 10-9 places an aliased spectral component at 800 Hz. Why? Symmetry about 1,000 Hz demands that the 1,200-Hz component fold to 200 Hz below 1,000 Hz, which is 800 Hz.

Don't worry: I haven't violated the principle alias formula. From the given information, $f_s = 2 \times 1,000 = 2,000\,\text{Hz}$, which indicates that a signal at 1,200 Hz has a principle alias frequency of $2,000 - 1,200 = 800$ Hz. The folded spectrum is just another way of looking at the inner workings of sampling theory.

Applying the Low-Pass Sampling Theorem

When $f_s > 2W$ (as it is in Figure 10-6c), the band of frequencies $-f_s/2 < f < f_s/2$ contains only the principle ($m = 0$) translate. This translate is simply an amplitude scaled version of the original spectrum $X(f)$. To *recover* $X(f)$ from

$X_s(f)$, all you need to do is pass $x_s(t)$ through an ideal low-pass filter (see Chapter 9) that keeps the principle translate and discards everything else. In mathematical terms,

$$X_r(f) = X_s(f) \cdot H_r(f)$$
$$x_r(t) = \mathcal{F}^{-1}\{X_s(f) \cdot H_r(f)\} = x_s(t) * h_r(t)$$

Here, $h_r(t)$ is the impulse response of an ideal low-pass filter having bandwidth (or cutoff frequency) $W < f_c < f_s - W$. Note that $H_r(f) = \mathcal{F}\{h_r(t)\}$. If $f_s < 2W$ recovery isn't possible because of the presence of aliasing inside the band of frequencies, then $-W \le f \le W$.

You can now formally state the low-pass sampling theorem:

- ✔ Let $x(t)$ be a bandlimited signal such that $X(f) = 0$ for $|f| > W$.

    Recover $x(t)$ from its samples $x(nT) = x[n]$, provided that $f_s = \dfrac{\omega_s}{2\pi} > 2W$.

- ✔ W is referred to as the *Nyquist frequency*.

- ✔ $2W$ is the *Nyquist rate*.

To find out how to use the antialising filter (the very first block in Figure 10-1) to block signals above the folding frequency, $f_s/2$, from entering the ADC and thus avoid aliasing, check out www.dummies.com/extras/signals andsystems.

Reconstructing a Bandlimited Signal from Its Samples: The DAC

I've said a lot about sampling a continuous-time signal. Now it's time to focus on how to reconstruct a continuous-time signal from a sequence of samples. When you listen to a CD, talk on a cellphone, or stream music from Internet radio, a sequence of signal values converts back to an intact continuous-time signal to reach the speaker. This happens in two steps:

1. The device places the signal samples back on the physical time axis.

2. The device interpolates a continuum of signal values between the sample values.

Take a system, or *behavioral,* level view of these two steps. A circuit designer would treat this as a *mixed signal* design problem, which means that the circuit is composed of both digital logic and analog circuit (continuous-time) building blocks.

The zoom inside the DAC at the bottom right of Figure 10-1 shows you that the first step for converting the discrete-time signal $y[n]$ to the continuous-time signal, $y(t)$, is placing the sequence values on the time axis using time-shifted impulse functions: $y_s(t) = \sum_{n=-\infty}^{\infty} y[n]\delta(t-nT)$. The DAC sampling rate clock, at frequency $f_s = 1/T$, is used to establish the time spacing.

Next, the impulse train signal passes through a zero-order hold (ZOH) operation. The ZOH models the output of a *realizable* DAC (a working design in electronic circuit form), where the output of each conversion cycle is held constant from one sample period to the next. This makes the output waveform resemble a set of stairs. The ZOH provides a conversion between the impulse train and the stairstep waveform. Viewed as a linear filter, the impulse response of the ZOH is $h_{ZOH}(t) = \Pi([t-T/2]/T)$.

The corresponding frequency response of the ZOH filter is $H_{ZOH}(f) = T\text{sinc}(Tf) \cdot e^{-j2\pi f \cdot T/2}$. The ZOH output is the convolution of $y_s(t)$ with $h_{ZOH}(t)$:

$$y_{ZOH}(t) = h_{ZOH}(t) * \sum_{n=-\infty}^{\infty} y[n]\delta(t-nT) = \sum_{n=-\infty}^{\infty} y[n]h_{ZOH}(t-nT)$$

Here, the sample values are placed every T seconds. Figure 10-10 shows the relation between the samples $y[n]$, the ZOH output, and the recovered/reconstructed output $y(t)$.

Figure 10-10: The DAC ZOH stairstep waveform and the associated input samples and perfectly reconstructed output.

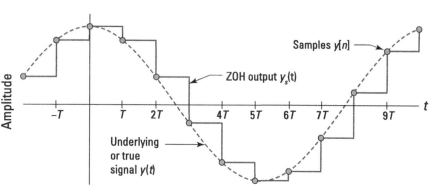

The output of the ZOH needs additional filtering for $y_s(t)$ to approximate $y(t)$, the true signal. The reconstruction filter has frequency response $H_r(f)$ and an impulse response $h_r(t)$. Getting the reconstructed output requires one more convolution:

$$y(t) = y_s(t) * \underbrace{h_{ZOH}(t) * h_r(t)}_{h_{r'}(t)} = \sum_{n=-\infty}^{\infty} y[n]h_{r'}(t-nT)$$

Note that $h_{r'}(t) = h_{ZOH}(t) * h_r(t)$. The combination of both filters smoothly interpolates signal values between the original samples spaced at T. The low-pass filter is $h_r(t)$. I explore the details of $h_r(t)$ in the next two sections.

Interpolating with an ideal low-pass filter

In the absence of a DAC ZOH filter, the ideal reconstruction filter can be an ideal low-pass filter with gain T (to compensate for the $f_s = 1/T$ gain from the sampling operation) and cutoff frequency $f_c = W$. The ideal low-pass filter has impulse response $h_r(t) = \mathcal{F}\left\{T\Pi\left[f/(2W)\right]\right\} = 2TW\text{sinc}(2Wt)$ (see Chapter 9).

You must deal with the impact of the ZOH filter before going any further. The ZOH sinc function frequency response introduces frequency droop over the frequency band $-f_s/2 \le f \le f_s/2$. *Frequency droop* means that the frequency response magnitude rounds downward instead of being flat for frequencies between 0 and $f_s/2$. See the plot in Figure 10-11.

Figure 10-11:
The frequency response of the ZOH filter and how to compensate for frequency droop with a compensator incorporated into $H_r(f)$.

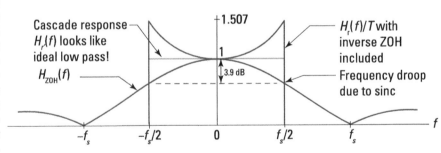

To make the cascade response $H_{r'}(f) = H_{ZOH}(f) \cdot H_r(f)$ look like an ideal low-pass filter, with constant frequency response magnitude out to $f_s/2$, you need to incorporate an inverse sinc function frequency response into $H_r(f)$. Figure 10-11 shows how you can modify $H_r(f)$ so the cascade response looks like an ideal low-pass filter. The compensated $H_r(f)$ is an ideal low-pass with the addition of a cup-shape amplitude response for $-f_s/2 \le f \le f_s/2$.

In practice, you can place the compensator in the discrete-time domain by using a simple digital filter or incorporate it into the reconstruction filter $H_r(f)$ (see Chapter 15 for a digital filter design example).

From this point forward, I assume that the compensation of the ZOH is already addressed. Keep in mind that some DAC chips make inverse sinc compensation totally transparent to the system designer.

To begin the ideal low-pass modeling process, write the time-domain formula for getting the recovered signal (assuming the ZOH is compensated):

$$y(t) = \sum_{n=-\infty}^{\infty} y[n] \cdot \text{sinc}\left[2W(t - nT)\right]$$

The sinc function serves as an ideal interpolation function in this case because the impulse response is non-causal.

Example 10-5: In this example, I show you how the sinc interpolation function (the ideal low-pass reconstruction filter) creates $y(t)$ from a finite set of sample values $y[n]$. In particular, I assume that $y[n]$ is found by sampling $\cos(2\pi t/5)$ with $f_s = 1$ Hz. The available samples are limited to $n = 0, 1, ..., 10$.

To show you the reconstruction process step by step, I first create an overlay plot in Figure 10-12a of the individual waveforms $y[n] \cdot \text{sinc}[2W(t - nT)]$, corresponding to each available sample. The sample values are filled circles, and the original continuous-time sinusoid is a dashed line.

In Figure 10-12b, the individual waveform contributions are added together to form the composite or reconstructed output $y(t)$.

Once inside the interpolation interval, the reconstruction is good, as evidenced by the two curves lying on top of each other.

Example 10-6: Consider an end-to-system that employs ideal sampling and ideal low-pass reconstruction. Assume that the discrete-time system placed between the ADC and DAC is a *signal pass through,* meaning $y[n] = x[n]$.

The analog input signal is $x(t) = 10\cos(2\pi \cdot 10^4 \cdot t) + 5\cos(2\pi \cdot 36 \times 10^3 \cdot t)$, and the sampling rate is 48 kHz. CD digital audio uses $f_s = 44.1$ kHz. The 48-kHz rate, and inter multiples thereof, are popular in recording studio audio processing.

Sketch the frequency spectra $X(f)$, $X_s(f) = X(e^{j\hat{\omega}}) = Y(e^{j\hat{\omega}})$, and $Y(f)$. The system block diagram is available in Figure 10-13.

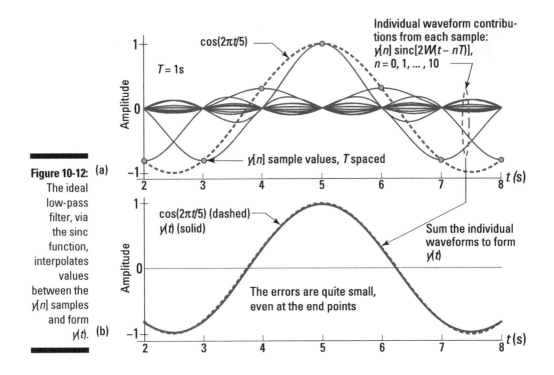

Figure 10-12: (a) The ideal low-pass filter, via the sinc function, interpolates values between the $y[n]$ samples and form $y(t)$. (b)

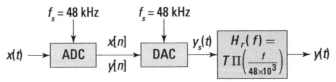

Figure 10-13: System block diagram for Example 10-6.

Before you jump into plotting anything, find out whether aliasing is present. Aliasing occurs for input frequencies above 24 kHz. The 10-kHz signal is fine, but the 36-kHz signal aliases to $48 - 36 = 12$ kHz. After it's inside the system, the 36 kHz signal loses its identity and becomes forever a 12-kHz signal. To find $X(f)$, use Fourier transform pair $\cos(2\pi f_0 t) \xleftrightarrow{\;\mathcal{F}\;} [\delta(f - f_0) + \delta(f + f_0)]/2$ to get the following:

$$X(f) = \frac{10}{2}\Big[\delta(f - 10^4) + \delta(f + 10^4)\Big]$$
$$+ \frac{5}{2}\Big[\delta(f - 36 \times 10^3) + \delta(f + 36 \times 10^3)\Big]$$

The spectrum $X(f)$ is sketched in Figure 10-14a. The spectrum $X_s(f)$, shown in Figure 10-14b, consists of $X(f)$ and translates $X(f \pm mf_s)$ scaled by f_s. When you use an ideal low-pass reconstruction filter (see overlay in Figure 10-14b), the output spectrum $Y(f)$ consists of the spectrum content of $X_s(f)$ that lies in the two-sided principle alias band -24 kHz $< f < 24$ kHz. The spectrum $Y(f)$ in Figure 10-14c shows that the 36-kHz sinusoid is aliased to 12 kHz, as expected.

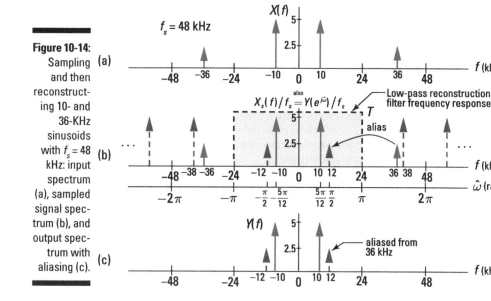

Figure 10-14:
Sampling (a) and then reconstructing 10- and 36-KHz sinusoids with $f_s = 48$ kHz: input spectrum (a), sampled signal spectrum (b), and output spectrum with aliasing (c).

Using a realizable low-pass filter for interpolation

Assuming $x(t)$ is bandlimited and f_s is chosen to satisfy the low-pass sampling theorem, the filtering action of an ideal low-pass filter (shaded region in Figure 10-6b) perfectly extracts the principle spectral translate of $X_s(f)$. In the time domain, the interpolation is perfect, too, because only $x(t)$ is returned at the filter output.

When using a realizable filter (see Chapter 9), the filter magnitude response doesn't jump instantly to 0 for $f > f_s/2$. The magnitude response becomes very small only gradually as frequency increases, depending on the filter order you choose. This means that spectral translates for $m \neq 0$ are present in the filter output. These translates are weak, but they're still present, so the reconstructed output contains distortion — imperfect interpolation.

You can increase the filter order but only at the expense of increased complexity. To ease the filter design requirements, you can use *oversampling*, which involves setting f_s to some integer factor greater than $2W$, to provide an effective guard band of frequencies between the principle translate and its nearest neighbors. The guard band allows the filter magnitude response to roll off more before encountering the first spectral translate. To visualize the impact of oversampling, think of moving the $m \neq 0$ spectral translates in Figure 10-6b much farther away from the $m = 0$ translate.

Oversampling in this fashion increases complexity. A popular oversampling technique is to increase the sampling rate (upsampling) in the discrete-time domain, just prior to the DAC. This way, low-pass reconstruction filtering can begin in the discrete-time domain at the oversampled rate by using an efficient digital filter.

The final stage of reconstruction filtering that takes place in the continuous-time domain then needs to be only a low-order analog filter. The downside is that the DAC must be clocked at a higher sampling rate. For audio signal processing used in consumer electronics, oversampling DACs are viable and thus used in devices you likely own.

Chapter 11

The Discrete-Time Fourier Transform for Discrete-Time Signals

- -

In This Chapter

▶ Checking out the Fourier transform of sequences

▶ Getting familiar with the characteristics and properties specific to the DTFT

▶ Working with LTI system relationships in the frequency domain

▶ Using the convolution theorem

- -

*I*f you're hoping to find out how the discrete-time Fourier transform (DTFT) operates on discrete-time signals and systems to produce spectra and frequency response representations with units of radians/sample, you're in the right place! And I hope it wasn't too hard to find; Fourier theory is covered in four different chapters.

The Fourier transform (FT) (explored in Chapter 9) has the same capabilities as the DTFT, but it applies to the continuous-time cousins in the lands of continuous frequency. If you're looking for a Fourier transform that's discrete in both time and frequency, you need the discrete Fourier transform (DFT), which is the subject of Chapter 12. The Fourier series (covered in Chapter 8) applies to continuous-time periodic signals with a discrete frequency-domain representation.

Trig functions are integral to all forms of Fourier theory. In signals and systems, the trig functions are usually hidden inside a complex exponential, but Euler's formula (see Chapter 2) tells you that a complex sinusoid is composed of a cosine on the real axis and a sine on the imaginary axis.

In terms of the DTFT, the forward transform (which moves a signal from the time to frequency domain) requires a summation; the inverse discrete-time

Fourier transform (IDTFT) (which takes a signal from the frequency domain back to the world of time) requires an integral. This asymmetry may be unexpected if you've worked only with the continuous-time Fourier transform. But the DTFT produces a spectral function of a continuous frequency variable from a discrete-time signal or sequence.

The forward transform takes a discrete-time signal (sequence) as input, so a sum is required (not an integral, as in the continuous-time case). Inverse transforming requires integration over the frequency variable to return to the discrete-time domain. As a bonus, the summation of the forward transform often involves only the use of the geometric series (see Chapter 2).

The DTFT has properties and theorems that are similar to the Fourier transform (FT). But, unlike the FT, the DTFT spectrum is a periodic function of the frequency variable, which may seem rudimentary if you're familiar with sampling theory (Chapter 10). Like the FT, the DTFT is a complex function of frequency, so magnitude and phase spectra appear.

In this chapter, I formally define the frequency response of linear time-invariant (LTI) systems. (Check out Chapter 7 for a peek at the frequency response for linear constant coefficient difference equations.) I also show you the full utility of the frequency response by using the convolution theorem for the DTFT.

Getting to Know DTFT

In this section, I define the DTFT and IDFT and point out when these tools apply to specific types of signals and systems work. I also cover basic DTFT properties for both signals and systems and show you the mathematical link between the spectrum of a continuous-time signal and the spectrum of the corresponding discrete-time signal via uniform sampling. My intent is to show you how nicely the frequency-domain view of sampling theory from Chapter 10 fits with the spectrum of a discrete-time signal and to help you get comfortable working with the DTFT/IDFT.

The DTFT of sequence $x[n]$ is defined by this summation:

$$X(e^{j\hat{\omega}}) \equiv \sum_{n=-\infty}^{\infty} x[n]e^{-j\hat{\omega}n}$$

Here, $\hat{\omega}$ is the discrete-time frequency variable. The quantity $X(e^{j\hat{\omega}})$ is the *signal spectrum* of $x[n]$. The synthesis formula for getting $x[n]$ from $X(e^{j\hat{\omega}})$ is the IDTFT equation:

$$x[n] \equiv \frac{1}{2\pi} \int_{-\pi}^{\pi} X(e^{j\hat{\omega}})e^{j\hat{\omega}n} \, d\hat{\omega}$$

The integration can be performed over any 2π interval. Formally, the DTFT exists if $x[n]$ is *absolutely summable,* meaning

$$\sum_{n=-\infty}^{\infty} |x[n]| < \infty$$

Explore signals that violate this condition in the section "The DTFT of Special Signals," later in this chapter.

Example 11-1: Find the DTFT of the exponential sequence $x[n] = a^n u[n]$. As a cautious first step, find out what conditions on a ensure that $x[n]$ is absolutely summable. Next, find $X(e^{j\hat{\omega}})$ and then follow these steps:

1. **Compute the absolute sum of $x[n]$ by using infinite geometric series results from Chapter 2:**

$$\sum_{n=-\infty}^{\infty} |a^n u[n]| = \sum_{n=0}^{\infty} |a^n| = \sum_{n=0}^{\infty} |a|^n = \frac{1}{1-|a|}, |a| < 1$$

   This first step reveals that the DTFT of $x[n]$ exists only for $|a| < 1$.

2. **Find the DTFT, which involves similar series-manipulation skills:**

$$X(e^{j\hat{\omega}}) = \mathcal{F}\{x[n]\} = \sum_{n=-\infty}^{\infty} a^n u[n] e^{-j\hat{\omega}n} = \sum_{n=0}^{\infty} a^n e^{-j\hat{\omega}n}$$

$$= \sum_{n=0}^{\infty} (a \cdot e^{-j\hat{\omega}})^n = \frac{1}{1-ae^{-j\hat{\omega}}}, |ae^{-j\hat{\omega}}| = |a| < 1$$

As expected, $X(e^{j\hat{\omega}})$ is a complex quantity. You now have your first DTFT transform pair:

$$a^n u[n] \xleftrightarrow{\ \mathcal{F}\ } \frac{1}{1-ae^{-j\hat{\omega}}}, |a| < 1$$

Not too bad! Hold on to your geometric series skills for calculating the DTFT sum throughout this chapter.

Checking out DTFT properties

When discrete-time signals are real, some useful symmetry properties fall into place in the frequency domain. Parallel results exist for the Fourier series of Chapter 8 and the Fourier transform of Chapter 10, but when you're working with a periodic spectrum for the DTFT, results are a bit different.

In general, the DTFT of $x[n]$ is a complex function of $\hat{\omega}$, so viewing it in polar form (magnitude and angle) is convenient:

- $\left|X(e^{j\hat{\omega}})\right|$ is the magnitude or amplitude spectrum, which parallels $\left|X(f)\right|$ for the FT.
- $\angle X(e^{j\hat{\omega}})$ is the *phase spectrum,* which parallels $\angle X(f)$ for the FT.

The fact that the frequency variable $\hat{\omega}$ always appears wrapped inside a complex exponential, such as $e^{j\hat{\omega}}$, means that $X(e^{j\hat{\omega}})$ is periodic, with period 2π. Why? Note that $X(e^{j(\hat{\omega}+2\pi)}) = X(e^{j\hat{\omega}}e^{j2\pi}) = X(e^{j\hat{\omega}})$ because $e^{j2\pi} = 1$.

The 2π periodicity of the discrete-time spectrum is in sharp contrast to the FT, because the spectrum of a sampled continuous-time signal is repeated at multiples of the sampling rate (see Chapter 10). For a discrete-time signal, the sampling period is pretty much once per integer, meaning the sampling rate in radians per sample is $1 \cdot 2\pi = 2\pi$. This justifies using any 2π interval in the IDTFT integral.

For LTI systems, the frequency response $H(e^{j\hat{\omega}})$ is the DTFT of the impulse response, $h[n]$. So returning to the impulse response is just a matter of taking the IDTFT of $H(e^{j\hat{\omega}})$. Find more on the impulse response and frequency response in the section "LTI Systems in the Frequency Domain," later in this chapter.

Relating the continuous-time spectrum to the discrete-time spectrum

Among the differences between the Fourier transform for discrete-time signals and the Fourier transform for continuous-time signals are the nature of sequences and the connection to uniform sampling a continuous-time signal. In practice, you frequently find the sequence values $x[n]$ by uniform sampling of the continuous-time signal $x(t)$. An impulse train with amplitude weights, the sample values $x[n] = x(nT)$ represent this sampled continuous-time waveform:

$$x_s(t) = \sum_{n=-\infty}^{\infty} x[n]\delta(t - nT)$$

The FT of $x_s(t)$ is $X_s(f) = f_s \sum_{m=-\infty}^{\infty} X(f - mf_s)$, where $X(f)$ is the FT of $x(t)$, and $f_s = 1/T$. Now, relate $X(e^{j\hat{\omega}}) = \mathcal{F}\{x[n]\}$ to $X_s(f)$ and ultimately $X(f)$.

Here's a three-step process for developing the $X(e^{j\hat{\omega}})$ to $X(f)$ relationship:

1. **Find $X_s(f)$.**

You can use the FT pair $\delta(t-t_0)\xleftrightarrow{\mathcal{F}}\exp(-j2\pi f\,t_0)$ and the linearity theorem to get a term-by-term solution:

$$X_s(f) = \mathcal{F}\left\{\sum_{n=-\infty}^{\infty}x[n]\delta(t-nT)\right\} = \sum_{n=-\infty}^{\infty}x[n]e^{-j2\pi fnT}$$

Notice that $X_s(f)$ looks very much like the definition of the DTFT of $x[n]$, except $2\pi fT$ appears instead of $\hat{\omega}$.

2. **Take the expression for $X_s(f)$ from Step 1 and substitute $f = \hat{\omega}/(2\pi T)$ to relate $X_s(f)$ to $X(e^{j\hat{\omega}})$:**

$$X_s(f)\big|_{f=\hat{\omega}/(2\pi T)} = \sum_{n=-\infty}^{\infty}x[n]e^{-j\hat{\omega}n} \overset{\text{DTFT}}{=} \mathcal{F}\{x[n]\} \overset{\text{also}}{=} X(e^{j\hat{\omega}})$$

3. **Pair the results of Step 2 with the FT of the sampled continuous-time signal $X_s(f)$, as developed in Chapter 9:**

$$X(e^{j\hat{\omega}}) = X_s(f)\big|_{f=\hat{\omega}/(2\pi T)} = f_s\sum_{m=-\infty}^{\infty}X\left(f-\tfrac{m}{T}\right)\bigg|_{f=\hat{\omega}/(2\pi T)}$$

$$= f_s\sum_{m=-\infty}^{\infty}X\left(\tfrac{\hat{\omega}}{2\pi T}-\tfrac{m}{T}\right) = f_s\sum_{m=-\infty}^{\infty}X_\omega\left(\tfrac{\hat{\omega}}{T}-m\tfrac{2\pi}{T}\right)$$

where in the last line, $X_\omega(\omega)$ is the FT of $x(t)$, using the $\omega = 2\pi f$ based FT.

For $x(t)$ *bandlimited* to $f_s/2$ — or $X(f) = 0$ for $|f| \geq f_s/2$, $X(e^{j\hat{\omega}}) = f_s X(\hat{\omega}/[2\pi T])$ for $-\pi \leq \hat{\omega} \leq \pi$ — $X(f)$ is the FT of $x(t)$. This result is significant because it links the frequency domain to continuous- and discrete-time systems, and it tells you — via the frequency axis mapping of f in hertz to $\hat{\omega}$ in radians/sample $(f \to \hat{\omega} = 2\pi\cdot f/f_s)$ — exactly how the continuous-time spectrum becomes the corresponding discrete-time spectrum.

When you use ideal reconstruction (see Chapter 10) to convert $y[n]$ back to $y(t)$, a similar result holds: $Y(f) = T\cdot Y(e^{j2\pi f/f_s})$, where $Y(f) = \mathcal{F}\{y(t)\}$ and $Y(e^{j\hat{\omega}}) = \mathcal{F}\{y[n]\}$. The reverse frequency axis mapping $\hat{\omega}$ to f is $\hat{\omega} \to f = \hat{\omega}\cdot f_s/(2\pi)$.

In Chapter 15, I apply the relationship between the continuous- and discrete-time spectrums for modeling across domains. The time-domain connection is simply $x[n] = x(nT)$.

Getting even (or odd) symmetry properties for real signals

A real signal may be an even or odd function, and these conditions leave their mark in the frequency domain.

For $x[n]$, a real sequence, the spectrum $X(e^{j\hat{\omega}})$ is *conjugate symmetric*, which means that $X(e^{-j\hat{\omega}}) = X^*(e^{j\hat{\omega}})$. To demonstrate this property, I work from the DTFT definition:

$$X(e^{-j\hat{\omega}}) = \sum_{n=-\infty}^{\infty} x[n]e^{-j(-\hat{\omega})n} = \left[\sum_{n=-\infty}^{\infty} x[n]e^{-j\hat{\omega}n}\right]^* = X^*(e^{j\hat{\omega}})$$

I rely on the fact that $(x[n])^* = x[n]$ and $\left(e^{-j\hat{\omega}}\right)^* = e^{j\hat{\omega}}$.

Here are the important consequences of conjugate symmetry:

- ✔ $\text{Re}\{X(e^{j\hat{\omega}})\}$ is even in $\hat{\omega}$, which implies that $\text{Re}\{X(e^{-j\hat{\omega}})\} = \text{Re}\{X(e^{j\hat{\omega}})\}$.
- ✔ $\text{Im}\{X(e^{j\hat{\omega}})\}$ is odd in $\hat{\omega}$, which implies that $\text{Im}\{X(e^{-j\hat{\omega}})\} = -\text{Im}\{X(e^{j\hat{\omega}})\}$.
- ✔ $\left|X(e^{j\hat{\omega}})\right|$ is even in $\hat{\omega}$, which implies that $\left|X(e^{-j\hat{\omega}})\right| = \left|X(e^{j\hat{\omega}})\right|$.
- ✔ $\angle X(e^{j\hat{\omega}})$ is odd in $\hat{\omega}$, which implies that $\angle X(e^{-j\hat{\omega}}) = -\angle X(e^{j\hat{\omega}})$.

These observations also point out that $X(e^{j\hat{\omega}})$ is unique on a π length interval, such as $[0, \pi]$. Consider $X(e^{j\hat{\omega}})$ over one 2π period spanning $[-\pi, \pi]$. Conjugate symmetry tells you that $X(e^{j\hat{\omega}})$ on the interval $[-\pi, 0]$ is $X^*(e^{j\hat{\omega}})$ on the interval $[0, \pi]$. So given $X(e^{j\hat{\omega}})$ on the $[0, \pi]$ interval, you can extend to the $[-\pi, 0]$.

From the periodicity of $X(e^{j\hat{\omega}})$, you can replicate any other π length interval knowing $X(e^{j\hat{\omega}})$ on the $[0, \pi]$ interval. Again, from periodicity, knowing any π length interval is sufficient.

If $x[n]$ is an even sequence $(x[-n] = x[n])$, it can be shown that $X(e^{j\hat{\omega}})$ is real, or $\text{Im}\{X(e^{j\hat{\omega}})\} = 0$. Similarly, if $x[n]$ is an odd sequence $(x[-n] = -x[n])$, it can be shown that $X(e^{j\hat{\omega}})$ is imaginary, or $\text{Re}\{X(e^{j\hat{\omega}})\} = 0$. (Find details on even and odd functions in Chapter 3.)

Verifying even and odd sequence properties

Begin this proof by expanding the complex exponential by using Euler's formula:

$$X(e^{j\hat{\omega}}) = \sum_{n=-\infty}^{\infty} x[n]e^{-j\hat{\omega}n} = \underbrace{\sum_{n=-\infty}^{\infty} x[n]\cos(\hat{\omega}n)}_{0 \text{ if } x[n] \text{ is odd}} - j\underbrace{\sum_{n=-\infty}^{\infty} x[n]\sin(\hat{\omega}n)}_{0 \text{ if } x[n] \text{ is even}}$$

In the first term, $x[n]\cos(\hat{\omega}n)$ is odd if $x[n]$ is odd, so the sum is 0, making the real part 0. In the second term, $x[n]\sin(\hat{\omega}n)$ is odd if $x[n]$ is even, so the sum is 0. This proof relies on cosine being even and sine being odd, and symmetric sums over an odd function is 0.

Example 11-2: Plot the spectra of $x[n] = a^n u[n]$ for $a = 0.8$ and comment on the observed symmetry properties. Example 11-1 reveals that $X(e^{j\hat{\omega}}) = \dfrac{1}{1 - 0.8e^{-j\hat{\omega}}}$.

You can use Python and Pylab to make the plots. Figure 11-1 shows plots of the real, imaginary, magnitude, and phase for $-\pi < \hat{\omega} < \pi$.

```
In [101]: w = arange(-pi,pi,pi/500.)
In [102]: X = 1/(1 - 0.8*exp(-1j*w))
In [105]: plot(w,real(X))
In [110]: plot(w,imag(X))
In [115]: plot(w,abs(X))
In [121]: plot(w,angle(X))
```

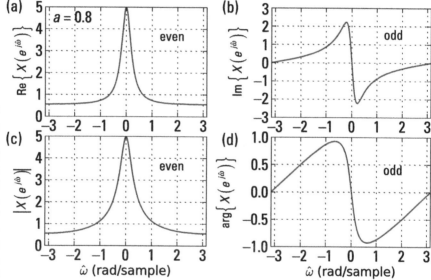

Figure 11-1: The spectrum of $0.8^n u[n]$ in terms of the real (a), imaginary (b), magnitude (c), and phase (d).

The conjugate symmetry of $X(e^{j\hat{\omega}})$ is visible in Figure 11-1 in both the rectangular and polar forms. Because $x[n]$ is neither even nor odd, $X(e^{j\hat{\omega}})$ contains both real and imaginary parts.

Example 11-3: Find the DTFT of the even sequence $x[n] = a^{|n|}$. From the start, you may be thinking that a two-sided exponential should be only twice as hard as the one-sided exponential $a^n u[n]$. Here, you can find out.

There are four steps to solving this problem:

1. **To successfully tackle a signal with an absolute value, break it into two pieces.**

   The convenient split point here is between $n = -1$ and $n = 0$. The DTFT of $x[n]$ is the sum of the DTFT of each piece:

$$X(e^{j\hat{\omega}}) = \sum_{n=-\infty}^{\infty} a^{|n|} e^{-j\hat{\omega}n} = \underbrace{\sum_{n=-\infty}^{-1} a^{-n} e^{-j\hat{\omega}n}}_{S_1} + \underbrace{\sum_{n=0}^{\infty} a^n e^{-j\hat{\omega}n}}_{S_2}$$

2. **In the first sum, S_1, change variables (let $m = -n$) so the sum runs from 1 to ∞, and then re-index the sum to start at 0 and subtract 1, which is what the $m = 0$ term contributes:**

$$S_1 = \sum_{m=1}^{\infty} a^m e^{j\hat{\omega}m} = -1 + \sum_{m=0}^{\infty} a^m e^{j\hat{\omega}m}$$

$$= -1 + \frac{1}{1 - ae^{j\hat{\omega}}} = \frac{ae^{j\hat{\omega}}}{1 - ae^{j\hat{\omega}}}, \left|ae^{j\hat{\omega}}\right| = |a| < 1$$

3. **Evaluate the second sum, S_2, by noting that it's already in standard infinite geometric series form (also see Example 11-1):**

$$S_2 = \sum_{n=0}^{\infty} a^n e^{-j\hat{\omega}n} = \sum_{n=0}^{\infty} \left(a \cdot e^{-j\hat{\omega}}\right)^n = \frac{1}{1 - ae^{-j\hat{\omega}}}, \left|ae^{-j\hat{\omega}}\right| = |a| < 1$$

4. **Combine the terms over a common denominator:**

$$X(e^{j\hat{\omega}}) = S_1 + S_2 = \frac{(ae^{j\hat{\omega}} - a^2) + (1 - ae^{j\hat{\omega}})}{1 - a(e^{j\hat{\omega}} + e^{-j\hat{\omega}}) + a^2}$$

$$= \frac{1 - a^2}{1 - 2a\cos(\hat{\omega}) + a^2}, |a| < 1$$

This last step is optional, but the final form is compact and clean.

As a check on the hand calculations, you can use a CAS like Maxima:

```
(%i1)    assume(cabs(a) < 1);
(%o1)    [ |a| < 1 ]

(%i2)    declare(w,real);
(%o2)    done

(%i3)    X:sum(a^-n*exp(-%i*w*n),n,minf,-1)+sum(a^n*exp(-%i*w*n),n,0,inf),simpsum;

              a %e^%i w          1
(%o3)    ─────────────── + ─────────────────
         1 - a %e^%i w      1 - a %e^-%i w

(%i4)    XX:demoivre(X),ratsimp;

              a² sin (w)² + a² cos (w)² - 1
(%o4)    - ─────────────────────────────────────
           a² sin (w)² + a² cos (w)² - 2 a cos (w) + 1

(%i5)    trigsimp(XX);

                  a² - 1
(%o5)    ──────────────────────
         2 a cos (w) - a² - 1
```

This calculation reveals $S_1 + S_2$ (line 3) and is a great help when you're trying to get comfortable with geometric series manipulation. Getting to the final form takes some finessing with the simplifying rules in Maxima. Notice that $X(e^{j\hat{\omega}})$ is indeed real, because the signal $x[n]$ is real and even.

Example 11-4: Find the DTFT of the odd sequence $x[n] = -2\delta[n+2] + \delta[n+1] - \delta[n-1] + 2\delta[n-2]$.

For short finite-length sequences, transforming term by term is best. Expand out the definition and include only the terms corresponding to nonzero values of $x[n]$:

$$X(e^{j\hat{\omega}}) = \sum_{n=-\infty}^{\infty} x[n]e^{-j\hat{\omega}n} = \cdots + x[-1]e^{j\hat{\omega}} + x[0]e^{j0\hat{\omega}} + x[1]e^{-j\hat{\omega}} + \cdots$$

Here, the only terms you need to include are n = –2, –1, 1, and 2. Simplify by using Euler's formula for sine:

$$X(e^{j\hat{\omega}}) = -2e^{j2\hat{\omega}} + e^{j\hat{\omega}} - e^{-j\hat{\omega}} + 2e^{-j2\hat{\omega}}$$
$$= -2\left(e^{j2\hat{\omega}} - e^{-j2\hat{\omega}}\right) + \left(e^{j\hat{\omega}} - e^{-j\hat{\omega}}\right)$$
$$= -4j \cdot \sin(2\hat{\omega}) + 2j \cdot \sin(\hat{\omega})$$

For $x[n]$ odd, $X(e^{j\hat{\omega}})$ is pure imaginary.

Studying transform theorems and pairs

Think of a DTFT theorem as a general purpose transform pair — or a catalog of frequency spectra corresponding to specific discrete-time signals — because a theorem considers the DTFT of one or more generic signals under some transformation, such as convolution. By taking full advantage of theorems and pairs, you can get fast and efficient in solving problems.

In this section, I provide tabular listings of DTFT theorems and pairs. I also provide short proofs of the most popular theorems and develop a couple of transform pairs.

Figure 11-2 offers a catalog of useful DTFT theorems.

These DTFT theorems are similar to the FT theorems in Chapter 9:

 ✔ **Linearity:** $ax_1[n] + bx_2[n] \xleftrightarrow{\ \mathcal{F}\ } aX_1(e^{j\hat{\omega}}) + bX_2(e^{j\hat{\omega}})$.

   The proof follows from the definition and the linearity of the sum operator itself.

✔ **Time shift:** $x[n-n_0] \xleftrightarrow{\mathcal{F}} e^{-j\hat{\omega}n_0} X(e^{j\hat{\omega}})$.

To prove, I start from the definition but change variables $m = n - n_0$:

$$\mathcal{F}\{x[n-n_0]\} = \sum_{m=-\infty}^{\infty} x[m]e^{-j\hat{\omega}(m+n_0)} = e^{-j\hat{\omega}n_0} \underbrace{\sum_{m=-\infty}^{\infty} x[m]e^{-j\hat{\omega}m}}_{X(e^{j\hat{\omega}})}$$

✔ **Frequency shift:** $x[n]e^{j\hat{\omega}_0 n} \xleftrightarrow{\mathcal{F}} X(e^{j(\hat{\omega}-\hat{\omega}_0)})$.

Using the DTFT definition,

$$\mathcal{F}\{x[n]e^{j\hat{\omega}_0 n}\} = \sum_{n=-\infty}^{\infty} x[n]e^{j\hat{\omega}_0 n}e^{-j\hat{\omega}n} = \underbrace{\sum_{n=-\infty}^{\infty} x[n]e^{-j(\hat{\omega}-\hat{\omega}_0)n}}_{X(e^{j(\hat{\omega}-\hat{\omega}_0)})}$$

✔ **Convolution:** The convolution of two sequences (see Chapter 6) is defined as $y[n] = x[n]*h[n] = \sum_{k=-\infty}^{\infty} x[k]h[n-k] \overset{also}{=} \sum_{k=-\infty}^{\infty} h[k]x[n-k] = h[n]*x[n]$.

It can be shown that convolving two sequences is equivalent to multiplying the respective DTFTs: $x[n]*h[n] \xleftrightarrow{\mathcal{F}} X(e^{j\hat{\omega}})H(e^{j\hat{\omega}})$.

	Property	Signal	Transform
1	Linearity	$ax_1[n] + bx_2[n]$	$aX_2(e^{j\hat{\omega}}) + bX_2(e^{j\hat{\omega}})$
2	Time shift	$x[n-n_0]$	$e^{-j\hat{\omega}n_0} X(e^{j\hat{\omega}})$
3	Frequency shift	$e^{j\hat{\omega}_0 n}x[n]$	$X(e^{j(\hat{\omega}-\hat{\omega}_0)})$
4	Convolution	$x[n]*h[n]$	$X(e^{j\hat{\omega}})H(e^{j\hat{\omega}})$
5	Multiplication	$x[n]w[n]$	$\frac{1}{2\pi}\int_{-\pi}^{\pi} X(e^{j\theta})W(e^{j(\hat{\omega}-\theta)})\,d\theta$
6	Reversal	$x[-n]$	$X^*(e^{j\hat{\omega}})$
7	Differentiation	$nx[n]$	$j\frac{dX(e^{j\hat{\omega}})}{d\hat{\omega}}$
8	First difference	$x[n] - x[n-1]$	$(1 - e^{-j\hat{\omega}})X(e^{j\hat{\omega}})$
9	Accumulation	$\sum_{k=-\infty}^{n} x[k]$	$\frac{1}{1-e^{-j\hat{\omega}}} X(e^{j\hat{\omega}})$
10	Downsample	$x[nM]$	$\frac{1}{M}\sum_{i=0}^{M-1} X(e^{j(\hat{\omega}/M - 2\pi k/M)})$
11	Upsample	$x[n/L]$ zero for $n/L \neq$ integer	$X(e^{j\hat{\omega}L})$
12	Parseval's theorem	$\sum_{n=-\infty}^{\infty} |x[n]|^2$	$\frac{1}{2\pi}\int_{-\pi}^{\pi} |X(e^{j\hat{\omega}})|^2\,d\hat{\omega}$

Figure 11-2:
Useful DTFT
theorems.

A table of DTFT pairs (provided in Figure 11-3) is invaluable when you work problems. Unless you're required to *prove* a particular pair, I see no sense in starting a problem empty-handed.

	Time	Frequency
1	$\delta[n]$	1
2	$\delta[n - n_0]$	$e^{-j\hat{\omega}n_0}$
3	$a^n u[n], (\lvert a \rvert < 1)$	$\dfrac{1}{1-ae^{-j\hat{\omega}}}$
4	$(n+1)a^n u[n], (\lvert a \rvert < 1)$	$\dfrac{1}{(1-ae^{-j\hat{\omega}})^2}$
5	$\dfrac{r^n \sin[\hat{\omega}_0(n-1)]}{\sin\hat{\omega}_0}u[n], (\lvert r \rvert < 1)$	$\dfrac{1}{1-2r\cos(\hat{\omega}_0)e^{-j\hat{\omega}}+r^2e^{-j2\hat{\omega}}}$
6	$u[n] - u[n - M]$	$\dfrac{\sin(\hat{\omega}M/2)}{\sin(\hat{\omega}/2)} \cdot e^{-j\hat{\omega}(M-1)/2}$
7	$\dfrac{\sin\hat{\omega}_c n}{\pi n}$	$\Pi\left(\dfrac{\hat{\omega}}{2\hat{\omega}_c}\right)$
8	$e^{j\hat{\omega}_0 n}$	$2\pi\sum_{m=-\infty}^{\infty}(\hat{\omega} - \hat{\omega}_0 - 2\pi m)$
9	$\cos(\hat{\omega}_0 n + \Phi)$	$\pi\sum_{m=-\infty}^{\infty}\begin{array}{l}[e^{j\phi}\delta(\hat{\omega} - \hat{\omega}_0 - 2\pi m)\\ + e^{-j\phi}\delta(\hat{\omega} - \hat{\omega}_0 - 2\pi m)]\end{array}$
10	1	$\pi\sum_{m=-\infty}^{\infty}(\hat{\omega} - 2\pi m)$
11	$u[n]$	$\dfrac{1}{1-e^{-j\hat{\omega}}} + \pi\sum_{m=-\infty}^{\infty}(\hat{\omega} - 2\pi m)$
12	$\sum_{k=-\infty}^{\infty}\delta[n - Mk]$	$\dfrac{2\pi}{M}\sum_{m=-\infty}^{\infty}\delta\left(\hat{\omega} - \dfrac{2\pi}{M}m\right)$

Figure 11-3: Useful DTFT pairs.

Here are a couple of transform pairs that you'll likely use when studying signals and systems.

✔ **Impulse sequence:** $\delta[n]\overset{\mathcal{F}}{\longleftrightarrow}1$.

This pair comes from the definition $\mathcal{F}\{\delta[n]\} = \sum_{n=-\infty}^{\infty}\delta[n]e^{-j\hat{\omega}n} = e^{-j\hat{\omega}\cdot 0} = 1$ from the sifting property of the impulse sequence.

✔ **Rectangular pulse (window) sequence:** The rectangular pulse or window sequence is defined as

$$w_M[n] = u[n] - u[n - M] = \begin{cases}1, & 0 \le n < M \\ 0, & \text{otherwise}\end{cases}$$

You can find the DTFT by direct evaluation, recognizing that the sum is a finite geometric series and factoring to form a ratio of sine functions:

$$W_M(e^{j\hat{\omega}}) = \sum_{n=0}^{M-1} e^{-j\hat{\omega}n} = \frac{1-e^{-j\hat{\omega}M}}{1-e^{-j\hat{\omega}}} = \frac{e^{j\hat{\omega}M/2}-e^{-j\hat{\omega}M/2}}{e^{j\hat{\omega}/2}-e^{-j\hat{\omega}/2}} \cdot \frac{e^{-j\hat{\omega}M/2}}{e^{-j\hat{\omega}/2}}$$

$$= \frac{\sin(\hat{\omega}M/2)}{\sin(\hat{\omega}/2)} \cdot e^{-j\hat{\omega}(M-1)/2}$$

The pair is $u[n]-u[n-M] \xleftrightarrow{\quad\mathcal{F}\quad} \dfrac{\sin(\hat{\omega}M/2)}{\sin(\hat{\omega}/2)} \cdot e^{-j\hat{\omega}(M-1)/2}$.

Example 11-5: If $x[n] = a^n u[n-5]$, find $\mathcal{F}\{x[n]\} = X(e^{j\hat{\omega}})$. The brute-force approach (plugging directly into the DTFT definition) works, but I recommend taking advantage of theorems and transform pairs to streamline your work:

1. **Rewrite $x[n]$ in a form that anticipates the use of certain theorems:**

$$x[n] = a^5\left(a^{n-5}u[n-5]\right)$$

2. **Apply the time shift theorem (Line 2 in Figure 11-2):**

$$X(e^{j\hat{\omega}}) = a^5 e^{-j5\hat{\omega}} \mathcal{F}\{a^n u[n]\}$$

3. **Apply transform pair, Line 3 in Figure 11-3, assuming $|a| < 1$:**

$$X(e^{j\hat{\omega}}) = \frac{a^5 e^{-j5\hat{\omega}}}{1-ae^{-j\hat{\omega}}}$$

Working with Special Signals

Some signals aren't absolutely summable, but you can find a meaningful DTFT for them. (I explore this kind of situation in Chapter 9, too, by using Fourier transforms in the limit to allow impulse functions in the frequency domain.) In this section, I describe mean-square convergence and Fourier transforms in the limit for the DTFT. By using mean-square convergence, I develop a transform pair for an ideal low-pass filter. Fourier transforms in the limit allow impulse functions in the frequency domain.

Getting mean-square convergence

A form of convergence that's weaker than absolute convergence is known as *mean-square convergence,* which requires *square summability* of $x[n]$:

$$\sum_{n=-\infty}^{\infty} |x[n]|^2 < \infty$$

This condition is easier to satisfy than absolute summability; but with mean-square convergence, the DTFT may not converge pointwise in the frequency domain. (Chapter 8 explores the trouble with getting the Fourier series of a square wave to converge.)

A rectangular or low-pass spectrum $X_{LP}(e^{j\hat{\omega}})$ is defined on the fundamental interval $[-\pi, \pi]$ to be

$$X_{LP}(e^{j\hat{\omega}}) = \Pi\left(\frac{\hat{\omega}}{2\hat{\omega}_c}\right) = \begin{cases} 1, & |\hat{\omega}| < \hat{\omega}_c \\ 0, & \hat{\omega}_c < |\hat{\omega}| \le \pi \end{cases}$$

where $\hat{\omega}_c$ is the spectrum bandwidth in rad/sample.

Here, I'm talking about a signal spectrum, but this definition also applies to the frequency response of an ideal low-pass filter. Being able to synthesize an ideal low-pass filter allows you to separate a desirable signal from an undesirable one, even when they're right next to each other spectrally.

Given the spectrum, you can work backward to get the signal by using the IDTFT:

$$x_{LP}[n] = \frac{1}{2\pi}\int_{-\hat{\omega}_c}^{\hat{\omega}_c} e^{j\hat{\omega}n}\, d\hat{\omega} = \frac{e^{j\hat{\omega}n}}{j2\pi n}\Big|_{-\hat{\omega}_c}^{\hat{\omega}_c} = \frac{e^{j\hat{\omega}_c n} - e^{-j\hat{\omega}_c n}}{2j\cdot\pi n}$$

$$= \frac{\sin\hat{\omega}_c n}{\pi n} = \frac{\hat{\omega}_c}{\pi}\operatorname{sinc}(\hat{\omega}_c n),\ -\infty < n < \infty$$

A rectangle in the frequency domain is a sampled sinc function in the time domain (find the continuous-time version in Chapter 9).

In Figure 11-4, I plot $x_{LP}[n]$ for $\hat{\omega}_c = \pi/2$ over the interval ± 50 to show how quickly the sinc function decays to 0 and to point out how important the small tail values are in the frequency domain when considering truncation.

The absolute sum of this sequence diverges because the terms are of the form $1/n$, which is the *harmonic series,* and known to diverge. The sum of $1/n^2$ converges so $x_{LP}[n]$ is square summable.

It would be nice if I could form a $2M+1$ term approximation to $X_{LP}(e^{j\hat{\omega}})$ and arrive at a likeable approximation to the ideal low-pass spectrum (read: filter), too. Luckily, I can! The approximation to the spectrum/frequency response takes the form

$$X_{M,LP}(e^{j\hat{\omega}}) = \sum_{n=-M}^{M} \frac{\sin(\hat{\omega}_c n)}{\pi n}\, e^{-j\hat{\omega}n}$$

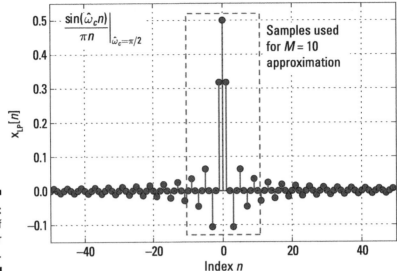

Figure 11-4:
A plot of
$x_{LP}[n]$ for
$\hat{\omega}_c = \pi/2.$

For the case of a filter, this means you can use a $2M + 1$-tap finite impulse response (FIR) filter to approximate an ideal low-pass filter. But to make the filter causal, you also need a time delay of M samples to the right. Your intuition may say that by increasing M, the filter approximation gets better — and it does get better in the sense of being more rectangular shaped (like the ideal rectangular spectrum definition) — but the ripples on both sides of $\hat{\omega}_c$ remain fixed in amplitude.

I use Python to check by writing a loop inside the IPython environment to numerically calculate the spectrum and then plot results (see Figure 11-5).

```
In [138]: w = arange(0,pi,pi/500.)
In [139]: X_LP = zeros(len(w))+1j*zeros(len(w))
In [140]: for n in range(-10,10+1):
     ...: X_LP += pi/2./pi*sinc(pi/2.*n/pi)*exp(-1j*w*n)
     ...:
In [140]: plot(w,20*log10(abs(X_LP)))
```

This exercise reveals that at the band edges, where the spectrum transitions from 1 to 0, quite a bit of ringing occurs. The peak passband (0 dB spectrum level) ripple is 0.75 dB, and the peak side lobe level is only 21 dB below the passband; both values are independent of $M!$ The passband to stopband transition occurs faster (narrower band of frequencies) as M increases, so your intuition is partially confirmed. The ripple isn't good.

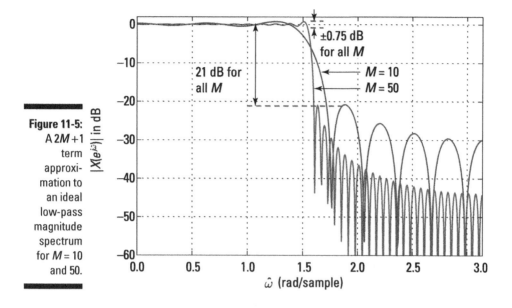

Figure 11-5:
A $2M+1$ term approximation to an ideal low-pass magnitude spectrum for $M = 10$ and 50.

Window functions to the rescue. To reduce the ripple level, you can employ window functions (described in Chapter 4). With windowing, you multiply the truncated sinc function signal by a nonconstant shaping function, $w[n]$, which smoothly transitions from 1 to 0 at the window edges. The DTFT of the smooth $w[n]$ has a much smaller ripple level, and passes that on to the overall spectrum of $x[n]w[n]$. The ripple level is made small at the expense of a wider transition frequency band. You can, however, narrow the transition band by increasing M. This story does indeed have a happy ending.

Finding Fourier transforms in the limit

The Fourier transform in the limit approach allows impulse functions to exist in the frequency domain. As a specific case, suppose you have a sequence $x_1[n]$, which has DTFT $\delta(\hat{\omega} - \hat{\omega}_0)$ for $-\pi \leq \hat{\omega} \leq \pi$. Because $X_1(e^{j\hat{\omega}})$ is always periodic, the complete representation is

$$X_1(e^{j\hat{\omega}}) = \sum_{m=-\infty}^{\infty} 2\pi\delta(\hat{\omega} - \hat{\omega}_0 - 2\pi m)$$

The spectrum in the discrete-time domain is always periodic with period 2π. When a spectrum involves impulse functions as opposed to a function of $e^{j\hat{\omega}}$, the periodicity isn't automatic, so you need the doubly infinite sum. Don't let this apparent complexity confuse you. For $-\pi \leq \hat{\omega} \leq \pi$, the only impulse function is $\delta(\hat{\omega} - \hat{\omega}_0)$.

To find $x_1[n]$, operate with the IDTFT $x_1[n] = \dfrac{1}{2\pi}\displaystyle\int_{-\pi}^{\pi} 2\pi\delta(\hat{\omega} - \hat{\omega}_0)e^{j\hat{\omega}n}\,d\hat{\omega} = e^{j\hat{\omega}_0 n}$, which establishes the following DTFT pair (through the sifting property of the impulse function):

$$e^{j\hat{\omega}_0 n} \overset{\mathcal{F}}{\longleftrightarrow} \sum_{m=-\infty}^{\infty} 2\pi\delta(\hat{\omega} - \hat{\omega}_0 - 2\pi m)$$

If $\hat{\omega}_0 = 0$, then $x[n]$ is just a constant, so A (for all n)$\overset{\mathcal{F}}{\longleftrightarrow} A\displaystyle\sum_{m=-\infty}^{\infty} 2\pi\delta(\hat{\omega} - 2\pi m)$.

Now, suppose $x_2[n] = A\cos(\hat{\omega}_0 n)$. Using Euler's formula for cosine, find that

$$X_2(e^{j\hat{\omega}}) = \mathcal{F}\{A\cos(\hat{\omega}_0 n)\} = \mathcal{F}\left\{\frac{A}{2}\left[e^{j\hat{\omega}_0 n} + e^{-j\hat{\omega}_0 n}\right]\right\}$$

$$= \sum_{m=-\infty}^{\infty} A\pi\left[\delta(\hat{\omega} - \hat{\omega}_0 - 2\pi m) + \delta(\hat{\omega} + \hat{\omega}_0 - 2\pi m)\right]$$

See the spectrum of $A\cos(\hat{\omega}_0 n)$ in Figure 11-6.

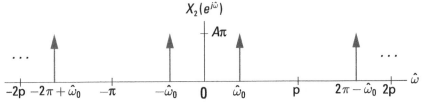

Figure 11-6: The spectrum of $A\cos(\hat{\omega}_0 n)$.

Example 11-6: Say a continuous-time sinusoid is sampled over a finite time interval to produce the discrete-time signal $x_3[n] = x_2[n]\cdot w[n] = A\cos(\hat{\omega}_0 n)\cdot w[n]$, $0 < \hat{\omega}_0 < \pi$, where $w[n] = u[n] - u[n - M]$ is a rectangular windowing function corresponding to the capture interval. The signal, both continuous-time and discrete-time, before and after windowing, is plotted in Figure 11-7.

A transform pair and transform theorem make short work of this problem.

1. **Use the transform pair from Line 6 of Figure 11-3:**

$$W(e^{j\hat{\omega}}) = \frac{\sin\left[\hat{\omega}M/2\right]}{\sin(\hat{\omega}/2)}\cdot e^{-j\hat{\omega}(M-1)/2}$$

2. **To get $X_3(e^{j\hat{\omega}})$, use the multiplication, or windowing, theorem (Line 5 from Figure 11-2).**

In the end, the calculation requires a convolution in the frequency domain between $X_2(e^{j\hat{\omega}})$ and $W(e^{j\hat{\omega}})$:

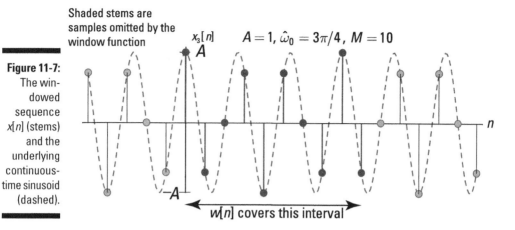

Shaded stems are samples omitted by the window function

$x_3[n]$ $A = 1$, $\hat{\omega}_0 = 3\pi/4$, $M = 10$

$w[n]$ covers this interval

Figure 11-7:
The windowed sequence $x[n]$ (stems) and the underlying continuous-time sinusoid (dashed).

Step 2 appears messy, but only one pair of the cosine impulse functions lies on the $[-\pi, \pi]$ interval that you integrate over. And the integration itself is straightforward because the impulse function just sifts out the integrand sampled at $\pm\hat{\omega}_0$.

I plot the spectrum of the windowed cosine in Figure 11-8, using Python. I compute the DTFT with the function `freqz()` from the SciPy module `signal`.

```
In [206]: n = arange(0,10)
In [207]: x = cos(3*pi/4.*n)
In [208]: w = arange(-pi,pi,pi/500.)
In [209]: w_in = arange(-pi,pi,pi/500.)
In [210]: w,X = signal.freqz(x,1,w_in)
In [211]: plot(w,abs(X))
```

Impressed? Only ten samples of the cosine signal results in the spectral *blobs* at $\pm\hat{\omega}_0$. Not too bad, but this is a far cry from impulse functions, which is what you'd see if the windowing wasn't present.

I increased M to 50 to show how much improvement a five-times the window length provides. In the end, M must be large enough to provide a reasonable estimate of the sinusoid parameters A and $\hat{\omega}_0$. As M increases the spectrum of the window, $W(e^{j\hat{\omega}})$, gets narrower, because the spectrum width is proportional to $1/M$. When the window spectrum is convolved with the frequency domain impulse functions of the sinusoid, the result is a more compact spectrum shape. Find a more detailed example of spectral estimation using windows online at www.dummies.com/extras/signalsandsystems.

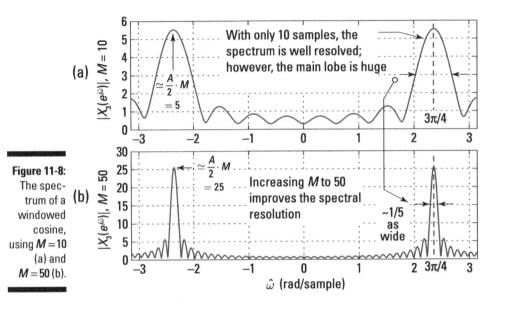

Figure 11-8:
The spectrum of a windowed cosine, using $M=10$ (a) and $M=50$ (b).

LTI Systems in the Frequency Domain

For LTI systems in the time domain (see Chapter 6), a fundamental result is that the output, $y[n]$, is the input, $x[n]$, convolved with the system impulse response, $h[n]$: $y[n] = x[n] * h[n]$.

To carry this result to the frequency domain, simply take the DTFT of both sides:

$$Y(e^{j\hat{\omega}}) = \mathcal{F}\{y[n]\} = \mathcal{F}\{x[n]*h[n]\} = \mathcal{F}\{x[n]\}\cdot\mathcal{F}\{h[n]\}$$
$$= X(e^{j\hat{\omega}})H(e^{j\hat{\omega}})$$

This is the convolution theorem for the DTFT.

The quantity $H(e^{j\hat{\omega}}) = \mathcal{F}\{h[n]\}$ is known as the *transfer function* or *frequency response* of the system having impulse response $h[n]$. This is a special FT. If you want to get $y[n]$ via multiplication in the frequency domain, you just need to compute the inverse DTFT of the product: $y[n] = \mathcal{F}^{-1}\{X(e^{j\hat{\omega}})H(e^{j\hat{\omega}})\}$.

You can write this input/output relationship as the ratio of the output spectrum to the input spectrum, $H(e^{j\hat{\omega}}) = Y(e^{j\hat{\omega}})/X(e^{j\hat{\omega}})$. If $x[n] = \delta[n]$, then $X(e^{j\hat{\omega}}) = 1$, and the output spectrum takes its shape entirely from $H(e^{j\hat{\omega}})$ because $Y(e^{j\hat{\omega}}) = 1\cdot H(e^{j\hat{\omega}})$.

Considering the properties of the frequency response, keep in mind that for $h[n]$ real, $\left|H(e^{-j\hat{\omega}})\right| = \left|H(e^{j\hat{\omega}})\right|$ and $\angle H(e^{-j\hat{\omega}}) = -\angle H(e^{j\hat{\omega}})$.

Also, the output energy spectral density is related to the input energy spectral density and the frequency response because the convolution theorem says $Y(e^{j\hat{\omega}}) = X(e^{j\hat{\omega}}) \cdot H(e^{j\hat{\omega}})$:

$$\left|Y(e^{j\hat{\omega}})\right|^2 = \left|X(e^{j\hat{\omega}})H(e^{j\hat{\omega}})\right|^2 = \left|X(e^{j\hat{\omega}})\right|^2 \cdot \left|H(e^{j\hat{\omega}})\right|^2$$

As a result of the FT convolution theorem, you can develop the cascade relationship from the block diagram of Figure 11-9. (See Chapter 5 for cascading LTI systems in the time domain.)

Figure 11-9: Cascade of LTI systems in the frequency domain.

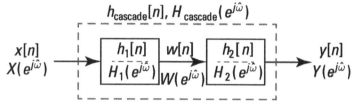

In the frequency domain, $W(e^{j\hat{\omega}}) = X(e^{j\hat{\omega}})H_1(e^{j\hat{\omega}})$ and $Y(e^{j\hat{\omega}}) = W(e^{j\hat{\omega}})H_2(e^{j\hat{\omega}})$ so, upon linking the two equations, you have the following:

$$Y(e^{j\hat{\omega}}) = X(e^{j\hat{\omega}}) \cdot H_1(e^{j\hat{\omega}})H_2(e^{j\hat{\omega}})$$

$$\text{and } H_{cascade}(e^{j\hat{\omega}}) = H_1(e^{j\hat{\omega}})H_2(e^{j\hat{\omega}})$$

As a result of the FT convolution theorem, you can develop the parallel connection relationship from the block diagram of Figure 11-10.

Figure 11-10: Parallel connection of LTI systems in the frequency domain.

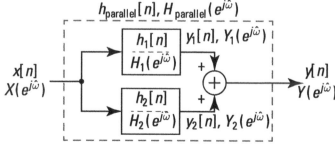

In the frequency domain, the following is true:

$$Y_1(e^{j\hat{\omega}}) = X(e^{j\hat{\omega}})H_1(e^{j\hat{\omega}}), \ Y_2(e^{j\hat{\omega}}) = X(e^{j\hat{\omega}})H_2(e^{j\hat{\omega}}),$$

and $Y(e^{j\hat{\omega}}) = Y_1(e^{j\hat{\omega}}) + Y_2(e^{j\hat{\omega}})$

So

$$Y(e^{j\hat{\omega}}) = X(e^{j\hat{\omega}})H_1(e^{j\hat{\omega}}) + X(e^{j\hat{\omega}})H_2(e^{j\hat{\omega}})$$
$$= X(e^{j\hat{\omega}}) \cdot \left[H_1(e^{j\hat{\omega}}) + H_2(e^{j\hat{\omega}}) \right]$$

and $H_{\text{parallel}}(e^{j\hat{\omega}}) = H_1(e^{j\hat{\omega}}) + H_2(e^{j\hat{\omega}})$

Taking Advantage of the Convolution Theorem

By working in the frequency domain, you can avoid the tedious details of the convolution integral (covered in Chapter 6). In particular, you can find time-domain signals at the output of a system by multiplying the input spectrum with the frequency response and then using the inverse transform to return to the time domain. The output you seek may be in response to an impulse or a step or a very specialized signal, and along the way, you may be interested in the output spectrum, too.

Example 11-7: Use the DTFT to determine both the impulse response and frequency response directly from a linear constant coefficient (LCC) difference equation. Consider the following:

$$y[n] - \frac{1}{2}y[n-1] = x[n] - \frac{1}{4}x[n-1]$$

Find $H(e^{j\hat{\omega}})$ and $h[n]$. The problem solution breaks down into four steps:

1. **Take the DTFT of both sides of the difference equation, using the time shift theorem (Line 2 of Figure 11-2):**

$$Y(e^{j\hat{\omega}}) - \frac{1}{2}e^{-j\hat{\omega}}Y(e^{j\hat{\omega}}) = X(e^{j\hat{\omega}}) - \frac{1}{4}e^{-j\hat{\omega}}X(e^{j\hat{\omega}})$$

Because $y[n] = x[n] * h[n], \ Y(e^{j\hat{\omega}}) = X(e^{j\hat{\omega}})H(e^{j\hat{\omega}})$.

2. **Solve for $H(e^{j\hat{\omega}}) = Y(e^{j\hat{\omega}}) / X(e^{j\hat{\omega}})$ by using simple algebra on the results of Step 1:**

$$H(e^{j\hat{\omega}}) = \frac{Y(e^{j\hat{\omega}})}{X(e^{j\hat{\omega}})} = \frac{1 - \frac{1}{4}e^{-j\hat{\omega}}}{1 - \frac{1}{2}e^{-j\hat{\omega}}}$$

3. To get **h[n]**, use the inverse transform $H(e^{j\hat{\omega}})$ by breaking the expression into two terms:

$$H(e^{j\hat{\omega}}) = \underbrace{\frac{1}{1-\frac{1}{2}e^{-j\hat{\omega}}}}_{\text{from } a^n u[n]} - \underbrace{\frac{\frac{1}{4}}{1-\frac{1}{2}e^{-j\hat{\omega}}}}_{\text{from } a^n u[n]} \cdot \underbrace{e^{-j\hat{\omega}}}_{\text{time shift by 1}}$$

4. Use transform pair Line 3 in Figure 11-3 on both terms and use the time shift theorem on the second term:

$$h[n] = \mathcal{F}^{-1}\{H[n]\} = \left(\tfrac{1}{2}\right)^n u[n] - \underbrace{\left(\tfrac{1}{4}\right)^{n-1} u[n-1]}_{n-1 \text{ from time shift}}$$

The time shift factor $n-1$ replaces n in all locations where it occurs in the second term.

Chapter 12

The Discrete Fourier Transform and Fast Fourier Transform Algorithms

. . . .

In This Chapter

▶ Looking at the DFT as uniformly spaced samples of the DTFT

▶ Working with the DFT/IDFT pair

▶ Exploring DFT theorems

▶ Implementing the fast Fourier transform algorithm

▶ Checking out an application example of transform domain filtering

. . . .

*F*ourier's name is associated with four different chapters of this book — nearly all the chapters in Part III. So what's special about the discrete Fourier transform (DFT)? Computer computation! Under the right conditions the DFT returns a sample version of the discrete-time Fourier transform (DTFT) but using a systematic computer computation.

That's right; the DFT allows you to move to and from the spectral representation of a signal with fast and efficient computation algorithms. You won't find any integrals here! Complex number multiplication and addition is all there is to the mathematical definition of the DFT.

In this chapter, I introduce efficient methods to calculate the DFT. Arguably, the most efficient methods or algorithms are known collectively as the fast Fourier transform (FFT). A popular FFT algorithm divides the DFT calculation into a set of two-point DFTs, and the results of the DFTs are combined, thereby reducing the number of complex multiplications and additions you need to perform during spectrum analysis and frequency response calculations. Yes, I tell you which algorithm it is in this chapter. Read on!

Establishing the Discrete Fourier Transform

In this section, I point out the connection between the discrete-time Fourier transform (DTFT) (covered in Chapter 11) and the discrete Fourier transform (DFT), which is the focus of this chapter. Knowing how these two concepts work together enables you to begin using the DFT to find the spectrum of discrete-time signals.

Unlike the discrete-time Fourier transform (DTFT), which finds the spectrum of a discrete-time signal in terms of a continuous-frequency variable, the DFT operates on an N-point signal and produces an N-point frequency spectrum. Here, I show you that as long as the DFT length, N, encompasses the entire discrete-time signal, the N spectrum values found with the DFT are equal to sample values of the DTFT.

Consider a finite duration signal $x[n]$ that's assumed to be 0 everywhere except possibly for $n = 0, 1, ..., N-1$. See a sketch of the finite duration signal in Figure 12-1.

Figure 12-1:
The finite duration signal $x[n]$ used for establishing the DFT from the DTFT.

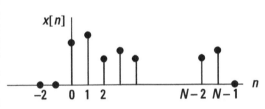

The DTFT of $x[n]$ is the periodic function $X\left(e^{j\hat{\omega}}\right) = \sum_{n=-\infty}^{\infty} x[n]e^{-j\hat{\omega}n}$.

Consider $X\left(e^{j\hat{\omega}}\right)$ at discrete frequencies $\hat{\omega}_k = 2\pi k/N$, $k = 0, 1, ..., N-1$:

$$X\left(e^{j\hat{\omega}_k}\right) = X[k] = \sum_{n=0}^{N-1} x[n]e^{-j\hat{\omega}}\bigg|_{\hat{\omega}=2\pi k/N} = \sum_{n=0}^{N-1} x[n]W_N^{kn}$$

where $W_N \equiv e^{-j2\pi/N}$. The sampling of $X\left(e^{j\hat{\omega}}\right)$ to produce $X[k]$ is shown graphically in Figure 12-2a.

WARNING!

The use of W_N here is an abbreviation for $e^{-j2\pi/N}$. In Chapter 11, $W(e^{j\hat{\omega}})$ represents the Fourier transform of a window function $w[n]$. Don't let this dual use confuse you. W_N is the standard notation for the complex weights, or *twiddle factors,* found in the DFT and FFT.

Figure 12-2:
The DFT as
sampling
the DTFT (a)
and sam-
pling the
z-transform
(b).

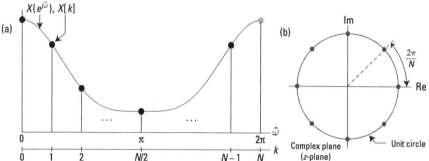

From a z-transform perspective (see Chapter 14), the DFT also corresponds to sampling $X(z)$, the z-transform $x[n]$, at uniformly spaced sample locations on the unit circle as shown in Figure 12-2b. Setting $z = e^{j\hat{\omega}}$ in $X(z)$ reveals the DTFT, and sampling the DTFT at $\hat{\omega} = 2\pi k/N$, $k = 0, 1, ..., N-1$ gives the DFT.

The DFT/IDFT Pair

The DFT is composed of a forward transform and an inverse transform, just like the Fourier transform (see Chapter 9) and the discrete-time Fourier transform (see Chapter 11). The forward transform is also known as *analysis,* because it takes in a discrete-time signal and performs spectral analysis. Here is the mathematical definition for the forward transform:

$$X[k] \equiv \sum_{n=0}^{N-1} x[n]e^{-j2\pi kn/N} \overset{also}{=} \sum_{n=0}^{N-1} x[n]W_N^{kn}, \ n = 0, 1, ..., N-1$$

The complex weights W_N^{kn} are known as the twiddle factors. And the inverse discrete Fourier transform (IDFT) is sometimes called *synthesis,* because it synthesizes $x[n]$ from the spectrum values, and is given by

$$x[n] \equiv \frac{1}{N}\sum_{k=0}^{N-1} X[k]e^{j2\pi kn/N} \overset{also}{=} \frac{1}{N}\sum_{k=0}^{N-1} X[k]W_N^{-kn}, \ n = 0, 1, ..., N-1$$

The analysis and synthesis are often referred to as the DFT/IDFT pair. The transform is *computable* in both directions. This means that each sum is composed of just N terms. The number of complex multiplications and additions to compute all N points is thus finite, allowing for computer implementation. The same can't be said about the continuous-time and discrete-time Fourier transform pairs, though, because they both involve integration over a continuous variable. This makes the DFT/IDFT very special.

Does the IDFT really undo the action of the DFT in exact mathematical terms? The best way to find out is to see for yourself.

Begin the proof by writing the definition of the IDFT and inserting the DFT in place of $X[k]$. Interchange sum orders and simplify:

$$\frac{1}{N}\sum_{k=0}^{N-1} X[k]W_N^{-kn} = \frac{1}{N}\sum_{k=0}^{N-1}\left[\sum_{m=0}^{N-1}x[m]W_N^{km}\right]W_N^{-kn}$$

$$(\text{swaps sums}) = \sum_{m=0}^{N-1}x[m]\cdot\left[\frac{1}{N}\sum_{k=0}^{N-1}W_N^{k(m-n)}\right]$$

To complete the proof, investigate the term in square brackets. You can use the finite geometric series (described in Chapter 2) to characterize this term

$$\frac{1}{N}\sum_{k=0}^{N-1}W_N^{k(m-n)} = \frac{1}{N}\cdot\frac{1-e^{-j2\pi(m-n)}}{1-e^{-j2\pi(m-n)/N}} = \delta[m-n]$$

because for $m \neq n\, e^{j2\pi(m-n)} = 1$ and for $m = n$, $W_N^{k(m-n)} = W_N^0 = 1$.

Throughout this chapter, I use the shorthand notation $x[n]\xleftarrow{\;DFT\;}X[k]$ to indicate the relationship between $x[n]$ and $X[k]$.

In the synthesis of $x[n]$ from the $X[k]$'s, you're actually representing $x[n]$ in terms of a Fourier series — or, more specifically, a *discrete Fourier series* (DFS) — with coefficients $X[k]$. The DFS point of view parallels the Fourier series analysis and synthesis pair of Chapter 8; only now you're working in the discrete-time domain. Still, when one period of a signal is represented (or synthesized) by using its Fourier series coefficients, all the periods of the waveform are nicely sitting side by side along the time axis.

In the case of the DFT, you start with a finite length signal. But in the eyes of the DFT, all the adjoining periods are also present, at least mathematically. The signal periods on each side of the original N-point sequence are called the *periodic extensions* of the signal. Here's the notation for the composite signal (a sequence of period N):

$$\tilde{x}[n] = \sum_{r=-\infty}^{\infty} x[n+rN] = x\big[(n\bmod N)\big] = x\Big[\big((n)\big)_N\Big]$$

A compact notation for $n \bmod N$ is $((n))_N$. If you're not familiar with the mod operator, think about any experience you may have with programming languages or digital electronics. The mod$N$ operator wraps the n value onto the interval $0 \le n \le N - 1$ by adding or subtracting integer multiples of N as needed. For example, 4 mod 10 = 4, 13 mod 10 = 3, and –5 mod 10 = 5.

On the frequency domain side, $X[k]$ is also periodic, because the DTFT itself has period 2π (see Chapter 11). With N samples of $X(e^{j\hat{\omega}})$ per 2π period, the period of $X[k]$ is also N. Formally, to maintain a duality relationship in the DFT/IDFT pair, you can also write this equation:

$$\tilde{X}[k] = \sum_{r=-\infty}^{\infty} X[k + rN] = X\big[(k \bmod N)\big] = X\big((k)\big)_N]$$

Example 12-1: To find the DFT of the four-point sequence $\{2, 2, 1, 1\}$, first assume, unless told otherwise, that a sequence being entered into the DFT starts at 0 and ends at $N - 1$. Here, $N = 4$. Working from the definition of the DFT, insert the sequence values in the sum formula:

$$X[k] = \sum_{n=0}^{N-1} x[n] W_N^{kn}$$
$$= 2 \cdot W_4^0 + 2 \cdot W_4^k + 1 \cdot W_4^{2k} + 1 \cdot W_4^{3k}$$
$$= 2 + 2e^{-j\pi k/2} + e^{-j\pi k} + e^{-j3\pi k/2}$$

Next, plug in $k = 0, 1, 2$, and 3 to find $X[0] = 2 + 2 + 1 + 1 = 6$, $X[1] = 2 - 2j - 1 + j = 1 - j$, $X[2] = 2 - 2 + 1 - 1 = 0$, and $X[3] = 2 + 2j - 1 - j = 1 + j$.

Using Python and PyLab, you can access FFT functions via the `fft` package (no need to use `import` in the IPython environment). I use `fft.fft(x,N)` and `fft.ifft(X,N)` throughout this chapter to check hand calculations. For this example, a Python check reveals the following:

```
In [272]: fft.fft([2,2,1,1],4)
Out [272]: array([6.+0.j, 1.-1.j, 0.+0.j, 1.+1.j]) # agree!
```

Example 12-2: Consider the DFT of the following pulse sequence, where $N \ge L$:

$$x[n] = u[n] - u[n - L] = \begin{cases} 1, & 0 \le n \le L - 1 \\ 0, & \text{otherwise} \end{cases}$$

In Chapter 11, the DTFT of $x[n]$ is shown to be

$$X(e^{j\hat{\omega}}) = \sum_{n=-\infty}^{L-1} e^{-j\hat{\omega}n} = \frac{1 - e^{-j\hat{\omega}L}}{1 - e^{-j\hat{\omega}}} = \frac{\sin(\hat{\omega}L/2)}{\sin(\hat{\omega}/2)} e^{-j\hat{\omega}(L-1)/2}$$

Check out the continuous function magnitude spectrum for $L = 10$ in Figure 12-3.

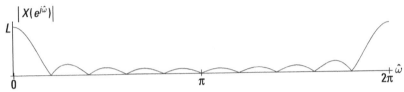

Figure 12-3: Magnitude spectrum plot of the DTFT for an $L = 10$ pulse.

The DFT of $x[n]$ is $X\left(e^{j\hat{\omega}}\right)$ evaluated at $\hat{\omega}_k = 2\pi k/N$, so

$$X[k] = \frac{\sin(\pi kL/N)}{\sin(\pi k/N)} \cdot e^{-j\pi k(L-1)/N}, \quad k = 0, 1, ..., N-1$$

$$X[k] = \begin{cases} L, & k = 0 \\ 0, & k = 1, 2, ..., L-1 \end{cases}$$

For the case of $N = L = 10$, shown in Figure 12-4, the results are somewhat unexpected because the most obvious shape resemblance to Figure 12-3 are the zero samples, which are correctly placed. There are no samples at the spectral peaks except at $k = 0$ and its periodic extensions.

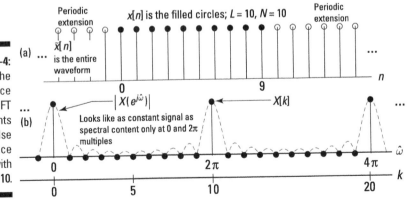

Figure 12-4: The sequence (a) and DFT (b) points for a pulse sequence with $L = N = 10$.

Figure 12-4 shows that sampling the DTFT with too few points results in a loss of spectral detail. The situation here is extreme. You'd hope that the DFT points would in some way represent the shape of the DTFT spectrum of Figure 12-3. Instead, spectral content is only at zero frequency (and 2π

multiples). The frequency domain samples at locations other than zero frequency hit the DTFT where it happens to null to zero.

Also notice that the $x[n]$ with its periodic extensions (making it $\tilde{x}[n]$) looks like a constant sequence. The input to the DFT is ten ones, so this sequence has absolutely no time variation. A constant sequence has all its spectral content located at zero frequency.

For the DFT to see some action, I pad the sequence with $N - L$ zeros, which makes the sequence I feed into the DFT look more like a pulse signal. Try $N = 20$ and see what happens. Figure 12-5 has the same layout as Figure 12-4 but with $N = 20$.

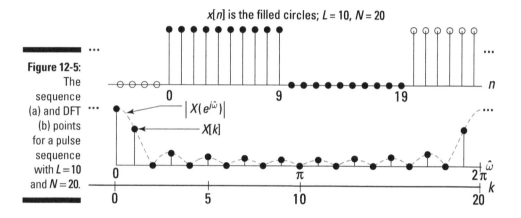

Figure 12-5: The sequence (a) and DFT (b) points for a pulse sequence with $L = 10$ and $N = 20$.

With twice as many samples, the DFT spectrum now has nonzero samples. The zero padding of the original $x[n]$ allows the DFT to more closely approximate the true DTFT of $x[n]$. The underlying DTFT hasn't changed because the nonzero points for the DFT are the same in both cases.

Figure 12-6 shows what happens when $N = 80$.

Here is PyLab code for generating the plot. Use the function `signal.freqz()` from the SciPy's `signal` package to approximate the DTFT of the sequence x. Inside `freqz`, you find the FFT is at work but with zero padding to 1,000 points.

```
In [288]: x = ones(10)
In [289]: w,X_DTFT = signal.freqz(x,1,arange(0,2*pi
              ,pi/500))
In [290]: k_DFT = arange(0,80)
In [291]: X_DFT = fft.fft(x,80)
In [293]: plot(w,abs(X_DTFT))
In [296]: stem(k_DFT*2*pi/80,abs(X_DFT),'r','ro')
```

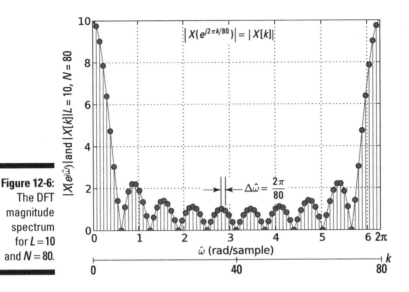

Figure 12-6:
The DFT
magnitude
spectrum
for $L=10$
and $N=80$.

DFT Theorems and Properties

The discrete Fourier transform (DFT) has theorems and properties that are similar to those of the DTFT (Chapter 11) and z-transform (Chapter 14). A major difference of the DFT is the implicit periodicity in both the time and frequency domains. The theorems and properties of this section are summarized in Figure 12-7 with the addition of a few more useful DFT theorems.

		Length N sequence	N-Point DFT
1	Linearity	$ax_1[n] + bx_2[n]$	$aX_1[k] + bX_2[k]$
2	Duality	$X[n]$	$N[((-k))_N]$
3	Circular shift	$x[((n-m))_N]$	$W_N^{km}X[k]$
4	Circular shift frequency	$W_N^{-mn}[x]n$	$X[((k-m))_N]$
5	Circular convolution	$\displaystyle\sum_{m=0}^{N-1} x_1[m]x_2[((n-m))_N]$	$X_1[k]X_2[k]$
6	Multiplication	$x_1[n]x_2[n]$	$\displaystyle\frac{1}{N}\sum_{l=0}^{N-1} X_1[l]\,X_2[((k-l))_N]$
7	Symmetry	$x[n]$ real $0 \le k \le N-1$	$\begin{cases} X[k] = X^*[N-k] \\ |X[k]| = |X[N-k]| \\ \angle X[k] = -\angle X[N-k] \end{cases}$

Figure 12-7:
DFT theorems and properties.

The first pair of properties that I describe in this section, linearity and symmetry, are parallel those of the DTFT, except in discrete or sampled frequency-domain form.

Two theorems that are distinctly different in behavior from their DTFT counterparts are circular sequence shift and circular convolution. The word *circular* appears in the theorem names, by the way, because of the periodic extension that's present in the time domain when dealing with the IDFT. Circular sequence shift is the DFT version of the time delay theorem that I describe in Chapter 11, and circular convolution is the DFT version of the DTFT convolution theorem, also in Chapter 11.

The circular convolution theorem is powerful because it opens the door to the world of *transform domain signal processing.* In other words, the computational properties of the DFT/IDFT make it possible to implement a signal filtering system in the frequency domain by multiplying the signal spectrum samples times the filter frequency response and then calculating the inverse DFT to return to the time domain. (Find out how this works in the real world in the section "Application Example: Transform Domain Filtering," later in this chapter.)

Carrying on from the DTFT

Linearity and frequency domain symmetry for real signals are two properties of the DFT that closely resemble their DTFT counterparts. The only differences are the discrete nature of the frequency domain and the fact that the time-domain signal has fixed length N.

The linearity theorem, $ax_1[n] + bx_2[n] \xleftrightarrow{\ DFT\ } aX_1[k] + bX_2[k]$, holds for the DFT — just as it does for all other transforms described in this book. But for linearity to make sense, all the sequence lengths must be the same.

For $x[n]$ a real sequence, symmetry in the DFT domain holds as follows:

$$X[k] = \sum_{n=0}^{N-1} x[n] W_N^{kn} = \left[\sum_{n=0}^{N-1} x[n] W_N^{(-k)n} \right]^* = X^*\left[((k))_N \right]$$
$$= X^*[N-k], 0 \le k \le N-1$$

This equation represents conjugate symmetry, but with a circular index twist. The last line is the most useful in typical problem solving because it relates to the fundamental transform domain period, which lies on the interval $0 \le k \le N-1$. When you compute the N-point DFT of $x[n]$ by using a computer tool, such as Python, the output is $X[k]$ on the fundamental interval. Example 12-1 reveals that $X[1] = 1 - j$, so you may expect $X^*[4-1] = X[3] = (1-j)^* = 1 + j$. True!

In terms of magnitude and phase:

$$|X[k]| = |X[N-k]|, 0 \le k \le N-1 \text{ (magnitude even)}$$
$$\angle X[k] = -\angle X[N-k], 0 \le k \le N-1 \text{ (phase odd)}$$

For real sequences, the DFT is unique over just the first $N/2$ points, which parallels the DTFT for real signals being unique for $0 \le \hat{\omega} \le \pi$ (see Chapter 11). You can verify this directly from $X[k] = X^*[N-k], 0 \le k \le N-1$.

Consider the following listing of the $X[k]$ points, where I use $X[k] = X^*[N-k]$ to relate point indexes above $N/2$ to those less than or equal to $N/2$. The case of N even means the point at $N/2$ is equal to itself ($N - N/2 = N/2$), and for N odd, the point $N/2$ doesn't exist.

$$N \text{ even: } X[0], \underbrace{X[1]}_{\substack{X^*[N-1]}}, \underbrace{X[2]}_{\substack{X^*[N-2]}}, \dots, \underbrace{X[N/2-1]}_{\substack{X^*[N-(N/2-1)] \\ =X^*[N/2+1]}}, \underbrace{X[N/2]}_{\substack{X^*[N-N/2] \\ =X^*[N/2]}}, X[N/2+1], \dots, X[N-1]$$

$$N \text{ odd: } X[0], \underbrace{X[1]}_{\substack{X^*[N-1]}}, \underbrace{X[2]}_{\substack{X^*[N-2]}}, \dots, \underbrace{X[(N-1)/2]}_{\substack{X^*[N-(N-1)/2] \\ =X^*[(N+1)/2]}}, X[(N+1)/2], \dots, X[N-1]$$

Circular sequence shift

The time shift theorem for the DTFT says $x[n-m] \xleftrightarrow{\mathscr{F}} e^{-j\hat{\omega}m} X(e^{j\hat{\omega}})$ (see Chapter 11). For the DFT, things are different due to the mod N characteristics of the DFT. It can be shown that $x\left[((n-m))_N\right] \xleftrightarrow{DFT} W_N^{km} X[k]$.

Example 12-3: Sketch $x\left[((n-3))_8\right]$ for an arbitrary eight-point sequence. See Figure 12-8 for a series of four waveform plots, starting with $x[n]$ and ending with $x\left[((n-3))_8\right]$.

PyLab's `roll(x,m)` function makes checking the numbers easy:

```
In [313]: x
Out[313]: array([0, 1, 2, 3, 4, 5, 6, 7])
In [314]: roll(x,3) # -3 rolls left by 3
Out[314]: array([5, 6, 7, 0, 1, 2, 3, 4])
```

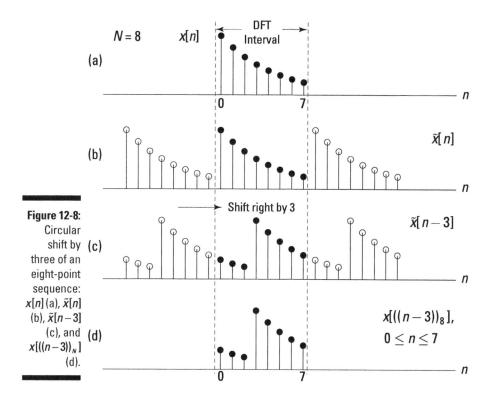

Figure 12-8: Circular shift by three of an eight-point sequence: $x[n]$ (a), $\tilde{x}[n]$ (b), $\tilde{x}[n-3]$ (c), and $x[((n-3))_N]$ (d).

Proving the DTFT time shift theorem

To prove the theorem, consider the relationship between $x[n]$ and $x_1[n]$ if the following equations hold:

$$x[n] \xleftarrow{\ DFT\ } X[k]$$
$$x_1[n] \xleftarrow{\ DFT\ } W_N^{mk} X[k]$$

The IDFT of $W_N^{mk} X[k]$ equals $x_1[n]$, so you can work with that to find a relationship with $x[n]$:

$$x_1[n] = \frac{1}{N}\sum_{k=0}^{N-1} W_N^{mk} X[k]\cdot W_N^{-nk} = \frac{1}{N}\sum_{k=0}^{N-1} X[k] W_N^{-(n-m)k}$$
$$= x[((n-m))_N],\ 0 \le n \le N-1$$

The last line follows from the apparent change of variables in the exponent of W_N from n to $n-m$. The mod N behavior is a result of $W_N = e^{-j2\pi/N}$ being a mod 2π function.

Circular convolution

Suppose that N-point sequences $x_1[n]$ and $x_2[n]$ have DFTs $X_1[k]$ and $X_2[k]$, respectively. Consider the product $X_3[k] = X_1[k] \cdot X_2[k]$. You may guess that $x_3[n]$, the inverse DFT of $X_3[k]$, is the convolution of $x_1[n]$ with $x_2[n]$ because convolution in the time domain is multiplication in the frequency domain (covered in Chapters 9 and 10). But it turns out that this assumption is only partially correct.

Because the IDFT of $X_1[k]$ and $X_2[k]$ produces periodic extensions of $x_1[n]$ and $x_2[n]$, the convolution sum, first developed in Chapter 6, is replaced by a *circular*, or periodic, convolution. The word *circular* is used because the periodic extensions show that time-domain signal values are imprinted on a cylinder (cross-section is a circle); as that cylinder rolls along the time axis, it imprints the same signal over and over again.

It can be shown that $x_3[n]$ is a circular convolution sum having mathematical form:

$$x_3[n] = \sum_{m=0}^{N-1} \tilde{x}_1[n]\tilde{x}_2[n-m], \ 0 \le n \le N-1$$

Notice the use of the periodic extension signals in the argument of the sum and the fact that the sum limits run over just the fundamental interval $0 \le m \le N-1$. The signal $x_3[n]$ can also be viewed with periodic extensions.

Circular convolution and the DFT

The circular convolution of two N-point sequences is the point-by-point product of the respective DFTs. You can formally establish this relationship in two steps starting from the product of DFTs $X_1[k]$ and $X_2[k]$:

1. Find the IDFT of $X_3[k] = X_1[k] \cdot X_2[k]$, but write $X_1[k]$ as the DFT of $x_1[n]$:

$$x_3[n] = \frac{1}{N}\sum_{k=0}^{N-1} \underbrace{X_1[k]}_{DFT\{x_1[n]\}} X_2[k]W_N^{-kn} = \frac{1}{N}\sum_{k=0}^{N-1}\left[\sum_{m=0}^{N-1} x_1[n]W_N^{mn}\right]X_2[k]W_N^{-kn}$$

2. Interchange the sum order to see that the inner sum is just $x_2[((n-m))_N]$ from the circular shift theorem. What remains is the circular convolution formula:

$$x_3[n] = \sum_{m=0}^{N-1} x_1[m]\underbrace{\left[\frac{1}{N}\sum_{k=0}^{N-1} X_2[k]W_N^{-k(n-m)}\right]}_{\text{circular sequence shift}} = \sum_{m=0}^{N-1} x_1[m]x_2[((n-m))_N]$$

A short-hand notation for circular convolution is $x_3[n] = x_1[n] *_N x_2[n] = x_2[n] *_N x_1[n]$ (also commutative).

Doing the actual circular convolution isn't the objective. The DFT/IDFT is a numerical algorithm in both directions. Do you see integrals anywhere? You can implement the convolution in the frequency domain as a multiplication of transforms (DFT-based frequency domain) and then return to the time domain via the IDFT. Figure 12-9 shows a block diagram that describes circular convolution in the frequency domain.

Figure 12-9:
N-point circular convolution as multiplication in the frequency (DFT) domain.

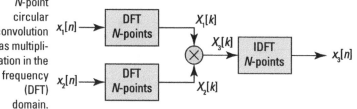

Practicing with short length circular convolutions is instructive, and it makes it easier to appreciate the utility of DFT/IDFT.

Example 12-4: Consider the four-point circular convolution of the sequences $x_1[n] = \{2, 2, 1, 1\}$ and $x_2[n] = \{1, 1, 0, 0\}$, shown in Figure 12-10.

Figure 12-10:
Two four-point sequences used for circular convolution.

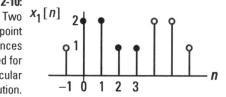
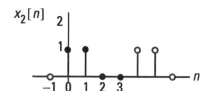

Find the output $x_3[n]$ by hand, calculating $x_3[n] = \sum_{m=0}^{N-1} x_1[m] x_2\big[((n-m))_N\big]$ with the help of the waveform sketches in Figure 12-11. The filled stems represent the sequence values that need to be pointwise multiplied over the interval $0 \le m \le 3$. For $n = 0$, the sum of products is $x_3[n] = 2 \cdot 1 + 2 \cdot 0 + 1 \cdot 0 + 1 \cdot 1 = 3$.

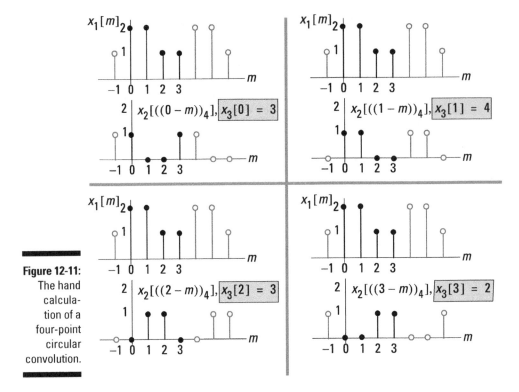

Figure 12-11:
The hand calcula-
tion of a
four-point
circular
convolution.

Now you can work the circular convolution in the DFT domain, using Pylab as the numerical engine for the DFT and IDFT:

```
In [316]: x1 = [2,2,1,1]
In [317]: x2 = [1,1,0,0]
In [318]: X3 = fft.fft(x1)*fft.fft(x2)
In [319]: X3
Out[319]: array([12.+0.j, 0.-2.j, 0.+0.j, 0.+2.j])
In [320]: x3 = fft.ifft(X3)
In [321]: x3
Out[321]: array([3.+0.j, 4.+0.j, 3.+0.j, 2.+0.j])
```

The Python results agree with the hand calculation.

A more rigorous check is to actually perform the DFT and IDFT calculations by hand. The four-point transforms aren't too difficult. The sequence $x[n]$ in Example 12-1 is $x_1[n]$, so you have $X_1[k]$. Next, you find $X_2[k]$ and then multiply the DFTs point by point. In the final step, you apply the IDFT to the four-point sequence to get $x_3[n]$.

Computing the DFT with the Fast Fourier Transform

The fast Fourier transform (FFT) is one of the pillars of modern signal processing because the DFT via an FFT algorithm is so efficient. You'll find it hard at work during the design, simulation, and implementation phase of all types of products.

For example, the FFT makes quick work of calculating the frequency response and input and output spectra of signals and systems. The functions available in Python's PyLab, as well as other popular commercial tools, routinely rely on this capability. And at the implementation phase of a design, you may use the FFT to perform transform domain filtering, get real-time spectral analysis, or lock to a GPS signal to get a position fix. The test equipment you use on the lab bench to verify the final design may use the FFT, and modern oscilloscopes use the FFT to display the spectrum of the waveform input to the instrument. The application list goes on and on, and the common thread is efficient computation by using computer hardware.

In this section, I point out the benefits of the FFT by looking at the computation burden that exists without it. I also describe the decimation-in-time (DIT) FFT algorithm to emphasize the computational performance gains that you can achieve with FFT algorithms. The DFT, or forward transform, is the focus here, but a modified version of the FFT algorithm can easily calculate the IDFT, too!

From the definition of the DFT, the number of complex multiplies and adds required for an N-point transform is N^2 and $N(N-1)$, respectively. The goal of the FFT is to significantly reduce these numbers, especially multiplication count.

Decimation-in-time FFT algorithm

Of the many FFT algorithms, the basic formula is the *radix-2 decimation-in-time* (DIT) algorithm, in which N must be a power of 2. If you let $N = 2^v$, where v is a positive integer, then $N = 8$ and $v = 3$. A property of the radix-2 FFT is that you can break down the DFT computation into $\log_2 N$ *stages* of $N/2$ two-point DFTs per stage. The fundamental computational unit of the radix-2 FFT is a two-point DFT.

Figure 12-12a is a *signal flowgraph* (SFG) of an eight-point FFT that illustrates
the complete algorithm. The SFG, a network of directed branches, is a way
to graphically depict the flow of input signal values being transformed to
output values. For the case of the FFT, the SFG shows you that N signal values
$x[0]$ through $x[N-1]$ enter from the left and exit at the far right as the transformed values $X[1]$ through $X[N-1]$.

Signal branches contain arrows to denote the direction of signal flow. When
a signal traverses a branch, it's multiplied by the *twiddle factor* (a term refer-
encing a complex constant) that's next to the branch arrow in Figure 12-12.
The absence of a constant next to an arrow means that you need to multiply
the signal by one.

The nodes of the SFG are the open circles where branches enter and depart.
When more than one branch enters a node, the signal value at that node is
the sum of branch values entering the node. For the case of the FFT SFG, the
nodes also delineate the stages of the FFT. A node with no entering branches
is called a *source node* (the input signal injection point), and a node with only
entering branches is called a *sink node* (the output signal extraction point).

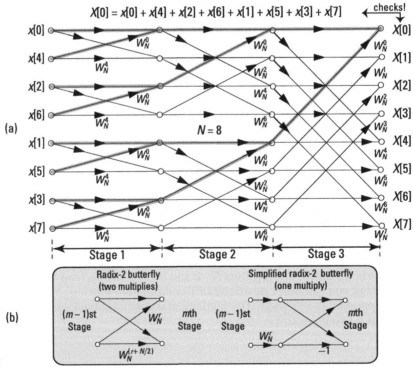

Figure 12-12:
An eight-
point radix-2
FFT signal
flowgraph
(a) and a
simplified **(b)**
radix-2
butterfly (b).

At each stage of the SFG are four *radix-2 butterflies,* similar to the two-multiplier version of Figure 12-12b (they're especially visible at Stage 1). The radix-2 butterfly is actually a two-point DFT; the radix-2 FFT is composed of two-point DFTs.

To demonstrate that the FFT is correct for the $N = 8$ case, consider $X[0]$. From the definition of the DFT

$$X[0] = \sum_{n=0}^{N-1} x[n]W_N^{kn}\Bigg|_{k=0} = \sum_{n=0}^{N-1} x[n]W_N^0 = x[0] + x[1] + \cdots + x[n-1]$$

In the SFG of Figure 12-12, I shaded the branches from each input $x[n]$ to the output $X[0]$. The branch gain applied to each input is unity. Branch values entering a node sum $X[0]$ is the sum of all the $x[n]$ values entering the SFG, so this agrees with the direct DFT calculation given earlier in this section.

Feel free to check the paths leading to any other output node and see that the DFT definition of the section "Establishing the Discrete Fourier Transform" is satisfied.

Referring again to Figure 12-12b, the radix-2 butterfly on the left may be replaced by a simplified butterfly that requires only one complex multiply. When using the simplified radix-2 butterfly, each stage requires $N/2$ complex multiplies. If there are $\log_2 N$ stages, then the total complex multiply count is $(N/2) \cdot \log_2(N)$. Find a comparison of DFT and FFT multiply complexity in Figure 12-13.

N	DFT Mult N^2	FFT Mult $(N/2)\log_2(N)$	Speed-up
8	64	12	5.3
32	1,024	80	12.8
64	4,096	192	21.3
256	65,526	1,024	64.0
512	262,144	2,304	113.8
1,024	1,048,576	5,120	204.8
4,096	16,777,216	24,576	682.7

Table 12-13: DFT versus radix-2 FFT complex multiplication count.

The speed-up factor of the FFT as N increases is impressive, I think. As a bonus, all computations of the FFT of Figure 12-12 are in-place. Two values enter each butterfly and are replaced by two new numbers; there's no need to create a working copy of the N-input signal values, hence the term *in-place computation.*

Working variables can be kept to a minimum. The in-place property comes at the expense of the input signal samples being scrambled. This scrambling is actually just a bit of reversing of the input index. For example:

$$x[6] = x\underbrace{[110]}_{\text{binary}} \overset{\text{is across from}}{\cdots\cdots} X\underbrace{[011]}_{\substack{\text{bit rev.} \\ \text{binary}}} = X[3]$$

Computing the inverse FFT

If you're wondering about efficient computation of the IDFT, rest assured that it's quite doable. To find the inverse FFT (IFFT), you follow these steps:

1. **Conjugate the twiddle factors.**

   Because the FFT implements the DFT, you can look at this mathematically by making the twiddle factor substitution in the definition of the DFT as a check: $X[k]\big|_{W_N^{kn} \to W_N^{-kn}} = \sum_{n=0}^{N-1} x[n] W_N^{-kn}, 0 \le k \le N-1$

2. **Scale the final output by N.**

   Make this substitution in the already modified DFT expression: $X[k]/N\big|_{W_N^{kn} \to W_N^{-kn}} = \frac{1}{N}\sum_{n=0}^{N-1} x[n] W_N^{-kn}, 0 \le k \le N-1$. On the right side is the definition of the IDFT (provided in the earlier section "The DFT/IDFT Pair" with $x[n]$ in place of $X[n]$).

The software tools fully support this, too. In the Python portion of Example 12-4, I use the function called `fft.fft(x,N)` for the FFT (and hence DFT) and `fft.ifft(X,N)` for the IFFT (the IDFT).

Application Example: Transform Domain Filtering

Filtering a signal $x[n]$ with finite impulse response $h[n]$ entirely in the frequency domain is possible by using the DFT/IDFT. Here are two questions you must answer to complete this process:

- ✔ How do I make circular convolution perform ordinary *linear* convolution, or the convolution sum (introduced in Chapter 6)?

- ✔ How do I handle a very long sequence $x[n]$ while keeping the DFT length manageable?

In this section, I point out how circular convolution can be made to act like linear convolution and take advantage of the computation efficiency offered by the FFT. I also tell you how to use the FFT/IFFT to continuously filter a discrete-time signal, using a technique known as *overlap and add*, which makes it possible to filter signals in the frequency domain. This isn't simply an analysis technique; it's a real implementation approach.

Making circular convolution perform linear convolution

The linear convolution of a length L signal $x[n]$ with a length P sequence $h[n]$ results in a signal $y[n]$ of length $L + P - 1$ (see Chapter 6). With circular convolution, both signal sequences need to have length N, and the circular convolution results in a length N sequence.

Can circular convolution emulate linear convolution? The answer is yes if you *zero pad* $x[n]$ and $h[n]$ to length $N \geq L + P - 1$.

Zero padding means you append zero signal values to the end of a signal: $\{x[0], x[1], \ldots,$ so $x[L-1]\}$ becomes $\{x[0], x[1], \ldots, x[L-1], 0, 0, \ldots, 0\}$, and the number of appended zero values is $N - L$. For the length P sequence $h[n]$, you zero pad by appending $N - P$ zero values.

Zero padding is a big deal because it allows you to use circular convolution to produce the same result as linear convolution. In practice, circular convolution happens in the frequency domain, using the block diagram of Figure 12-8.

Using overlap and add to continuously filter sequences

Assume that $x[n]$ is a sequence of possibly infinite duration and $h[n]$ is a length P FIR filter. Divide $x[n]$ into segments of finite length L and then filter each segment by using the DFT/IDFT.

You can complete the segment filtering by using zero padded length $N \geq L + P - 1$ transforms. Afterward, add the filtered sections together — with overlap because $N > L$. This process requires the following steps:

1. **Decompose $x[n]$, as shown in Figure 12-14.**

2. **Compute the DFT (actually use an FFT) of each subsequence $x_r[n] = x[n+rL]$ in Figure 12-15 as they're formed with increasing n.**

   Also compute $H[k] = \text{DFT}_N\{h[n]\}$, or you may specify $H[k]$ directly in the frequency domain.

3. **Multiply the frequency domain points together: $X_r[k]H[k]$ for each r.**

4. **Take the IDFT of the overlapping frequency domain subsequences $X_r[k]H[k]$ to produce $y_r[n]$, and then overlap and add the time domain signal sequences as shown in Figure 12-15.**

The function $y = OA\_filter(x,h,N)$ in the module $ssd.py$ implements overlap and add (OA). A variation on OA is overlap and save (OS), which you can also find in $ssd.py$ as $OS\_filter(x,h,N)$.

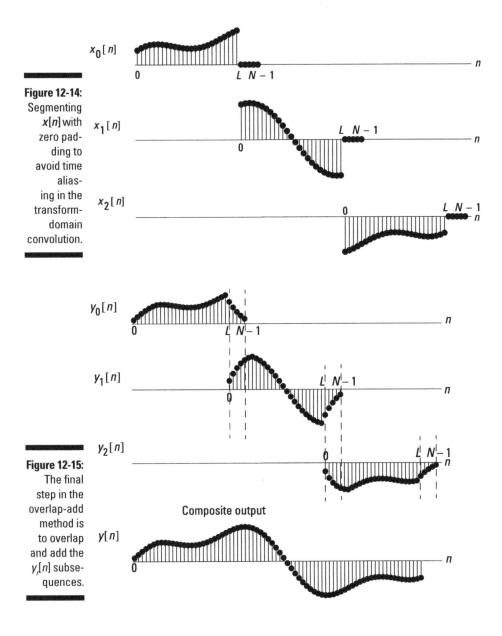

Figure 12-14: Segmenting $x[n]$ with zero padding to avoid time aliasing in the transform-domain convolution.

Figure 12-15: The final step in the overlap-add method is to overlap and add the $y_r[n]$ subsequences.

Part IV

Entering the *s*- and *z*-Domains

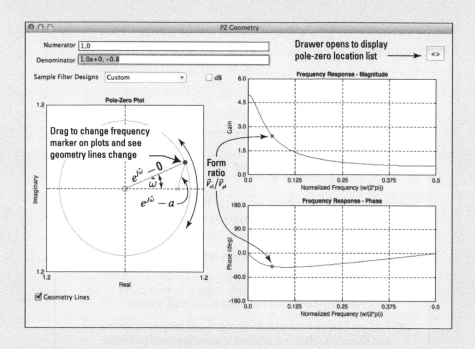

Part IV

Entering the s- and z-Domains

In this part . . .

- ✔ Generalize the Fourier transform to check out the s-domain for continuous-time signals and systems and the z-domain for the discrete-time counterpart.

- ✔ Let the Laplace transform help you solve differential equations with friendly algebra.

- ✔ See how to avoid the convolution integral by multiplying in the s- or z-domains and then returning to the time domain.

- ✔ Find out how to work across domains and between continuous and discrete signals and systems to solve practical problems.

Chapter 13

The Laplace Transform for Continuous-Time

In This Chapter

▶ Checking out the two-sided and one-sided Laplace transforms

▶ Getting to know the Laplace transform properties

▶ Inversing the Laplace transform

▶ Understanding the system function

The Laplace transform (LT) is a generalization of the Fourier transform (FT) and has a lot of nice features. For starters, the LT exists for a wider class of signals than FT. But the LT really shines when it's used to solve linear constant coefficient (LCC) differential equations (see Chapter 7) because it enables you to get the total solution (forced and transient) for LCC differential equations and manage nonzero initial conditions automatically with algebraic manipulation alone.

Unlike the frequency domain, which has real frequency variable f or ω, the LT transforms signals and linear time-invariant (LTI) impulse responses into the s-domain, where s is a complex variable. This means you can avoid using the convolution integral by simply multiplying the transformed quantities in the s-domain. The impulse response of an LCC differential equation is a ratio of polynomials in s (rational function). The denominator roots are the poles, and the numerator roots are the zeros. And a system's poles and zeros reveal a lot about the system, including whether it's stable and what the frequency response's shape is.

Returning from the s-domain requires an inverse LT procedure. In this chapter, I describe the two forms of the Laplace transform. I also tell you how to apply the Laplace transform by using partial fraction expansion and table lookup. Check out Chapter 2 for a math refresher if you think you need it.

Seeing Double: The Two-Sided Laplace Transform

Signals and systems use two forms of the Laplace transform (LT): the two-sided and the more specific one-sided. In this section, I introduce the basics of the two-sided form before diving into its one-sided counterpart.

The two-sided LT accepts *two-sided* signals, or signals whose extent is infinite for both $t < 0$ and $t > 0$. This form of the LT is closely related to the FT, which also accepts signals over the same time axis interval. But the two-sided LT can work with both causal and non-causal system models; the one-sided LT can't. The drawback of the two-sided form is that it's unable to deal with systems having nonzero initial conditions.

The two-sided LT takes the continuous-time signal $x(t)$ and turns it into the *s*-domain function:

$$X_{\mathrm{II}}(s) = \mathcal{L}_{\mathrm{II}}\{x(t)\} = \int_{-\infty}^{\infty} x(t)e^{-st}\, dt$$

where $s = \sigma + j\omega = \sigma + j2\pi f$ is a complex variable and the subscript in $X_{\mathrm{II}}(s)$ denotes the two-sided LT.

It's no accident that I chose σ as the real axis variable and ω as the imaginary axis variable. The use of σ for the real axis name is almost universal in the signals and systems community. The imaginary axis is $\omega = 2\pi f$ because of the connection to the FT.

You can find the relationship to the radian frequency FT by writing

$$X_{\mathrm{II}}(s)\Big|_{s=\sigma+j\omega} = \int_{-\infty}^{\infty} x(t)\cdot e^{-(\sigma+j\omega)t}\, dt = \int_{-\infty}^{\infty} \left[x(t)e^{-\sigma t}\right]\cdot e^{-j\omega t}\, dt = \mathcal{F}\left\{x(t)e^{-\sigma t}\right\}$$

The two-sided LT is always equivalent to the FT of signal $x(t)$ multiplied by an exponential weighting factor $e^{-\sigma t}$. This factor allows improved convergence of the LT. If, however, the FT of $x(t)$ exists, it's a simple matter to set $\sigma = 0$ in the equation and see that $X_{\mathrm{II}}(j\omega) = \mathcal{F}\{x(t)\} = X(\omega)$. This relationship is shown in Figure 13-1.

Finding direction with the ROC

The LT doesn't usually converge over the entire *s*-plane. The region in the *s*-plane for which the LT converges is known as the *region of convergence* (ROC). Uniform convergence requires that

$$\int_{-\infty}^{\infty} \left|x(t)e^{-st}\right| dt = \int_{-\infty}^{\infty} |x(t)|\left|e^{-(\sigma t+j\omega t)}\right| dt = \int_{-\infty}^{\infty} |x(t)|\, e^{-\sigma t}\, dt < \infty$$

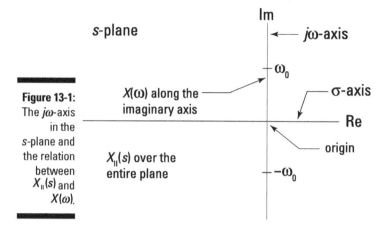

Figure 13-1:
The *jω*-axis
in the
s-plane and
the relation
between
$X_{II}(s)$ and
$X(\omega)$.

Convergence depends only on Re$\{s\} = \sigma$, so if the LT converges for $s = s_0$, then the ROC also contains the vertical line Re$\{s_0\} = \sigma_0$. It can be shown that the general ROC for a two-sided signal is the vertical strip $\sigma_1 < $ Re$\{s\} < \sigma_2$ in the *s*-plane, as shown in Figure 13-2.

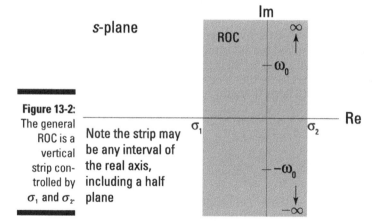

Figure 13-2:
The general
ROC is a   Note the strip may
vertical   be any interval of
strip con- the real axis,
trolled by including a half
σ_1 and σ_2. plane

You can determine the value of σ_1 and σ_2 by the nature of the signal or impulse response being transformed. That both $\sigma_1 \rightarrow -\infty$ and/or $\sigma_2 \rightarrow \infty$ is also possible.

To show that the ROC is a vertical strip in the *s*-plane for $x(t)$ two-sided, first consider $x(t)$ is *right-sided,* meaning $x(t) = 0$ for $t < t_1$. If the LT of $x(t)$ includes the vertical line $\sigma = \sigma_1$, it must be that $\int_{-\infty}^{\infty} |x(t)| e^{-\sigma_1 t}\, dt = \int_{t_1}^{\infty} |x(t)| e^{-\sigma_1 t}\, dt < \infty.$

The ROC must also include $\sigma_1^+ > \sigma_1$ because the integral $\int_{t_1}^{\infty} |x(t)| e^{-\sigma_1^+ t}\, dt =$ $\int_{t_1}^{\infty} |x(t)| e^{\sigma_1 t} \cdot e^{-(\sigma_1^+ - \sigma_1)t}\, dt \le e^{-(\sigma_1^+ - \sigma_1)t_1} \int_{t_1}^{\infty} |x(t)| e^{\sigma_1 t}\, dt$ is finite. What makes this true is the fact that $e^{-(\sigma_1^+ - \sigma_1)t} < 1$ for $t > 0$ and t_1 finite. You can generalize this result for σ_1^+ to say for $x(t)$ right-sided, the ROC is the half-plane $\mathrm{Re}\{s\} > \sigma_1$.

If $x(t)$ is *left-sided,* meaning $x(t) = 0$ for $t > t_2$ and the LT of $x(t)$ includes the vertical line $\sigma = \sigma_2$, you can show, through an argument similar to the right-sided case, that the ROC is the half-plane $\mathrm{Re}\{s\} < \sigma_2$.

Finally, for $x(t)$ two-sided, the ROC is the intersection of right and left half-planes, which is the vertical strip $\sigma_1 < \mathrm{Re}\{s\} < \sigma_2$, provided that $\sigma_2 > \sigma_1$; otherwise, the ROC is empty, meaning the LT isn't absolutely convergent anywhere.

Example 13-1: Find the two-sided LT of the right-sided signal $x_a(t) = e^{-at} u(t)$, where a may be real or complex. Note that *right-sided* means the signal is 0 for $t < t_0 < \infty$. A further specialization of right-sided is to say that a signal is *causal,* meaning $t_0 = 0$. Use the definition of the two-sided LT:

$$X_{\text{all}}(s) = \int_{-\infty}^{\infty} e^{-at} u(t) \cdot e^{-st}\, dt = \int_0^{\infty} e^{-at} \cdot e^{-st}\, dt = \lim_{T \to \infty} \frac{e^{-(a+s)t}}{-(a+s)}\bigg|_0^T$$

$$= \frac{1}{s+a} - \frac{1}{s+a} \cdot \lim_{T \to \infty} e^{-(a+s)T} = \frac{1}{s+a}, \ \text{ROC: } \mathrm{Re}\{s\} > -\mathrm{Re}\{a\}$$

Example 13-2: Find the two-sided LT of the left-sided signal $x_b(t) = -e^{-at} u(-t)$, where a may be real or complex. *Left-sided* means the signal is 0 for $t > t_0 > -\infty$. Another way to refer to a left-sided signal is to say it's *anticausal,* meaning $t_0 = 0$ (the opposite of causal). Note that a *non-causal* signal is two-sided. Use this definition to reveal ROC: $\mathrm{Re}\{s\} < -\mathrm{Re}\{a\}$.

$$X_{\text{all}}(s) = \int_{-\infty}^{\infty} -e^{-at} u(-t) \cdot e^{-st}\, dt = -\int_{-\infty}^0 e^{-at} \cdot e^{-st}\, dt = \lim_{T \to \infty} \frac{e^{-(a+s)t}}{(a+s)}\bigg|_{-T}^0$$

$$= \frac{1}{s+a} - \frac{1}{s+a} \cdot \lim_{T \to \infty} e^{(a+s)T} = \frac{1}{s+a}$$

Both $x_a(t)$ of Example 13-1 and $x_b(t)$ of Example 13-2 have the same LT! The ROCs, however, are distinct — they're complementary regions — making the LTs distinguishable only by the ROCs being different. The purpose of Example 13-2 is to point out that, without the ROC, you can't return to the time domain without ambiguity — is the signal left-sided or right-sided?

Locating poles and zeros

When the LT is a rational function, as in Examples 13-1 and 13-2, it takes the form $N(s)/D(s)$, where $N(s)$ and $D(s)$ are each polynomials in s. The roots of $N(s)$ — where $N(s) = 0$ — are the *zeros* of $X_{\text{II}}(s)$, and the roots of $D(s)$ are the *poles* of $X_2(s)$.

Think of the poles and zeros as the magnitude $|N(s)/D(s)|$ of a 3D stretchy surface placed over the *s*-plane. The surface height ranges from zero to infinity. At the zero locations, because $N(s) = 0$, the surface is literally tacked down to zero. At the pole locations, because $D(s) = 0$, you have $N(s)/0 = \infty$, which you can view as a tent pole pushing the stretchy surface up to infinity. It's not your average circus tent or camping tent, but when I plot the poles and zeros in the *s*-plane, I use the symbols × and O.

Take a look at the pole-zero plots of $X_{all}(s)$ and $X_{bll}(s)$, including the ROC, in Figure 13-3.

Figure 13-3: The pole-zero plot and ROC of $X_{all}(s)$ (a) and $X_{bll}(s)$ (b) of Examples 13-1 and 13-2, respectively.

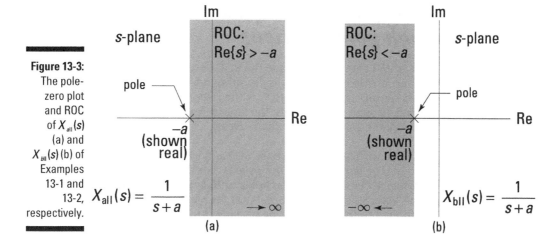

Checking stability for LTI systems with the ROC

An LTI system is bounded-input bounded-output (BIBO) stable (see Chapter 5) if $\int_{-\infty}^{\infty} |h(t)| \, dt < \infty$.

Transforming to the *s*-domain has its perks. Here's what I mean: The LT of the impulse response is called the *system function*. The Fourier transform of the impulse response is the frequency response (covered in Chapter 9), and the system function generalizes this result to the entire *s*-plane. The frequency domain is just the Laplace domain evaluated along the $j\omega$ axis. It can be shown that if the ROC of $H_{II}(s) = \mathcal{L}_{II}\{h(t)\}$ contains the $j\omega$-axis, then the system is BIBO stable. Cool, right?

The proof follows by expanding $\left|H_{\mathrm{II}}(s)\right|_{s=j\omega}$, using the *triangle inequality* — a general mathematical result that states that the magnitude of the sum of two quantities is less than or equal to the sum of the magnitudes:

$$\left|H_{\mathrm{II}}(s)\right|_{s=j\omega} = \left|\int_{-\infty}^{\infty} h(t)e^{-j\omega t}\, dt\right| \leq \int_{-\infty}^{\infty}\left|h(t)e^{-j\omega t}\right|\, dt = \int_{-\infty}^{\infty}\left|h(t)\right|\, dt < \infty$$

The triangle inequality reveals that the magnitude of a sum is bounded by the sum of the magnitudes.

Checking stability of causal systems through pole positions

The poles of $H_{\mathrm{II}}(s)$ must sit outside the ROC because the poles of $H_{\mathrm{II}}(s)$ are singularities. In other words, $\left|H_{\mathrm{II}}(s)\right|$ is unbounded at the poles. For right-sided (and causal) systems, the ROC is the region to the right of the vertical line passing through the pole with the largest real part. In Example 13-1, this represents all the values greater than $-\mathrm{Re}\{a\}$.

An LTI system with impulse response $h(t)$ and system function $H_{\mathrm{II}}(s)$ is BIBO stable if the ROC contains the $j\omega$-axis. For the special case of a causal system, when $h(t) = 0$ for $t < 0$, the poles of $H_{\mathrm{II}}(s)$ must lie in the left half of the s-plane, indicating a negative real part, because causal sequences have ROC positioned to the right of a line passing through the pole with the largest real part. In short, a causal LTI system is stable if the poles lie in the left-half plane (LHP).

Digging into the One-Sided Laplace Transform

The one-sided Laplace transform (LT) offers the capability to analyze causal systems with nonzero initial conditions and inputs applied, starting at $t = 0$. Unlike two-sided LT signals, the one-sided LT doesn't allow you to analyze signals that are nonzero for $t < 0$. The trade-off is acceptable because the transient analysis of causal LTI systems in the s-domain involves only algebraic manipulations.

The one-sided LT restricts the integration interval to $[0^-, \infty)$:

$$\mathcal{L}\{x(t)\} = X(s) = \int_{0^-}^{\infty} x(t)e^{-st}\, dt$$

Use the 0^- to accommodate signals such as the impulse $\delta(t)$ and the step $u(t)$ (see Chapter 5). To be clear, 0^- is $t = 0 - \varepsilon$ as $\varepsilon > 0$ approaches zero.

All the results developed for the two-sided LT in the previous section hold under the one-sided case as long as the signals and systems are causal — that is, both $x(t)$ and $h(t)$ are zero for $t < 0$. The poles of $H(s)$ must lie in the left-half plane for stability.

Example 13-3: To find the LT of $x(t) = \delta(t - t_0)$, $t_0 \geq 0$, use the definition and the sifting property of the following impulse function:

$$X(s) = \mathcal{L}\{\delta(t - t_0)\} = \int_{0^-}^{\infty} \delta(t - t_0) e^{-st} \, dt = e^{-st_0}$$

The ROC is $\left| e^{-st_0} \right| = \left| e^{-(\sigma + j\omega)t_0} \right| = \left| e^{-\sigma t_0} \right| < \infty$. In terms of $\sigma = \text{Re}\{s\}$, the ROC is just $\text{Re}\{s\} > -\infty$. This means that the ROC includes the entire s-plane. And when $t_0 = 0$, the 0^- is needed to properly handle the integral of $\delta(t)$.

Example 13-4: Find the LT of $x(t) = u(t)$. This signal is causal, so the one-sided and two-sided LTs are equal because the step function, $u(t)$, makes the integration limits identical.

Use Example 13-1 and set $a = 0$:

$$X(s) = \mathcal{L}\{u(t)\} = \frac{1}{s}, \text{ ROC: } \text{Re}\{s\} > 0$$

The ROC doesn't include the $j\omega$-axis because a pole is at zero.

Example 13-5: To find the system function corresponding to the impulse response $h(t) = 2e^{-3t}u(t) + 5e^{-2t}u(t)$, first recognize that linearity holds for the LT: $\mathcal{L}\{h_1(t) + h_2(t)\} = \mathcal{L}\{h_1(t)\} + \mathcal{L}\{h_2(t)\}$.

Again, use Example 13-1, where two-sided and one-sided signals are identical:

$$H(s) = \mathcal{L}\{h(t)\} = \frac{2}{s+3} + \frac{5}{s+2} = \frac{7s+19}{(s+3)(s+2)} = \frac{7(s+19/7)}{s^2 + 5s + 6}$$

One zero and two poles exist in this solution: $z_1 = -19/7$, $p_1 = -3$, and $p_2 = -2$. You can find the ROC for the solution as a whole by intersecting the ROC for each exponential term. Just use the ROC from Example 13-1: $\text{ROC}_h = (\text{Re}\{s\} > -3) \cap (\text{Re}\{s\} > -2) = \text{Re}\{s\} > -2$

The pole with the largest real part sets the ROC boundary. The system is stable because the ROC includes the $j\omega$-axis. But a more memorable finding is the fact that the poles are in the left-half s-plane, which indicates that this causal system is stable.

Checking Out LT Properties

Problem solving with the LT centers on the use of transform theorems and a reasonable catalog of transform pairs. Knowing key theorems and pairs can help you move quickly through problems, especially when you need an inverse transform. In this section, I point out some of the most common LT theorems and transform pairs.

The theorems and transform pairs developed in this section are valid for the one-sided LT. Those applicable to the two-sided LT are different in some cases.

Transform theorems

In this section of basic theorems, assume that the signals and systems of interest are causal.

Find a collection of significant one-sided Laplace transform theorems in Figure 13-4.

	Property	$x(t)$	$X(s)$	ROC
1	Linearity	$ax_1(t) + bx_2(t)$	$aX_1(s) + bX_2(s)$	at least $R_{x_1} \cap R_{x_2}$
2	Delay (time shift)	$x(t - t_0\ u)(t - t_0)$	$e^{-st_0}X(s)$	R_x
3	Differentiation	$\dfrac{d^n x(t)}{dt^n}$	$sX(s) - s^{n-1}x(0^-) - \cdots$ $- sx^{(n-2)}(0^-) - x^{(n-1)}(0^-)$	R_x
4	Integration	$\displaystyle\int_{0^-}^{\infty} x(\lambda)d\lambda$	$\dfrac{X(s)}{s}$	$R_x \cap [\mathrm{Re}(s) > 0]$
5	Convolution	$x_1(t) * x_2(t)$	$X_1(s)X_2(s)$	at least $R_{x_1} \cap R_{x_2}$
6	*s*-shift	$x(t)e^{-\alpha t}$	$X(s + \alpha)$	R_x shifted left by $\propto$
7	Initial value	$x(0^+) =$	$\displaystyle\lim_{s \to \infty} sX(s)$	if limit exists
8	Final value	$x(\infty) =$	$\displaystyle\lim_{s \to 0} sX(s)$	if limit exists
	R_x is the ROC	of $(x)t$	$x^{(k)}(0^-) = d^k\ (x)t/dt^k\big|_{t=0^-}$	

Figure 13-4: One-sided Laplace transform theorems.

Delay

The one-sided LT of a signal with time delay $t_0 > 0$ is an important modeling capability. To set up the theorem, consider $x(t - t_0)u(t - t_0)$ for $t_0 \geq 0$:

$$\mathcal{L}\{x(t - t_0)u(t - t_0)\} = \int_{t_0^-}^{\infty} x(t - t_0)e^{-st}\, dt = \int_{0^-}^{\infty} x(\tau)e^{-s(\tau + t_0)}\, d\tau$$

$$= e^{-st_0}X(s), \text{ ROC: } R_x$$

Example 13-6: Suppose that $x(t)$ is a *periodic-like* waveform with period T_0. A true periodic waveform has periods extending to $-\infty$ as well. Here, the signal is causal with the first period starting at $t = 0$:

$$x_0(t) = \begin{cases} x(t), & 0 \le t \le T_0 \\ 0, & \text{otherwise} \end{cases}$$

To find $\mathcal{L}\{x(t)\}$, write $x(t) = x_0(t) + x(t - T_0)u(t - T_0)$.

Taking the LT of both sides results in the following equation:

$$X(s) = X_0(s) + X(s)e^{sT_0}$$

$$\text{or } X(s) = \frac{X_0(s)}{1 - e^{sT_0}}, \text{ ROC: R}_{x_0}$$

Differentiation

The differentiation theorem is fundamental in the solution of LCC differential equations with nonzero initial conditions. If you start with $dx(t)/dt$, you ultimately get a result for $d^n x(t)/dt^n$. Using integration by parts, you can show that

$$\mathcal{L}\left\{\frac{dx(t)}{dt}\right\} = sX(s) - x(0^-)$$

Proving the differentiation theorem

To develop the proof for the differentiation theorem, start with the definition of the one-sided LT, and input $dx(t)/dt$. Use the integration by parts formula from calculus (find calculus review in Chapter 2):

$$\mathcal{L}\left\{\frac{dx(t)}{dt}\right\} = \int_{0^-}^{\infty} \frac{dx(t)}{dt}e^{-st}\, dt = e^{-st}x(t)\Big|_{0^-}^{\infty} + s\int_{0^-}^{\infty} x(t)e^{-st}\, dt$$

The parts formula is

$$\int_a^b u\, dv = uv\Big|_a^b - \int_a^b v\, du$$

In this formula, let $u = e^{-st}$ and $dv = dx(t)$. To complete the proof, you can reasonably assume that $\lim_{T \to \infty} e^{-sT}x(T) = 0$. Note that if the initial conditions are zero, the LT is just $sX(s)$, which is equivalent to the two-sided LT.

Using mathematical induction, you can then show the following, where $x^{(k)}(0^-) = d^k x(t)/dt^k \big|_{t=0^-}$:

$$\mathcal{L}\left\{\frac{d^n x(t)}{dt^n}\right\} = s^n X(s) - s^{n-1} x(0^-) - \cdots - s x^{(n-2)}(0^-) - x^{(n-1)}(0^-)$$

Integration

The integration theorem is the complement to the differentiation theorem. It can be shown that

$$\int_{0^-}^\infty x(\tau)\,d\tau \xleftrightarrow{\;\mathcal{L}\;} \frac{X(s)}{s}$$

Example 13-7: Find the LT of the *ramp* signal $tu(t)$. The study of singularity functions (see Chapter 3) reveals that the step function $u(t)$ is the integral of the impulse function. If you integrate the step function, you get the ramp function. In the s-domain, the LT of a step function is $1/s$, so applying the integration theorem shows that

$$\mathcal{L}\{tu(t)\} = \mathcal{L}\left\{\int_{0^-}^\infty u(\tau)\,d\tau\right\} = \frac{1}{s}\cdot\frac{1}{s} = \frac{1}{s^2}$$

This is a repeated pole at $s = 0$ with the ROC the right-half plane.

Convolution

Given causal $x(t)$ and $h(t)$, the convolution theorem states that $Y(s) = \mathcal{L}\{x(t)*h(t)\} = X(s)H(s)$, ROC: $R_x \cap R_h$. This theorem is fundamental to signals and systems modeling, because it allows you to study the action of passing a signal through a system in the s-domain. An even bigger deal is that convolution in the time domain is multiplication in the s-domain. The FT convolution theorem (see Chapter 9) provides this capability, too, but the LT version is superior, especially when your goal is to find a time-domain solution.

s-shift

Solving LCC differential equations frequently involves signals with an exponential decay. The s-shift theorem provides this information:

$$\mathcal{L}\{x(t)e^{-\alpha t}\} = \int_{0^-}^\infty x(t)e^{-\alpha t}\cdot e^{-st}\,dt = \int_{0^-}^\infty x(t)e^{-(s+\alpha)t}\,dt = X(s+\alpha)$$

The s-shift moves all the poles and zeros of $X(s)$ to the left by α. The left edge of the ROC also shifts to the left by α. Stability improves!

Developing the convolution theorem proof

The proof of the convolution theorem follows by direct substitution and an interchange of the order of integration:

$$Y(s) = \int_{0^-}^{\infty}\left[\int_{0^-}^{\infty} x(\lambda)h(t-\lambda)d\lambda\right]e^{-st}\,dt$$

$$= \int_{0^-}^{\infty} x(\lambda)\underbrace{\left[\int_{0^-}^{\infty} h(t-\lambda)e^{-s(t-\lambda)}\,dt\right]}_{H(s)} e^{-s\lambda}\,d\lambda$$

$$= \int_{0^-}^{\infty} x(\lambda)e^{-s\lambda}\,d\lambda \cdot H(s) = X(s)H(s)$$

If pole-zero cancellation occurs, the ROC may become larger.

Example 13-8: To find $\mathcal{L}\left\{e^{-at}\cos(\omega_0 t)\right\}$ and $\mathcal{L}\left\{e^{-at}\sin(\omega_0 t)\right\}$ with $u(t)$ implied as a result of the one-sided LT, start with $\mathcal{L}\{\cos\omega_0 t\}$ and $\mathcal{L}\{\sin\omega_0 t\}$. Using linearity, Euler's inverse formulas for sine and cosine, and the transform pair $e^{-at}u(t)\xleftrightarrow{\mathcal{L}} 1/(s+a)$ developed in Example 13-1, solve for the LT:

$$\mathcal{L}\{\cos\omega_0 t\} = \mathcal{L}\left\{\frac{e^{j\omega_0 t}+e^{-j\omega_0 t}}{2}\right\} = \frac{1}{2}\cdot\left[\frac{1}{s-j\omega_0}+\frac{1}{s+j\omega_0}\right]$$

$$= \frac{s}{s^2+\omega_0^2}, \text{ ROC: Re}\{s\} > 0$$

$$\mathcal{L}\{\sin\omega_0 t\} = \mathcal{L}\left\{\frac{e^{j\omega_0 t}-e^{-j\omega_0 t}}{2j}\right\} = \frac{1}{2j}\cdot\left[\frac{1}{s-j\omega_0}-\frac{1}{s+j\omega_0}\right]$$

$$= \frac{\omega_0}{s^2+\omega_0^2}, \text{ ROC: Re}\{s\} > 0$$

Finally, let $s \to s+\alpha$:

$$\mathcal{L}\left\{e^{-at}\cos(\omega_0 t)\right\} = \frac{s+\alpha}{(s+\alpha)^2+\omega_0^2} \text{ and } \mathcal{L}\left\{e^{-at}\sin(\omega_0 t)\right\} = \frac{\omega_0}{(s+\alpha)^2+\omega_0^2}$$

The poles that were at $\pm j\omega_0$ shift to $-\alpha\pm j\omega_0$. The ROC for both transforms is Re$\{s\} > -\alpha$.

Proving the final value theorem

To prove the final value theorem, start with the differentiation theorem and take the limit as $s \to 0$ on both sides. On the left side, move the limit inside the LT integral, which makes $\lim_{s \to 0} e^{-st} = e^0 = 1$, so the integral of $x'(t)$ is just $x(t)$ evaluated at the upper and lower limits:

$$\mathcal{L}\{x'(t)\} = sX(s) - x(0^-)$$

$$\lim_{s \to 0} \int_{0^-}^{\infty} x'(t) \cdot e^{-st}\, dt = \lim_{s \to 0} sX(s) - x(0^-)$$

$$x(\infty) - x(0^-) = \lim_{s \to 0} sX(s) - x(0^-)$$

Final value

The final value theorem states that $\lim_{t \to \infty} x(t) = x(\infty) = \lim_{s \to 0} sX(s)$ *if the limit exists* — or $x(t)$ reaches a bounded value only if the poles of $X(s)$ lie in the left-half plane (LHP). This theorem is particularly useful for determining steady-state error in control systems (covered in Chapter 18). You don't need a complete inverse LT just to get $x(\infty)$.

Example 13-9: Apply the final value theorem to $H_1(s) = \dfrac{5}{s+1}$ and $H_2(s) = \dfrac{4}{s^2+4}$:

$$\lim_{s \to 0} sH_1(s) = \lim_{s \to 0} s\,\frac{5}{s+1} = 0,\ \text{looks good as } h_1(t) = 5e^{-t}u(t) \xrightarrow{\ t \to \infty\ } 0$$

$$\lim_{s \to 0} sH_2(s) = \lim_{s \to 0} s\,\frac{4}{s^2+4} = 0,\ \text{not good as } h_2(t) = 4\sin(2t)u(t)\ \text{oscillates}$$

What's wrong with the analysis of $H_2(s)$? Poles on the $j\omega$-axis ($s^2 = -4 \to s = \pm 2j$) make $h_2(t)$ oscillate forever. The limit doesn't exist because the poles of $H_2(s)$ don't lie in the LHP, so the theorem can't be applied. The moral of this story is to study the problem to be sure the theorem applies!

Transform pairs

Transform pairs play a starring role in LT action. After you see the transform for a few signal types, you can use that result the next time you encounter it in a problem. This approach holds for both forward and inverse transforms (see the section "Getting Back to the Time Domain," later in this chapter). Figure 13-5 highlights some LT pairs.

	$x(t)$	$X(s)$	ROC
1	$\delta(t)$	1	all s
2	$\delta^{(n)}(t) = d^n\delta(t)/dt^n$	s^n	all s
3	$e^{-at}u(t)$	$\dfrac{1}{s+a}$	$\mathrm{Re}\{s\} > -\mathrm{Re}\{a\}$
4	$\dfrac{t^n \exp(-at)u(t)}{n!}$	$\dfrac{1}{(s+a)^{n+1}}$	$\mathrm{Re}\{s\} > -\mathrm{Re}\{a\}$
5	$\cos\omega_0 t\, u(t)$	$\dfrac{s}{s^2+\omega_0^2}$	$\mathrm{Re}\{s\} > 0$
6	$\sin\omega_0 t\, u(t)$	$\dfrac{\omega_0}{s^2+\omega_0^2}$	$\mathrm{Re}\{s\} > 0$
7	$e^{-\alpha t}\cos\omega_0 t\, u(t)$	$\dfrac{s+\alpha}{(s+\alpha)^2+\omega_0^2}$	$\mathrm{Re}\{s\} > \alpha$, α real pos.
8	$e^{-\alpha t}\sin\omega_0 t\, u(t)$	$\dfrac{\omega_0}{(s+\alpha)^2+\omega_0^2}$	$\mathrm{Re}\{s\} > \alpha$, α real pos.

Figure 13-5:
Laplace
transform
pairs.

Example 13-10: Develop a transform pair where the time-domain signal is a linear combination of exponentially damped cosine and sine terms. You need this pair when finding the inverse Laplace transform (ILT) of a rational function $X(s)$ containing a complex conjugate pole pair (see Example 13-12 for a full example).

The starting point from the s-domain side is $X(s) = (b_1 s + b_0)/(s^2 + a_1 s + a_0)$, with $4a_0 > a_1^2$ to ensure complex conjugate poles. You can write the time-domain solution as a linear combination of the time-domain side of transform pairs in Lines 7 and 8 of Figure 13-5: $x(t) = e^{-\alpha t}[a\cos(\omega_0 t) + b\sin(\omega_0 t)]u(t)$. Your objective is to find a and b in terms of the coefficients of $X(s)$. You can do so with the following steps:

1. **Equate $X(s)$ with a linear combination of the s-domain side of transform pairs in Lines 7 and 8 of Figure 13-5 by first matching up the denominators in**

$$X(s) = \frac{b_1 s + b_0}{s^2 + a_1 s + a_0} = \frac{a(s+\alpha) + b\omega_0}{(s+\alpha)^2 + \omega_0^2}$$

Recall the old algebra trick called *completing the square,* in which, given a quadratic polynomial $ax^2 + bx + c$, you rewrite it as $a(x - c_1)^2 + c_2$, where c_1 and c_2 are constants. You then equate like terms:

$$s^2 + a_1 s + a_0 = (s + a_1/2)^2 + \left[a_0 - (a_1/2)^2\right] = (s+\alpha)^2 + \omega_0^2$$

So $\alpha = a_1/2$ and $\omega_0^2 = [a_0 - (a_1/2)^2]$ or $\omega_0 = \sqrt{a_0 - (a_1/2)^2}$.

2. **Find a and b by matching like terms in the numerator:** $a = b_1$ and $b_0 = a\alpha + b\omega_0 = b_1\alpha + b\omega_0$ or $b = (b_0 - b_1\alpha)/\omega_0$.

As a test case, consider $X(s) = (s+12)/(s^2+6s+15)$. Using the equations for α, ω_0, a, and b in Steps 1 and 2, you find that $\alpha = 3$, $\omega_0 = \sqrt{6}$, $a = 1$, and $b = 9/\sqrt{6}$; for example:

$$s^2 + 6s + 15 = (s+3)^2 + \left[15 - (3)^2\right] = (s+3)^2 + \left(\sqrt{6}\right)^2 = (s+\alpha)^2 + \omega_0^2$$

On the time-domain side, you have

$$x(t) = e^{-3t}\left[\cos\left(\sqrt{6}\,t\right) + \frac{9}{\sqrt{6}}\sin\left(\sqrt{6}\;t\right)\right]u(t)$$

To check this with Maxima you use the ILT function `ilt()` and find agreement when using the original $X(s)$ function:

```
(%i1) ilt((s+12)/(s^2+6*s+15),s,t);

(%o1) %e^{-3 t} ( 9 sin(√6 t) / √6 + cos(√6 t) )
```

Getting Back to the Time Domain

Working in the *s*-domain is more of a means to an end than an end itself. In particular, you may bring signals to the *s*-domain by using the LT and then performing some operations and/or getting an *s*-domain function; but in the end, you need to work with the corresponding time-domain signal.

Returning to the time domain requires the inverse Laplace transform (ILT). Formally, the ILT requires *contour integration,* which is a line integral over a closed path in the *s*-plane. This approach relies on a complex variable theory background.

In this section, I describe how to achieve contour integration by using partial fraction expansion (PFE) and table lookup. I also point out a few PFE considerations and include examples that describe how these considerations play out in the real world.

The general formula you need to complete the ILT is

$$X(s) = \frac{N(s)}{D(s)} = \frac{\sum_{k=0}^{M} b_k s^k}{\sum_{k=0}^{N} a_k s^k}$$

The first step is to ensure that $X(s)$ is proper rational, that $N > M$. If it isn't, you need to use long division to reduce the order of the denominator. After long division, you end up with this:

$$X(s) = \frac{N(s)}{D(s)} = \sum_{r=0}^{M-N} K_r s^r + \frac{N_1(s)}{D(s)}$$

$N_1(s)$ is the remainder, having polynomial degree less than N.

Dealing with distinct poles

If the poles are simple (unrepeated) and $X(s)$ is proper rational, you can write $X(s) = \sum_{k=1}^{N} R_k / (s - p_k)$, where the p_k represents the nonzero poles of $X(s)$. Find the PFE coefficients by using the *residue* formula: $R_k = (s - p_k) X(s)\big|_{s=p_k}$ (see Chapter 2).

If you need to perform long division (such as when the order of the numerator polynomial is greater than or equal to the order of the denominator polynomial), be sure to augment your final solution with the long division terms $K_r s^r, r = 0, 1, \ldots, M - N$.

Working double time with twin poles

When $X(s)$ has the pole p_i repeated once (a multiplicity of two), the expansion form is augmented as follows:

$$X(s) = \sum_{k=1, k \neq i}^{N} \frac{R_k}{s - p_k} + \frac{S_1}{s - p_i} + \frac{S_2}{(s - p_i)^2}$$

You can find the coefficients R_k by using the residue formula. The formula for S_2 is $S_2 = (s - p_i)^2 X(s)\big|_{s=p_i}$ and S_1 is found last by substitution. See Example 13-11.

Completing inversion

For the general ILT scenario outlined at the start of the section, invert three transform types via table lookup. The first manages the long division result; the second, distinct poles; and the third, poles of multiplicity two:

$$K_r \cdot \delta^{(n)}(t) \xleftarrow{\mathcal{L}} K_r s^r$$

$$R_k \cdot p_k u(t) \xleftarrow{\mathcal{L}} \frac{R_k}{s - p_k}$$

$$S_2 \cdot t e^{-at} u(t) \xleftarrow{\mathcal{L}} \frac{S_2}{(s - p_k)^2}$$

Using tables to complete the inverse Laplace transform

This section contains two ILT examples that describe how to complete PFE with the table-lookup approach. The first example, Example 13-11, considers a repeated real pole; the second, Example 13-12, considers a complex conjugate pole pair. Both examples follow this step-by-step process:

1. **Find out whether $X(s)$ is a proper rational function; if not, perform long division to make it so.**

2. **Find the roots of the denominator polynomial and make note of any repeated poles (roots) and/or complex conjugate poles (roots).**

3. **Write the general PFE of $X(s)$ in terms of undetermined coefficients.**

   For the case of complex conjugate pole pairs, use the special transform pair developed in Example 13-9 so the resulting time domain is conveniently expressed in terms of exponentially damped cosine and sine functions.

4. **Solve for the coefficients by using a combination of the residue formula and substitution techniques as needed.**

5. **Apply the ILT to the individual PFE terms, using table lookup.**

6. **Check your solution by using a computer tool, such as Maxima or Python.**

Example 13-11: Find the impulse response corresponding to a system function containing three poles: one at $s = -2$ and two at $s = -1$, and one zero at $s = 0$. Specifically, $H(s)$ is the proper rational function $H(s) = (5s)/[(s+2)(s+1)^2]$, ROC: Re{$s$} > 1.

1. **Use the repeated roots form for PFE to expand $H(s)$ into three terms:**

$$H(s) = \frac{5s}{(s+2)(s+1)^2} = \frac{R_1}{s+2} + \frac{S_1}{s+1} + \frac{S_2}{(s+1)^2}$$

2. **Find the coefficients R_1 and S_2, using the residue formula as described in section "Working double time with twin poles":**

$$R_1 = \frac{5s}{(s+1)^2}\bigg|_{s=-2} = \frac{-10}{(-1)^2} = -10, \quad S_2 = \frac{5s}{s+2}\bigg|_{s=-1} = \frac{-5}{1} = -5$$

3. **Solve for S_1 by direct substitution. Let $s = 0$:**

$$\frac{5s}{(s+2)(s+1)^2}\bigg|_{s=0} = 0 = \left[\frac{-10}{s+2} + \frac{S_1}{s+1} + \frac{-5}{(s+1)^2}\right]_{s=0} = -10 + S_1 \Rightarrow S_1 = 10$$

4. **Place the values you found for R_1, S_1, and S_2 into the expanded form for $H(s)$ of Step 1, and then apply the ILT term by term:**

$$X(s) = \frac{-10}{s+2} + \frac{10}{s+1} - \frac{5}{(s+1)^2} \Rightarrow x(t) = \left(-10e^{-2t} + 10^{-t} - 5te^{-t}\right)u(t)$$

Check the results with Maxima to see that all is well:

```
(%i1) ilt(5*s/((s+2)*(s+1)^2),s,t);
(%o1)  -5 t %e^-t + 10 %e^-t - 10 %e^-2 t
```

Example 13-12: Find the impulse response corresponding to a system function that has one real pole and one complex conjugate pair. Here's the equation for a third-order system function with pole at $s = -1$ and complex conjugate poles at $s = -1 \pm j$:

$$H(s) = (s^2 + 3s + 1)/[(s+1)(s^2 + 2s + 2)], \text{ ROC: } \text{Re}\{s\} > 1$$

All the poles are distinct, so you can use the residue formula to find the PFE coefficients and the impulse response. This approach results in real and complex exponential time-domain terms — not a clean solution without further algebraic manipulation. But the impulse response is real because all the poles are either real or complex conjugate pairs, so the complex exponentials do reduce to real sine and cosine terms.

For this scenario, I recommend using the s-shift theorem and the associated transform pairs found in the earlier section "s-shift," because this approach takes you to a usable time-domain solution involving damped sine and cosine terms. Example 13-10 boiled all of this down to a nice set of coefficient equations. The PFE expansion for the conjugate pole term is $[a(s+\alpha) + b\omega_0]/[(s+\alpha)^2 + \omega_0^2]$. You can find the equations for α, ω_0, a, and b in Example 13-10.

Now it's time for action. Here's the process:

1. **For the three-pole system, complete the square of the complex conjugate pole term.**

   Use the equations from Step 1 of Example 13-10, noting that, here, $a_1 = 2$ and $a_0 = 2$: $\alpha = a_1/2 = 2/2 = 1$ and $\omega_0 = \sqrt{a_0 - (a_1/2)^2} = \sqrt{2 - (1/1)^2} = 1$. Write the PFE as

   $$H(s) = \frac{s^2 + 3s + 1}{(s+1)\left[(s+1)^2 + 1\right]} = \frac{R_1}{s+1} + \frac{a(s+1) + b \cdot 1}{(s+1)^2 + 1}, \alpha = 1, \omega_0 = 1$$

2. **Solve for R_1 by using the residue formula; then use variable substitution to find a and b.**

$$R_1 = \frac{s^2 + 3s + 1}{s^2 + 2s + 2}\bigg|_{s=-1} = \frac{1 - 3 + 1}{1 - 2 + 2} - 1$$

$$s = 0: \frac{s^2 + 3s + 1}{s^2 + 2s + 2}\bigg|_{s=0} = \boxed{\frac{1}{2}} = \frac{-1}{s+1} + \frac{a(s+1) + b \cdot 1}{(s+1)^2 + 1}\bigg|_{s=0} = \boxed{-1 + \frac{a+b}{2}}$$

$$s = 1: \frac{s^2 + 3s + 1}{s^2 + 2s + 2}\bigg|_{s=1} = \boxed{\frac{5}{10}} = \frac{-1}{s+1} + \frac{a(s+1) + b \cdot 1}{(s+1)^2 + 1}\bigg|_{s=1} = \boxed{\frac{-1}{2} + \frac{2a+b}{5}}$$

3. **Solve the two equations for the unknowns to find $a = 2$ and $b = 1$.**

   Use the transform pairs Lines 3, 7, and 8 in Figure 13-5 to get the inverse transform of the three terms:

$$h(t) = -e^{-t}u(t) + 2e^{-t}\cos(t)u(t) + e^{-t}\sin(t)u(t)$$

Checking with Maxima shows agreement:

```
(%i1) ilt((s^2+3*s+1)/((s+1)*(s^2+2*s+2)),s,t);
(%o1) %e^-t (sin(t)+2 cos(t)) -%e^-t
```

Working with the System Function

The general LCC differential equation (see Chapter 7) is defined as

$$\sum_{k=0}^{N} a_k \frac{d^k y(t)}{dt^k} = \sum_{k=0}^{M} b_k \frac{d^k x(t)}{dt^k}$$

By taking the LT of both sides of this equation, using linearity and the differentiation theorem (described in the section "Transform theorems," earlier in this chapter), you can find the system function $H(s)$ as the ratio of $Y(s)/X(s)$.

$$H(s) = \frac{Y(s)}{X(s)} = \frac{\sum_{k=0}^{M} b_k s^k}{\sum_{k=0}^{N} a_k s^k}$$

This works because of the convolution theorem (see Figure 13-4),
$y(t) = x(t) * h(t) \Rightarrow Y(s) = X(s)H(s)$ or $H(z) = Y(z)/X(z)$.

Assuming zero initial conditions, you use the differentiation theorem (see Figure 13-4) to find the LT of each term on the left and right sides of the general LCC differential equation. You then factor out $Y(s)$ and $X(s)$ and form the ratio $Y(s)/X(s) = H(s)$ on the left side, and what remains on the right side is the system function:

$$\mathcal{L}\left\{\sum_{k=0}^{N} a_k \frac{d^k y(t)}{dt^k}\right\} = \underbrace{\sum_{k=0}^{N} a_k s^k Y(s) = \sum_{k=0}^{M} b_k s^k X(s)}_{\text{Form the ratio } \frac{Y(s)}{X(s)}, \text{ done!}} = \mathcal{L}\left\{\sum_{k=0}^{M} b_k \frac{d^k x(t)}{dt^k}\right\}$$

After you solve for $H(s)$, you can perform analysis of the system function alone or in combination with signals.

✔ Plot the poles and zeros of $H(s)$ and see whether the poles are in the left-half plane, making the system stable.

✔ Assuming the system is stable, plot the frequency response magnitude and phase by letting $s = j2\pi f$ in $H(s)$ to characterize the system filtering properties.

✔ Find the inverse LT $\mathcal{L}^{-1}\{H(s)\}$ to get the impulse response, $h(t)$ — the core time-domain characterization of the system.

✔ Find the inverse LT $\mathcal{L}^{-1}\{\mathcal{L}\{u(t)\}\cdot H(s)\} = \mathcal{L}^{-1}\{H(s)/s\}$ to get the step response.

✔ Find the inverse LT $\mathcal{L}^{-1}\{X(s)H(s)\}$ to get the output $y(t)$ in response to a specific input $x(t)$.

You can also work initial conditions into the solution by *not* assuming zero initial conditions in the earlier system function analysis.

Managing nonzero initial conditions

A feature of the one-sided LT is that you can handle nonzero initial conditions when solving the general LCC differential equation. The key to this is the differentiation theorem. The lead term $s^n X(s)$ is the core result of the theorem, but the nonzero initial conditions are carried by the terms $-s^{n-(k+1)} d^k x(t)/dt^k\big|_{t=0^-}$, $k = 0, \dots, n-1$.

Example 13-13: The input to an LTI system is given by $x(t) = 2e^{-5t}u(t)$, and the system LCC differential equation is $y'(t) + 3y(t) = x(t)$, $y(0^-) = 2$. To find the system output $y(t)$ with nonzero initial conditions, follow these steps:

1. **Apply the differentiation theorem (see Figure 13-4) to both sides of the LCC differential equation:**

$$\left[sY(s) - y(0^-) \right] + 3Y(s) = X(s)$$

2. **Solve for $Y(s)$, making substitutions for $X(s)$ and $y(0^-)$:**

$$Y(s) = \frac{X(s) + y(0^-)}{s+3} = \frac{2}{(s+5)(s+3)} + \frac{2}{s+3} = \frac{R_1}{s+5} + \frac{R_2}{s+3} + \frac{2}{s+3}$$

Use Line 3 of Figure 13-5: $\mathcal{L}\left\{ 2e^{-5t}u(t) \right\} = 2/(s+5)$.

3. **Solve for the partial fraction coefficients R_1 and R_2:**

$$R_1 = \left. \frac{2}{s+3} \right|_{s=-5} = -1 \text{ and } R_2 = \left. \frac{2}{s+5} \right|_{s=-3} = 1$$

4. **Insert the coefficients into the PFE for $Y(s)$ and then apply the inverse Laplace transform to each term:**

$$y(t) = \underbrace{-e^{-5t}u(t)}_{\text{forced resp.}} + \underbrace{e^{-3t}u(t) + 2e^{-3t}u(t)}_{\text{transient response}}$$

Check Maxima to see whether it agrees:

```
(%i1) ilt(2/(s^2+8*s+15)+2/(s+3),s,t);
(%o1) 3 %e^-3 t -%e^-5 t
```

Checking the frequency response with pole-zero location

The location of the poles and zeros of a system controls the frequency response $H(\omega) - H(f)$. Consider the following:

$$H(\omega) = \left. \frac{\sum_{k=0}^{M} b_k s^k}{\sum_{k=0}^{N} a_k s^k} \right|_{s=j\omega} = \left. \frac{\prod_{k=1}^{M} (s - z_k)}{\prod_{k=1}^{N} (s - p_k)} \right|_{s=j\omega} = \frac{\prod_{k=1}^{M} \vec{V}_{zk}(\omega)}{\prod_{k=1}^{N} \vec{V}_{pk}(\omega)}$$

p_k and z_k are the poles and zeros of $H(s)$ and $\vec{V}_{pk}(\omega) = j\omega - p_k$ and $\vec{V}_{zk}(\omega) = j\omega - z_k$ are the frequency response contributions of each pole and zero. The vectors $\vec{V}_{zk}$ and $\vec{V}_{pk}$ point from the zero and pole locations to $j\omega$ on the imaginary axis. This provides a connection between the pole-zero geometry and the frequency response. In particular, the ratio of vector magnitudes, $|\vec{V}_{z1}(\omega)\cdots\vec{V}_{zM}(\omega)|/|\vec{V}_{p1}(\omega)\cdots\vec{V}_{pN}(\omega)|$, is the magnitude response. The phase response is $[\angle\vec{V}_{z1}(\omega)+\cdots+\angle\vec{V}_{zM}(\omega)]-[\angle\vec{V}_{p1}(\omega)+\cdots+\angle\vec{V}_{pN}(\omega)]$.

Example 13-14: Consider the simple first-order system:

$$H(s) = \frac{s}{s+a} \Rightarrow H(\omega) = \frac{j\omega}{j\omega-(-a)} = \frac{\vec{V}_{z1}(\omega)}{\vec{V}_{p1}(\omega)}$$

The single numerator and denominator vectors indicates that the geometry is rather simple. You can use the software app PZ_Geom_S to further explore the relationship between frequency response and pole-zero location. Check out the screen shot of PZ_geom_S in Figure 13-6.

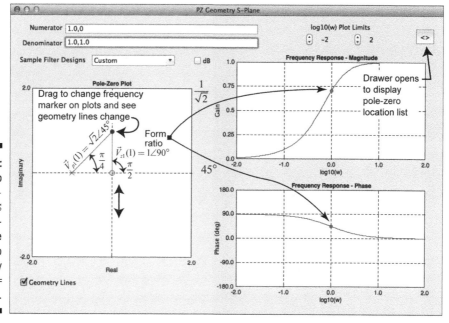

Figure 13-6: The app PZ-Geom_S displaying the pole-zero geometry for $H(s) = s/(s+1)$.

Chapter 14

The *z*-Transform for Discrete-Time Signals

- -

In This Chapter

▶ Traveling to the *z*-domain with the two-sided *z*-transform

▶ Appreciating the ROC for left- and right-sided sequences

▶ Using partial fraction expansion to return to the time domain

▶ Checking out *z*-transform theorems and pairs

▶ Visualizing how pole-zero geometry controls the frequency response

- -

*T*he *z*-transform (ZT) is a generalization of the discrete-time Fourier transform (DTFT) (covered in Chapter 11) for discrete-time signals, but the ZT applies to a broader class of signals than the DTFT. The ZT notation is also more user friendly than the DTFT. For example, linear constant coefficient (LCC) difference equations (see Chapter 7) can be solved by using just algebraic manipulation when the ZT is involved.

Here's the deal: When a discrete-time signal is transformed with the ZT, it becomes a function of a complex variable; the DTFT creates a function of a real frequency variable only. The transformed signal is said to be in the *z*-domain.

The ZT closely parallels the Laplace transform (see Chapter 13) for continuous-time signals, and you can use the ZT to analyze signals and the impulse response of linear time-invariant (LTI) systems.

In this chapter, I introduce the two-sided ZT, which allows you to transform signals that cover the entire time axis. When computing the ZT, you immediately bump into the region of convergence (ROC), which tells you where the transform exists. This chapter also describes important ROC properties that offer insight into the nature of a signal or system.

The Two-Sided z-Transform

The *two-sided* or *bilateral* z-transform (ZT) of sequence $x[n]$ is defined as $X(z) = \mathcal{Z}\{x[n]\} = \sum_{n=-\infty}^{\infty} x[n]z^{-n}$, where z is a complex variable. The ZT operator transforms the sequence $x[n]$ to $X(z)$, a function of the continuous complex variable z. The relationship between a sequence and its transform is denoted as $x[n] \xleftrightarrow{\ z\ } X(z)$.

You can establish the connection between the discrete-time Fourier transform (DTFT) and the ZT by first writing $z = r \cdot e^{j\hat{\omega}}$ (polar form):

$$X(z)\big|_{z=re^{j\hat{\omega}}} = X\left(re^{j\hat{\omega}}\right) = \sum_{n=-\infty}^{\infty} x[n]\left(re^{j\hat{\omega}}\right)^{-n}$$

$$= \sum_{n=-\infty}^{\infty} \left(x[n]r^{-n}\right)e^{-j\hat{\omega}n} = \mathcal{F}\left\{x[n]r^{-n}\right\}$$

The special case of $r = 1$ evaluates $X(z)$ over the unit circle — $z = e^{j\hat{\omega}} = 1\angle\hat{\omega}$ sweeps around the unit circle as $\hat{\omega}$ varies — and is represented as $X\left(e^{j\hat{\omega}}\right)$ — the DTFT of $x[n]$ (see Chapter 11). This result holds as long as the DTFT is absolutely summable (read: impulse functions not allowed).

The view that $X(e^{j\hat{\omega}})$ is $X(z)$ sampled around the unit circle in the z-plane $(z \to e^{j\hat{\omega}})$ shows that the DTFT has period 2π because $e^{j(\hat{\omega}+2\pi k)} = e^{j\hat{\omega}}e^{j2\pi k} = e^{j\hat{\omega}}$ for k an integer. When $z = e^{j\hat{\omega}}$, $z = 1 \leftrightarrow \hat{\omega} = 0$, $z = j \leftrightarrow \hat{\omega} = \pi/2$, $z = -1 \leftrightarrow \hat{\omega} = \pm\pi$, and $z = -j \leftrightarrow \hat{\omega} = -\pi/2$. See this solution in Figure 14-1.

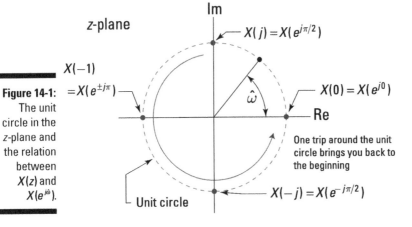

Figure 14-1: The unit circle in the z-plane and the relation between $X(z)$ and $X(e^{j\hat{\omega}})$.

The Region of Convergence

The ZT doesn't converge for all sequences. When it does converge, it's only over a region of the z-plane. The values in the z-plane for which the ZT converges are known as the *region of convergence* (ROC).

Convergence of the ZT requires that

$$\left|X(z)\right| \le \sum_{n=-\infty}^{\infty}\left|x[n]z^{-n}\right| = \sum_{n=-\infty}^{\infty}\left|x[n]\right|\left|z\right|^{-n} = \sum_{n=-\infty}^{\infty}\left|x[n]r^{-n}\right| < \infty$$

The right side of this equation shows that $x[n]r^n$ is *absolutely summable* (the sum of all terms $\left|x[n]r^{-n}\right|$ is less than infinity). This condition is consistent with the absolute summability condition for the DTFT to converge to a continuous function of $\hat{\omega}$ (see Chapter 11).

Convergence depends only on $\left|z\right| = r$, so if the series converges for $z = z_1$, then the ROC also contains the circle $\left|z\right| = \left|z_1\right|$. In this case, the general ROC is an annular region in the z-plane, as shown in Figure 14-2. If the ROC contains the unit circle, the DTFT exists because the DTFT is the ZT evaluated on the unit circle.

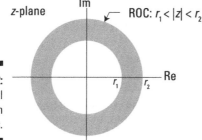

Figure 14-2: The general ROC is an annulus.

The significance of the ROC

The ROC has important implications when you're working with the ZT, especially the two-sided ZT. When the ZT produces a rational function, for instance, the roots of the denominator polynomial are related to the ROC. And for LTI systems having a rational ZT, the ROC is related to a system's bounded-input bounded-output (BIBO) stability. The uniqueness of the ZT is also ensured by the ROC.

Example 14-1: Consider the right-sided sequence $x_a[n] = a^n u[n]$. The term *right-sided* means that the sequence is 0 for $n < n_0 < \infty$, and the nonzero values extend from n_0 to infinity. The value of n_0 may be positive or negative.

To find $X(z)$ and the ROC, follow these steps:

1. **Reference the definition to determine the sum:**

$$X_a(z) = \sum_{n=-\infty}^{\infty} a^n u[n] z^{-n} = \sum_{n=0}^{\infty} \left(az^{-1}\right)^n$$

2. **Find the condition for convergence by summing the infinite geometric series:**

$$\sum_{n=0}^{\infty} \left|az^{-1}\right| = \frac{1}{1-\left|az^{-1}\right|} < \infty \text{ provided } |a/z| < 1 \text{ or } |z| > |a|$$

Thus, the ROC is $|z| > |a|$.

3. **To find the sum of Step 1 in closed form, use the finite geometric series sum formula (see Chapter 2):**

$$X_a(z) = \underbrace{\frac{1}{1-az^{-1}}}_{\text{neg. powers of } z \text{ form}} \cdot \frac{z}{z} = \underbrace{\frac{z}{z-a}}_{\text{pos. powers of } z \text{ form}} , \text{ROC: } |z| > |a|$$

The geometric series convergence condition corresponds with the ROC. In Examples 14-2 and 14-3, I find the ROC from the geometric series convergence condition.

Example 14-2: Consider the left-sided sequence $x_b[n] = -a^n u[-n-1]$. The term *left-sided* means that the sequence is 0 for $n > n_0 > -\infty$. To find $X(z)$ and the ROC, write the definition and the infinite geometric series:

$$X_b(z) = -\sum_{n=-\infty}^{\infty} a^n u[-n-1] z^{-n} = \sum_{n=-\infty}^{-1} \left(az^{-1}\right)^n$$

If the sum looks unfamiliar in its present form, you can change variables in the sum, an action known as *re-indexing* the sum, by letting $m = -n$, then in the limits $n = -1 \rightarrow m = 1$, $n = -\infty \rightarrow m = \infty$, and in the sum itself $n \rightarrow -m$:

$$X_b(z) = -\sum_{n=-1}^{\infty} a^{-m} z^m = 1 - \sum_{n=0}^{\infty} \left(z/a\right)^m = 1 - \frac{1}{1-a^{-1}z}$$

$$= \frac{1}{1-az^{-1}} \cdot \frac{z}{z} = \frac{z}{z-a}, \text{ROC: } |z| < |a|$$

Note: Both $x_a[n]$ from Example 14-1 and $x_b[n]$ from Example 14-2 have the same ZT. The ROCs, however, are complements; the ZTs are thus distinguishable only by the ROCs being different, demonstrating that you need the ROC to return to the time domain without ambiguity — is it left-sided or right-sided? The rest of this chapter focuses exclusively on right-sided sequences, because they correspond to the most common scenarios.

Plotting poles and zeros

For rational functions, such as the $X(z)$ of Examples 14-1 and 14-2, you want to plot the *pole* and *zero* locations in the z-plane as well as the ROC. A rational $X(z)$ takes the form $N(z)/D(z)$, where $N(z)$ and $D(z)$ are both polynomials in either z or z^{-1}. The roots of $N(z)$, where $N(z)=0$, are the zeros of $X(z)$, and the roots of $D(z)$ are the poles of $X(z)$.

I recommend placing $X(z)$ in positive powers of z when determining the poles and zeros. You can do this by multiplying the numerator and denominator by the largest positive power of z to eliminate any negative powers of z.

Check out the pole-zero plot of $X(z)$, including the ROC, for $X_a(z)$ and $X_b(z)$ in Figure 14-3.

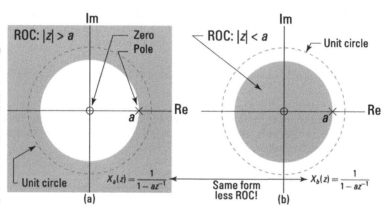

Figure 14-3: The pole-zero plot and ROC of $X_a(z)$ (a) and $X_b(z)$ (b), of Examples 14-1 and 14-2.

The ROC and stability for LTI systems

An LTI system is BIBO stable if the impulse response $h[n]$ is absolutely *summable*, which means that the sum of all terms $|h[n]|$ is less than infinity (see Chapter 6). To relate this to the ROC, consider the corresponding ZT $H(z) = \mathcal{Z}\{h[n]\}$, also known as the *system function*. You can find out whether the ROC of $H(z)$ includes the unit circle by setting $z = e^{j\hat{\omega}}$ and then checking for absolute summability:

$$\sum_{n=-\infty}^{\infty}\left|h[n]z^{-n}\right|\Bigg|_{z=e^{j\hat{\omega}}} = \sum_{n=-\infty}^{\infty}|h[n]|\underbrace{\left|e^{j\hat{\omega}}\right|}_{1} = \sum_{n=-\infty}^{\infty}|h[n]| \overset{\text{if BIBO stable}}{<} \infty$$

Here, you can conclude that, yes, the ROC of a BIBO stable system must include the unit circle.

Carrying the analysis a bit further, the poles of $H(z)$ must sit outside the ROC because the poles of $H(z)$ sit at *singularities,* points where $|H(z)|$ is unbounded. For right-sided sequences, the ROC is the exterior of a circle with a radius corresponding to the largest pole radius of $H(z)$. This is the case in Example 14-1, where $X_a(z)$ has a pole at $z = a$.

An LTI system with impulse response $h[n]$ and system function $H(z)$ is BIBO stable if the ROC contains the unit circle. The system function is the ZT version of frequency response $H(e^{j\hat{\omega}})$.

For the special case of a causal system, in which $h[n] = 0$ for $n < 0$, the poles of $H(z)$ *must* lie inside the unit circle because right-sided sequences, where causal sequences are a special case, have an ROC that's the exterior of the largest pole radius of $H(z)$. In short, poles inside the unit circle ensure stability of causal discrete-time LTI systems.

Example 14-3: To find $H(z)$ given $h[n] = (1/3)^n u[n] + (1/2)^n u[n]$, use the transform pair developed in Example 14-1 with $a = 1/3$ and $1/2$ in the first and second terms, respectively, and make use of linearity:

$$\mathcal{Z}\left\{\left(\tfrac{1}{3}\right)^n u[n]\right\} = \frac{1}{1 - \tfrac{1}{3}z^{-1}}, |z| > \frac{1}{3} \quad \text{and} \quad \mathcal{Z}\left\{\left(\tfrac{1}{2}\right)^n u[n]\right\} = \frac{1}{1 - \tfrac{1}{2}z^{-1}}, |z| > \frac{1}{2}$$

The ROC is the intersection of the two ROCs unless a pole-zero cancellation occurs. So, tentatively, the ROC is $\max\{1/3, 1/2\} = 1/2$. When placed over a common denominator, $H(z)$ is

$$H(z) = \frac{1}{1 - \tfrac{1}{3}z^{-1}} + \frac{1}{1 - \tfrac{1}{2}z^{-1}} = \frac{2\left(1 - \tfrac{5}{12}z^{-1}\right)}{\left(1 - \tfrac{1}{3}z^{-1}\right)\left(1 - \tfrac{1}{2}z^{-1}\right)}$$

$$= \frac{2z\left(z - \tfrac{5}{12}\right)}{\left(z - \tfrac{1}{3}\right)\left(z - \tfrac{1}{2}\right)}, |z| > \frac{1}{2}$$

The system pole-zero plot and ROC are shown in Figure 14-4.

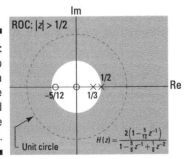

Figure 14-4: Pole-zero plot of a two-pole causal and stable system.

Finite length sequences

If $h[n]$ has finite length (finite impulse response [FIR] in this case), then $H(z)$ will converge everywhere as long as $|h[n]|$ is finite for all nonzero values. If $h[n]$ is nonzero only on $N_1 \leq n \leq N_2$, then the ZT of $h[n]$ is simply $H(z) = x[N_1]z^{-N_1} + \cdots + x[N_2]z^{-N_2}$, ROC at least: $0 < |z| < \infty$. Here, the ROC is the entire z-plane, excluding perhaps $|z| \rightarrow \infty$ and/or $|z| = 0$.

If $H(z)$ contains negative powers of z, then the ROC must exclude $|z| = 0$ (ROC: $|z| > 0$) to avoid the $1/z$ singularity. Similarly, if $H(z)$ contains positive powers of z, then the ROC must exclude $|z| \rightarrow \infty$ (ROC: $|z| < \infty$) to remain bounded.

Example 14-4: Find the ZT of the finite length sequence $x[n] = a^n\big(u[n] - u[n-N]\big)$. Note that this sequence is a rectangular window with exponential weighting a^n. Use the definition of the ZT and the finite geometric series sum formula:

$$X(z) = \sum_{n=0}^{N-1}\left(az^{-1}\right)^n = \frac{1-\left(az^{-1}\right)^N}{1-az^{-1}} \cdot \frac{z^N}{z^N} = \frac{1}{z^{N-1}} \cdot \frac{z^N - a^N}{z-a}$$

Because $X(z)$ includes only negative powers of z, ROC: $|z| > 0$. What about the poles and zeros? The denominator (poles) is clear, but you may be stuck on the numerator (zeros). The roots of $z^N - a^N$ are given by $z_k = ae^{j\pi k/N}$, $k = 0, \ldots, N-1$ (see Chapter 2 for details on the N roots of unity). The pole at $z = a$ cancels the zero at $z = a$. Putting it all together, you get the following:

$$X(z) = \frac{(z-a)\cdot\prod_{k=1}^{N-1}\left(z - ae^{j2\pi k/N}\right)}{z^{N-1}\cdot(z-a)} = \frac{\prod_{k=1}^{N-1}\left(z - ae^{j2\pi k/N}\right)}{z^{N-1}}$$

This equation shows $N-1$ zeros equally spaced around a circle of radius a, excluding $z = 0$ and $N = 16$ poles at $z = 0$. The pole-zero plot is shown in Figure 14-5 for $N = 16$.

Figure 14-5: Pole-zero plot for an exponentially weighted window having length $N=16$.

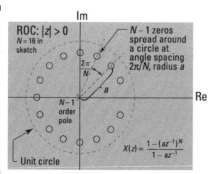

You may have noticed that the number of poles and zeros plotted is always equal to each other. This is a property of $X(z)$ when it's a ratio of polynomials in z or z^{-1}. Figure 14-5 shows $N-1$ poles at $z = 0$ and $N-1$ zeros spread around a circle of radius a.

In some instances, a pole or zero may be at infinity! Consider the one-zero system $(z-a)$. As $|z| \rightarrow \infty$, $|z-a| \rightarrow \infty$, which is the behavior of a pole at infinity. Likewise, consider the one-pole system $1/(z-a)$. As $|z| \rightarrow \infty$, $|1/(z-a)| \rightarrow 0$, which is the behavior of a zero at infinity. The next example investigates a pole at infinity following sequence shifting.

Example 14-5: Consider $x[n] = \delta[n] - 3/2\delta[n-1] - \delta[n-2]$. To find $H(z)$, the ROC, and the poles and zeros, follow this process:

1. **Apply the definition of the ZT:**

$$X(z) = 1 - (3/2)z^{-1} - z^{-2} = \frac{z^2 - \frac{3}{2}z - 1}{z^2} = \frac{(z-2)(z+1/2)}{z^2}$$

2. **From $X(z)$, find that the zeros are $z_1 = 2$ and $z_2 = -1/2$, the two poles are at $z = 0$, and ROC is $0 < |z| < \infty$.**

3. **Generate the pole-zero plot in Figure 14-6 by using PyLab and the custom function ssd.zplane(b,a,radius=2).**

   Then check this on your own, using the available ssd.py Python code module.

   ```
   In [17]: ssd.zplane([1,-3/2.,-1],[1],2.25)
   ```

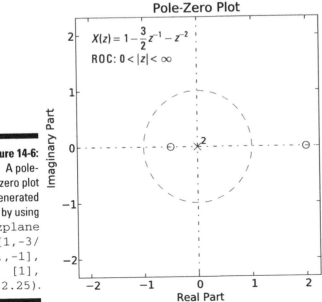

Figure 14-6:
A pole-zero plot generated by using zplane ([1,-3/2.,-1], [1], 2.25).

If you shift $x[n]$ to the left one sample (let $n \to n+1$), $x_{m1}[n] = x[n+1]$, then the ZT becomes

$$X_{m1}(z) = z - (3/2) - z^{-1} = \frac{z^2 - \frac{3}{2}z - 1}{z} = \frac{(z-2)(z+1/2)}{z}$$

which still has zeros at 2 and –1/2. Now the poles are at zero and infinity. You need the pole at infinity to equalize the pole-zero count and account for the fact that $|X_{m1}(z)| \to \infty$ as $|z| \to \infty$, which is the behavior of a pole at infinity. Shifting $x[n]$ to the right one sample (let $n \to n-1$) introduces a zero at infinity; the ZT goes to zero as $|z| \to \infty$.

Returning to the Time Domain

After working in the *z*-domain, you need to get your signal back to the time domain to work with it as a true signal once again, following some *z*-domain operations, such as filtering. The quick way back is through the inverse *z*-transform.

Formally, the inverse *z*-transform requires *contour integration,* a function of a complex variable. With a contour integral, you integrate on a closed curve in the *z*-plane instead of using an ordinary integral along a line. Note that I said *formally.* The most practical way to return from the *z*-domain is to form a partial fraction expansion (PFE) (see Chapter 2) and then apply the inverse *z*-transform to each term of the expansion by using a table of *z*-transform pairs.

If you can't directly find $X(z)$ from a list of transform pairs, such as you can find in Figure 14-8, then you can use PFE to decompose $X(z)$ into a sum of terms that each have an inverse transform readily available. In this book, assume that $X(z)$ is expressed in negative powers of *z*. Start here:

$$X(z) = \frac{N(z)}{D(z)} = \frac{\sum_{k=0}^{M} b_k z^{-k}}{\sum_{k=0}^{N} a_k z^{-k}}$$

Ensure that $X(z)$ is proper rational, or $N > M$. If it isn't, you need to use long division to reduce the order of the denominator. Example 14-6 demonstrates the long division process for polynomials written in terms of negative powers of *z*.

When $M \geq N$, $X(z)$ is an improper rational function. Following long division, you end up with

$$X(z) = \frac{N(z)}{D(z)} = \sum_{r=0}^{M-N} K_r z^{-r} + \frac{N_1(z)}{D(z)}$$

Here, $N_1(z)$ is the remainder polynomial from long division, having degree less than N.

In this section, I run through a few PFE considerations and include examples that point out how these considerations play out in the real world.

Working with distinct poles

For the case distinct poles, if you assume that $X(z)$ is proper rational and $D(z)$ contains no repeated roots (poles), then you can write $X(z) = \sum_{k=1}^{N} R_k / (1 - p_k z^{-1})$ where the p_k's are the nonzero poles of $X(z)$ and the R_k's are the PFE coefficients.

Some textbooks and references use positive powers of z for PFE, but I use negative powers of z for this technique. Don't attempt to mix the two approaches. Using negative powers of z is the most common approach, and it fits the very nature of the signals being transformed. Signals are most often delayed in time, which corresponds to a negative power of z in the z-domain.

For the case of negative powers of z, the PFE coefficients are given by the *residue* formula:

$$R_k = \left(1 - p_k z^{-1}\right) X(z) \Big|_{z^{-1} = p_k^{-1}}$$

If you need to perform long division, be sure to augment your final solution with the long division quotient terms $K_r z^{-r}, r = 0, 1, ..., M - N$.

Managing twin poles

When the pole p_i for $X(z)$ is repeated once (a multiplicity of two), you augment the expansion form as follows:

$$X(z) = \sum_{k=1, k \neq i}^{N} \frac{R_k}{1 - p_k z^{-1}} + \frac{S_1}{1 - p_i z^{-1}} + \frac{S_2}{\left(1 - p_i z^{-1}\right)^2}$$

For the distinct poles, you can find the coefficients R_k by using the residue formula from the previous section. For the S_2 coefficient, use a modified version of the residue formula:

$$S_2 = \left(1 - p_i z^{-1}\right)^2 X(z) \Big|_{z^{-1} = p_i^{-1}}$$

Find S_1 by substitution. (This step is covered in Example 14-7.)

Performing inversion

After you complete the PFE, move on to using the inverse transform of the form $K_r z^{-r}$, $R_k / (1 - p_k z^{-1})$, and $S_2 / (1 - p_k z^{-1})^2$ to get the signal back to the time domain, one by one. Unless stated otherwise, assume that the ROC is the exterior of a circle whose radius corresponds to the largest pole radius of $X(z)$. This assumption makes the corresponding time-domain signals right-sided.

Here's the table lookup for each of the three transform types:

$$K_r \cdot \delta[n-r] \xleftarrow{\ Z\ } K_r z^{-r}, \text{ROC: all } z \text{ except zero for } r > 0$$

$$R_k \cdot p_k^n u[n] \xleftarrow{\ Z\ } \frac{R_k}{1 - p_k z^{-1}}, \text{ROC: } |z| > |p_k|$$

$$S_2 \cdot (n+1) \cdot p_k^n u[n] \xleftarrow{\ Z\ } \frac{S_2}{\left(1 - p_k z^{-1}\right)^2}, \text{ROC: } |z| > |p_k|$$

Using the table-lookup approach

This section contains two inverse z-transform (IZT) examples that describe how to complete PFE with the table-lookup approach. The first example requires long division but has only distinct poles; the second doesn't require long division but has one pole repeated. In both examples, you follow these steps:

1. **Find out whether $X(z)$ is a proper rational function; if not, perform long division to make it so.**

2. **Find the roots of the denominator polynomial and make note of any repeated roots.**

3. **Write the general PFE of $X(z)$ in terms of undetermined coefficients.**

4. **Solve for the coefficients by using a combination of the residue formula and substitution techniques as needed.**

5. **Apply the IZT to the individual PFE terms by using table lookup.**

6. **Check your solution by using a computer tool, such as Python.**

Example 14-6: Find $x[n]$ given the following:

$$X(z) = \frac{1 + 2z^{-1} + z^{-2}}{\left(1 - \tfrac{1}{2}z^{-1}\right)\left(1 + \tfrac{1}{4}z^{-1}\right)} = \frac{1 + 2z^{-1} + z^{-2}}{1 - \tfrac{1}{4}z^{-1} - \tfrac{1}{8}z^{-2}}, \text{ROC: } |z| > \frac{1}{4}$$

Here's the process for returning this signal to the time domain:

1. **Because the numerator and denominator are both second degree, reduce the numerator order to one or less by using long division.**

   In this case, you want to reduce the numerator to order zero:

$$1-\tfrac{1}{4}z^{-1}-\tfrac{1}{8}z^{-2}\overline{)\,\dfrac{-8}{1-2z^{-1}+z^{-2}}}$$
$$\dfrac{-8+2z^{-1}+z^{-2}}{9}$$

   The numbers are deceptively simple in this example, so work carefully.

2. **Factor the denominator polynomial to find the roots.**

   Breathe easy, it's already done for you in this example, but you could have started with $1-(1/4)z^{-1}-(1/8)z^{-2}$. Notice that the roots, 1/2 and −1/4 are distinct.

3. **Using the results of long division and the roots identified in Step 2, write out the general PFE so you can see the coefficients you need to find.**

$$X(z)=-8+\underbrace{\dfrac{9}{\left(1-\tfrac{1}{2}z^{-1}\right)\left(1+\tfrac{1}{4}z^{-1}\right)}}_{\text{Partial fraction expand this!}}=-8+\dfrac{R_1}{\left(1-\tfrac{1}{2}z^{-1}\right)}+\dfrac{R_2}{\left(1+\tfrac{1}{4}z^{-1}\right)}$$

4. **Solve for R_1 and R_2 coefficients, using the residue formula:**

$$R_1=\dfrac{9\cdot\left(1-\tfrac{1}{2}z^{-1}\right)}{\left(1-\tfrac{1}{2}z^{-1}\right)\left(1+\tfrac{1}{4}z^{-1}\right)}\Bigg|_{z^{-1}=2}=\dfrac{9}{1+\tfrac{1}{4}z^{-1}}\Bigg|_{z^{-1}=2}=\dfrac{9}{1+\tfrac{1}{2}}=6$$

$$R_2=\dfrac{9\cdot\left(1+\tfrac{1}{4}z^{-1}\right)}{\left(1-\tfrac{1}{2}z^{-1}\right)\left(1+\tfrac{1}{4}z^{-1}\right)}\Bigg|_{z^{-1}=-4}=\dfrac{9}{1+\tfrac{1}{2}z^{-1}}\Bigg|_{z^{-1}=-4}=\dfrac{9}{1+2}=3$$

5. **With the PFE coefficients and long division quotient in hand, apply the inverse transform to the terms of the PFE:**

$$X(z)=-8+\dfrac{6}{1-\tfrac{1}{2}z^{-1}}+\dfrac{3}{1+\tfrac{1}{4}z^{-1}},\ \text{ROC: }|z|>\tfrac{1}{4}$$

   The ROC is the exterior of a circle of radius 1/4 so all the time-domain terms are right-sided. Carry out the inversion of each term:

$$x[n]=-8\delta[n]+6\left(\tfrac{1}{2}\right)^{n}u[n]+3\left(-\tfrac{1}{4}\right)^{n}u[n]$$

6. **Check your work with PyLab and the SciPy signal package function R,p,K = residuez(b,a).**

   The ndarray b contains the numerator coefficients, and a contains the denominator coefficients. The function returns a tuple of three ndarrays, R (the residues), p (the poles), and K (the results of long division):

```
In [7]: R,p,K = signal.residuez([1,2,1],[1,-.25,-.125])
In [8]: R
Out[8]: array([3., 6.])
In [9]: p
Out[9]: array([-0.25, 0.5 ])
In [10]: K
Out[10]: array([-8.])
```

The Python check agrees with the hand calculation!

Example 14-7: Find $x[n]$ given the following equation:

$$X(z) = \frac{1}{\left(1 - \frac{1}{2}z^{-1}\right)\left(1 - \frac{1}{4}z^{-1}\right)^2}, \text{ROC: } |z| > \frac{1}{2}$$

Now go through the six-step process:

1. **Examine $X(z)$; if you find it to be proper rational as $M = 0$ and $N = 2$, no long division is required.**

2. **Factor the denominator polynomial to find the roots.**

   Here, the denominator polynomial is already in factored form, and the roots are 1/2 and 1/4 with the 1/4 repeated, making this a multiplicity of two.

3. **Write the general PFE with one root repeated:**

$$X(z) = \frac{R_1}{1 - \frac{1}{2}z^{-1}} + \frac{S_1}{1 - \frac{1}{4}z^{-1}} + \frac{S_2}{\left(1 - \frac{1}{4}z^{-1}\right)^2}$$

4. **Using the residue formula, find that the coefficient R_1 is 4. Using the modified residue formula, find that the coefficient S_2 is –1.**

   At this point, you have the following:

$$X(z) = \frac{1}{\left(1 - \frac{1}{2}z^{-1}\right)\left(1 - \frac{1}{4}z^{-1}\right)^2} = \frac{4}{\left(1 - \frac{1}{2}z^{-1}\right)} + \frac{S_1}{1 - \frac{1}{4}z^{-1}} - \frac{1}{\left(1 - \frac{1}{4}z^{-1}\right)^2}$$

Use substitution to solve for S_1 in the expansion by choosing a convenient value for z^{-1} that doesn't place a zero in the denominator of any term. Choosing $z^{-1} = 0$ gives you the equation $1 = 4 + S_1 - 1 \Rightarrow S_1 = -2$. With all the coefficients solved for, write

$$X(z) = \frac{4}{\left(1 - \frac{1}{2}z^{-1}\right)} - \frac{2}{1 - \frac{1}{4}z^{-1}} - \frac{1}{\left(1 - \frac{1}{4}z^{-1}\right)^2}$$

5. **Because the ROC is the exterior of a circle of radius 1/2, all the time-domain terms are right-sided; carry out the inversion of each term to get this equation:**

$$x[n] = 4\left(\tfrac{1}{2}\right)^n u[n] - 2\left(\tfrac{1}{4}\right)^n u[n] - (n+1)\left(\tfrac{1}{4}\right)^n u[n]$$

6. **Multiply out the denominator polynomial and then check your work with Python:**

$$\left(1 - \tfrac{1}{2}z^{-1}\right)\left(1 - \tfrac{1}{4}z^{-1}\right)^2 = 1 - z^{-1} + \tfrac{5}{16}z^{-2} - \tfrac{1}{32}z^{-3}$$

Load the numerator and denominator coefficients into `residuez()`:

```
In [33]: R,p,K = signal.residuez([1],[1.,-1., 5/16.,-
         1/32.])
In [34]: R
Out[34]: array([-2.+0.j, -1.+0.j, 4.+0.j])
In [35]: p
Out[35]: array([0.25+0.j, 0.25+0.j, 0.50+0.j])
In [36]: K
Out[36]: array([0.])
```

Nice. Total agreement with the hand calculation.

Surveying z-Transform Properties

The ZT has many useful theorems and transform pairs. The transform theorems can be applied generally to any signal, but the transform pairs involve a specific signal and corresponding z-transform. The transform pairs in this section emphasize right-sided signals and causal systems because that's the emphasis of this chapter. This means that the ROC is generally the exterior of a circle having finite radius.

Transform theorems

You can find some useful z-transform theorems in Figure 14-7. All these theorems are valuable, but some are used more frequently than others. For instance, the linearity theorem (Line 1) is especially popular because it allows you to break down big problems into smaller pieces. The proof follows immediately from the definition of the ZT.

	Property	$x[n]$	$X(z)$	ROC
1	Linearity	$ax_1[n] + bx_2[n]$	$aX_1(z) + bX_2(z)$	at least $R_{x_1} \cap R_{x_2}$
2	Delay (time shift)	$x[n - n_0]$	$z^{-n_0} X(z)$	R_x, maybe exclude 0 or ∞
3	Differentiation	$nx[n]$	$-z\frac{dX(z)}{dz}$	R_x, maybe exclude 0 or ∞
4	Conjugation	$x^*[n]$	$X^*(z^*)$	R_x
5	Convolution	$x_1[n] * x_2[n]$	$X_1(z)X_2(z)$	at least $R_{x_1} \cap R_{x_2}$
6	First difference	$x[n] - x[n-1]$	$(1 - z^{-1}) X(z)$	at least $R_x \cap \{|z| > 0\}$
7	Accumulation	$\sum_{k=-\infty}^{n} x[k]$	$\frac{1}{1-z^{-1}} X(z)$	at least $R_x \cap \{|z| < \infty\}$
8	Initial value	$x[0] = \lim_{z \to \infty} X(z)$		

R_x is the ROC of $x[n]$, and so on

Figure 14-7: Useful ZT theorems.

In the following sections, I lavish some special attention on two theorems — the delay and convolution theorems — because they apply to so many of the problems you're likely to encounter as an electrical engineer.

Delay

One of the most widely used ZT theorems is delay (also sometimes called time shift): $\mathcal{Z}\{x[n - n_0]\} = z^{-n_0} X(z)$. Here, the ROC is the ROC of $X(z)$, denoted R_x with the possible exclusion of zero or infinity depending on both $x[n]$ and the sign of n_0.

The proof follows from the definition and the substitution $m = n - n_0$:

$$\mathcal{Z}\{x[n - n_0]\} = \sum_{n=-\infty}^{\infty} x[n - n_0]z^{-n} = z^{-n_0} \sum_{n=-\infty}^{\infty} x[m]z^{-n} = z^{-n_0} X(z)$$

Convolution

The convolution theorem of sequences is fundamental when working with LTI systems. It can be shown that $x[n] * h[n] \xleftarrow{z} X(z)H(z)$, where the ROC $= R_x \cap R_h$ or larger if pole-zero cancellation occurs.

You can develop the proof of the convolution in two steps:

1. **Insert the convolution sum into the ZT and then change the sum order:**

$$Y(z) = \sum_{n=-\infty}^{\infty} \left\{ \sum_{k=-\infty}^{\infty} x[k]h[n-k] \right\} z^{-n} = \sum_{k=-\infty}^{\infty} x[k] \left\{ \sum_{n=-\infty}^{\infty} h[n-k] \right\} z^{-n}$$

2. **Change variables on the inner sum by letting $m = n - k$.**

The inner sum becomes $H(z)$, and the outer sum becomes $X(z)$:

$$Y(z) = \sum_{k=-\infty}^{\infty} x[k] \underbrace{\left\{ \sum_{m=-\infty}^{\infty} h[m]z^{-m} \right\}}_{H(z)} z^{-k} = \underbrace{\sum_{k=-\infty}^{\infty} x[k]z^{-k}}_{X(z)} H(z)$$

Transform pairs

Transform pairs are the lifeblood of working with the ZT. If you need to transform a signal or inverse transform a *z*-domain function, start by consulting a transform pairs table. Combined with theorems, these pairs practically give you super powers. Some basic transform pairs are developed in Examples 14-1 and 14-2, earlier in this chapter.

Example 14-8: Determine the *z*-transform of the sequence $x[n] = n \cdot a^n u[n]$. From the differentiation theorem, you know that the following must be true:

$$X(z) = z \cdot \frac{d}{dz} \left(1 - az^{-1} \right)^{-1} = -z \cdot -1 \cdot \left(1 - az^{-1} \right)^{-2} \cdot (-a)(-1)z^{-2}$$

$$= \frac{az^{-1}}{\left(1 - az^{-1} \right)^{-2}}, \text{ROC: } |z| > |a|$$

The transform pair developed is

$$na^n u[n] \xleftarrow{\ z\ } \frac{az^{-1}}{\left(1 - az^{-1} \right)^2}$$

Working with the delay theorem and linearity makes this pair into Line 5 of Figure 14-8.

Find a short table of *z*-transform pairs in Figure 14-8, which emphasizes right-sided sequences because that's the focus of this chapter.

If you notice the absence of the step function, $u[n]$, all you need to do is let $a = 1$ in Line 3 of Figure 14-8, and you have that pair, too!

	$x[n]$	$X(z)$	ROC
1	$\delta[n]$	1	all z
2	$\delta[n-n_0]$	z^{-n_0}	all z except 0 or ∞ depending on n_0
3	$a^n u[n]$	$\dfrac{1}{1-az^{-1}}$	$|z|>|a|$
4	$-a^n u[-n-1]$	$\dfrac{1}{1-az^{-1}}$	$|z|<|a|$
5	$(n+1)a^n u[n]$	$\dfrac{1}{(1-az^{-1})^2}$	$|z|>|a|$
6	$r^n\cos(\hat\omega_0 n)u[n]$	$\dfrac{1-(r\cos\hat\omega_0)z^{-1}}{1-(2r\cos\hat\omega_0)z^{-1}+r^2z^{-2}}$	$|z|>r$
7	$r^n\sin(\hat\omega_0 n)u[n]$	$\dfrac{(r\sin\hat\omega_0)z^{-1}}{1-(2r\cos\hat\omega_0)z^{-1}+r^2z^{-2}}$	$|z|>r$
8	$a^n(u[n]-u[n-N])$	$\dfrac{1-a^N z^{-N}}{1-az^{-1}}$	$|z|>0$

Figure 14-8: Common ZT pairs.

Leveraging the System Function

The system function $H(z)$ for LTI discrete-time systems, like its counterpart for continuous-time signals $H(s)$ (see Chapter 13), plays an important role in the design and analysis of systems. Of special interest are LTI systems that are represented by a linear constant coefficient (LCC) difference equation.

When you couple the system function with the convolution theorem, you have the capability to model signals passing through systems entirely in the z-domain. In this section, I show you how the poles and zeros work together to shape the frequency response of a system.

The general LCC difference equation $\sum_{k=0}^{N} a_k y[n-k] = \sum_{k=0}^{M} b_k x[n-k]$ describes a causal system where the output $y[n]$ is related to input $x[n]$, M past inputs and N past outputs (see Chapter 7). You can find the system function by taking the z-transform of both sides of this equation and using linearity and the delay theorem, form the ratio of $Y(z)/X(z)$ to get

$$H(z)=\frac{Y(z)}{X(z)}=\frac{\sum_{k=0}^{M} b_k z^{-k}}{\sum_{k=0}^{N} a_k z^{-k}}$$

This works as a result of the convolution theorem (see Figure 14-7),
$y[n] = x[n] * h[n] \Rightarrow Y(z) = X(z)H(z)$ or $H(z) = Y(z)/X(z)$.

$$\mathcal{Z}\left\{\sum_{k=0}^{N} a_k y[n-k]\right\} = \underbrace{\sum_{k=0}^{N} a_k z^{-k} Y(z) = \sum_{k=0}^{M} b_k z^{-k} X(z)}_{\text{Form the ratio } \frac{Y(z)}{X(z)}, \text{ done!}} = \mathcal{Z}\left\{\sum_{k=0}^{M} b_k x[n-k]\right\}$$

After you solve for $H(z)$, you can go almost anywhere:

- $\mathcal{Z}^{-1}\{H(z)\}$ gives you the impulse response, $h[n]$.
- $\mathcal{Z}^{-1}\{\mathcal{Z}\{u[n]\} \cdot H(z)\} = \mathcal{Z}^{-1}\{H(z)/(1-z^{-1})\}$ reveals the step response.
- $\mathcal{Z}^{-1}\{X(z)H(z)\}$ provides the output $y[n]$ in response to input $x[n]$.

Applying the convolution theorem

The convolution theorem gives you some powerful analysis capabilities. When you write $Y(z) = X(z)H(z)$, the door opens to the time-domain, _z_-domain, and even frequency-domain results for the input, output, and the system itself. You can find $Y(z)$, $y[n]$ (via the inverse _z_-transform), and $Y(e^{j\hat{\omega}}) = Y(z)\big|_{z=e^{j\hat{\omega}}}$. You can also find $X(z)$ if you're given $Y(z)$ first as well as $H(z)$, because $X(z) = Y(z)/H(z)$.

In some cases, you may have the time-domain quantities $y[n]$ and/or $h[n]$ instead, but getting to the _z_-domain is no problem because $Y(z) = \mathcal{Z}\{y[n]\}$ and $H(z) = \mathcal{Z}\{h[n]\}$. Or, given $X(z)$ and $Y(z)$, you can discover the system function $H(z)$ with a starting point of the time-domain quantities.

Example 14-9: You're given the system input $x[n] = 4(1/2)^n u[n]$ and LTI system function

$$H(z) = \frac{1 + \frac{1}{3}z^{-1}}{1 - \frac{1}{4}z^{-1}}, \text{ ROC: } |z| > \frac{1}{4}$$

To find the system output $y[n] = x[n] * h[n]$, the most direct approach is to find $Y(z) = \mathcal{Z}\{x[n]\} \cdot H(z)$ and then calculate the inverse transform to get $y[n] = \mathcal{Z}^{-1}\{Y(z)\}$. Using transform pair Line 3 from Figure 14-8, you can write

$$Y(z) = \frac{4}{1 - \frac{1}{2}z^{-1}} \cdot \frac{1 + \frac{1}{3}z^{-1}}{1 - \frac{1}{4}z^{-1}} = \frac{4\left(1 + \frac{1}{3}z^{-1}\right)}{\left(1 - \frac{1}{2}z^{-1}\right)\left(1 - \frac{1}{4}z^{-1}\right)} = \frac{R_1}{1 - \frac{1}{2}z^{-1}} + \frac{R_2}{1 - \frac{1}{4}z^{-1}}$$

Solve for the coefficients:

$$R_1 = \left.\frac{4\left(1 + \frac{1}{3}z^{-1}\right)}{1 - \frac{1}{4}z^{-1}}\right|_{z^{-1}=2} = \frac{4\left(1 + 2/3\right)}{1 - 1/2} = \frac{40}{3} = 13.333$$

$$R_1 = \left.\frac{4\left(1 + \frac{1}{3}z^{-1}\right)}{1 - \frac{1}{2}z^{-1}}\right|_{z^{-1}=4} = \frac{4\left(1 + 4/3\right)}{1 - 2} = -\frac{28}{3} = -9.333$$

By calculating the inverse z-transform term by term, you get this equation:

$$y[n] = \frac{40}{3}\left(\tfrac{1}{2}\right)^n u[n] - \frac{28}{3}\left(\tfrac{1}{4}\right)^n u[n].$$

Check your results in PyLab or Maxima to verify that your hand calculations are correct.

Finding the frequency response with pole-zero geometry

The poles and zeros of a system function control the frequency response $H\left(e^{j\hat{\omega}}\right)$. Consider the following development:

$$H\left(e^{j\hat{\omega}}\right) = \left.\frac{\sum\limits_{k=0}^{M} b_k e^{j\hat{\omega}k}}{\sum\limits_{k=0}^{N} a_k e^{j\hat{\omega}k}}\right|_{z=e^{j\hat{\omega}}} = \left.\frac{e^{j\hat{\omega}N}}{e^{j\hat{\omega}M}} \cdot \frac{\prod\limits_{k=1}^{M}(z - z_k)}{\prod\limits_{k=1}^{N}(z - p_k)}\right|_{z=e^{j\hat{\omega}}} = \frac{\prod\limits_{k=1}^{M}\vec{V}_{zk}\left(e^{j\hat{\omega}}\right)}{\prod\limits_{k=1}^{N}\vec{V}_{pk}\left(e^{j\hat{\omega}}\right)}$$

where p_k and z_k are the poles and zeros of $H\left(e^{j\hat{\omega}}\right)$ and $\vec{V}_{pk}\left(e^{j\hat{\omega}}\right) = e^{j\hat{\omega}} - p_k$, and $\vec{V}_{zk}\left(e^{j\hat{\omega}}\right) = e^{j\hat{\omega}} - z_k$ are the frequency response contributions of each pole and zero. The vectors $\vec{V}_{zk}$ and $\vec{V}_{pk}$ point from the zero and pole locations to the point $e^{j\hat{\omega}}$ on the unit circle. This gives you a connection between the pole-zero geometry and the frequency response. In particular, the ratio of vector magnitudes, $\left|\vec{V}_{z1}\left(e^{j\hat{\omega}}\right)\cdots\vec{V}_{zM}\left(e^{j\hat{\omega}}\right)\right| / \left|\vec{V}_{p1}\left(e^{j\hat{\omega}}\right)\cdots\vec{V}_{pN}\left(e^{j\hat{\omega}}\right)\right|$, is the magnitude response. The phase response is $\left[\angle\vec{V}_{z1}\left(e^{j\hat{\omega}}\right) + \cdots + \angle\vec{V}_{zM}\left(e^{j\hat{\omega}}\right)\right] - \left[\angle\vec{V}_{p1}\left(e^{j\hat{\omega}}\right) + \cdots + \angle\vec{V}_{pN}\left(e^{j\hat{\omega}}\right)\right].$

Two custom software apps, PZ_Geom and PZ_Tool, at www.dummies.com/extras/signalsandsystems can help you further explore the relationship between frequency response and pole-zero location.

> ✔ **PZ_Geom app:** You can load a predefined system or filter or load custom coefficients and then see the pole-zero geometry with the geometry lines to represent the vectors $\vec{V}_z$ and $\vec{V}_p$ that change as you drag $e^{j\hat{\omega}}$ with the mouse. A marker on the magnitude and frequency response follows along in adjacent plots.

✔ **PZ_Tool app:** You can place and then drag poles and zeros arbitrarily over the *z*-plane to see the magnitude and phase response change interactively as poles and zeros are moved.

Example 14-10: Consider the simple first-order system:

$$H(z) = \frac{1}{1 - az^{-1}} = \frac{z - 0}{z - a} \qquad H\left(e^{j\hat{\omega}}\right) = \frac{e^{j\hat{\omega}} - 0}{e^{j\hat{\omega}} - a} = \frac{\overrightarrow{V_{z1}}(\hat{\omega})}{\overrightarrow{V_{p1}}(\hat{\omega})}$$

This system contains single numerator and denominator vectors, so the geometry is rather simple: $\overrightarrow{V_{z1}}(\hat{\omega})/\overrightarrow{V_{p1}}(\hat{\omega})$. The software app PZ_Geom allows you to see magnitude and angle of the vectors change as $\hat{\omega}$ is changed by dragging the mouse. The result is that you're able to explore the relationship between frequency response and pole-zero location. A screen shot of PZ_geom is shown Figure 14-9.

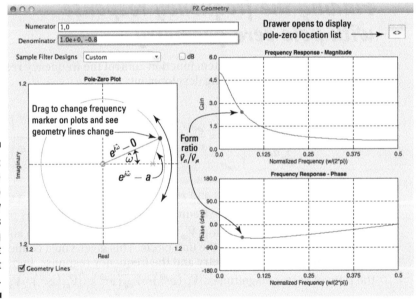

Figure 14-9:
The
pole-zero
geometry
app allows
you to drag
the $e^{j\hat{\omega}}$ point
on the unit
circle.

Chapter 15

Putting It All Together: Analysis and Modeling Across Domains

....

In This Chapter

▶ Looking at the big picture of signals and systems domains

▶ Working with LCCDEs across both continuous- and discrete-time domains

▶ Tackling signals and systems problems beyond the one-liner solution

....

*W*orking across domains is a fact of life as a computer and electronic engineer. Solving real computer and electrical engineering tasks requires you to assimilate the vast array of signals and systems concepts and techniques and apply them in a smart and efficient way. For many people, putting everything together and coming up with real solutions is tough. My goal in this chapter is to help you see how various concepts interact and function in actual situations.

When you receive design requirements for a new project, the system architecture will likely contain a variety of subsystems that cover both continuous- and discrete-time. A powerful subsystem building block is the linear constant coefficient (LCC) differential equation and the LCC difference equation. These subsystems are responsible for implementing what are commonly referred to as *filters*.

Filters allow some signals to pass and they block others. For design and analysis purposes, you can view filters in the time, frequency, and *s*- and *z*-domains (see Chapters 7, 9, 11, 13, and 14 for details on how filters function in the specific domains). Or system design constraints may result in a filter design that's distributed between the continuous and discrete domains.

This chapter explores two common scenarios that require you to work across domains t, f, and s for continuous-time signals and systems and across n, $\hat{\omega}$, and z-domains for discrete time as well as function between continuous and discrete domains as a whole. I point out how to approach these situations and use analysis techniques that draw upon powerful software tool sets.

The systems-level problems in this chapter require you to have an approach in mind before moving forward with specific action steps. Yet arriving at a solution approach gets a bit hairy when real problems are complex. Eventually, experience can guide you to the best approaches.

In the meantime, here are some general problem-solving guidelines:

- ✔ Create a system block diagram from the problem statement.
- ✔ Create detailed subsystem models for critical areas in the design.
- ✔ Start with simple behavioral level models for perceived non-critical subsystems.
- ✔ Through analysis, get a first-cut design of the critical subsystem(s).
- ✔ Verify composite system performance via a high-level analysis model and simulation.

Relating Domains

To successfully apply the various signals and systems concepts as part of practical engineering scenarios, you need to know what analysis tools are available. Figure 15-1 shows the mathematical relationships between time, frequency, and the *s*- and *z*-domains as described in Chapters 1 through 14 of this book. This portrayal offers perspective on a well-rounded study of signals and systems and reveals that you can establish relationships between domains in more than one way.

When working through engineering problems, I recommend staying on a single branch when possible — but sometimes you just can't. Fortunately, if you need to go from the time domain t to the discrete-time frequency domain $\hat{\omega}$, you can hop from t to n and then from n to $\hat{\omega}$ or go from t to f and then f to $\hat{\omega}$.

Some of the interconnects in Figure 15-1 are simply conceptual; they don't hold mathematically under all conditions. For example, Fourier transforms in the limit (see Chapters 9 and 11 for continuous- and discrete-time signals) make it invalid to sample the *s*- or *z*-domains by letting $s = j2\pi f$ and $z = e^{j\hat{\omega}}$. Figure 15-1 tells you that the region of convergence (ROCs) must include the frequency axis for this connection to be valid. Sampling theory offers the links between t and n, but a causal interpolation filter leads to less than perfect reconstruction, revealing that the idea of exact is conceptual only. An approximation is used in real-world systems.

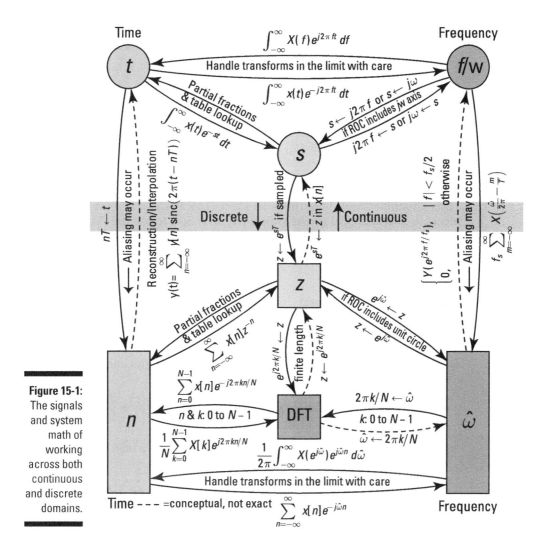

Figure 15-1:
The signals and system math of working across both continuous and discrete domains.

Using PyLab for LCC Differential and Difference Equations

Computer tools play a big part in modern signals and systems analysis and design. LCC differential and difference equations (covered in Chapters 7, 9,

11, 13, and 14) are a fundamental part of simple and highly complex systems. Fortunately, current software tools make it possible to work across domains with these LCC equations without too much pain.

LCC differential and difference equations are completely characterized by the $\{a_k\}$ and $\{b_k\}$ coefficient sets (find details on coefficient sets in Chapter 7). You can use such tools as Pylab with the SciPy `signal` package to design high performance filters, particularly in the discrete-time domain. The filter design functions of `signal` give you the $\{a_k\}$ and $\{b_k\}$ coefficients in response to the design requirements you input. You can then use the filter designs in the simulation of larger systems.

Throughout this book, I use the open-source tool PyLab and its various pieces. In this section, I provide information on how to use the PyLab tools to design and analyze LCC differential and difference equations systems across the continuous- and discrete-time domains.

Continuous time

Three representations of the LCC differential equation system are the time, frequency, and *s*-domains, and the same coefficient sets, $\{b_k\}$ and $\{a_k\}$, exist in all three representations. Here are the corresponding input and output relationships in these domains:

✔ Time domain (from differential equation): $\displaystyle\sum_{k=0}^{N} a_k \frac{d^k y(t)}{dt^k} = \sum_{k=0}^{M} b_k \frac{d^k x(t)}{dt^k}$

✔ Time domain (from impulse response): $\displaystyle y(t) = \int_{-\infty}^{\infty} x(\lambda) h(t-\lambda) d\lambda$

✔ Frequency domain: $Y(f) = X(f)H(f)$, where $\displaystyle H(f) = \frac{\sum_{k=0}^{M} b_k (j2\pi f)^k}{\sum_{k=0}^{N} a_k (j2\pi f)^k}$

✔ *s*-domain: $Y(s) = X(s)H(s)$, where $\displaystyle H(s) = \frac{\sum_{k=0}^{M} b_k s^k}{\sum_{k=0}^{N} a_k s^k}$

In the second line of the differential equation, the impulse response, $h(t)$, along with the convolution integral of Chapter 5 produce the output, $y(t)$, from the input, $x(t)$. In the third line, the convolution theorem for Fourier transforms (Chapter 9) produces the output spectrum, $Y(f)$, as the product of the input spectrum, $X(f)$, and the frequency response, $H(f)$ — which is the Fourier transform of the impulse response. In the fourth line, the convolution theorem for Laplace transforms (Chapter 13) produces the s-domain output, $Y(s)$, as the product of the input, $X(s)$, and the system function, $H(s)$ — which is the Laplace transform of the impulse response.

Figure 15-2 highlights the key functions in PyLab and the `ssd.py` code module you can use to work across continuous-time domains. Remember, these functions are at the top level. You can integrate many lower-level functions (such as math, array manipulation, and plotting library functions) with these top-level functions to carry out specific analysis tasks.

	PyLab with `scipy.signal` and the module `ssd` from the book	**Purpose / Description**
Time domain	`t, y, x_state=signal.lsim ((b,a),x,t)`	Numerical solution to differential equation.
	Recipe: `t =arange(0, tstop, tstep)` `x = 1*ones(len(t))` `t, y, x_state=signal.lsim ((b,a),x,t)` `plot(t, y)`	Find system output with input $u(t)$ via simulation. Other inputs are possible. You may also define the input x of your choice.
s-domain	`ssd.splane(b, a)`	s-plane pole-zero plot.
	`R ,p,K = signal.residue (b, a)`	Partial fraction expand $H(s)$ to get exact $h(t)$.
	`R, p, K= signal.residue (b, signal. convolve(a,[1,0]))`	Find exact step response with s included in the denominator.
Frequency domain	`w, H = signal freqs(b, a,2*pi*f)`	Frequency response given array of frequency values.
	Recipe: `f = linspace(f1,f2,Npt) or` `f = logspace(logf1,logf2,Npt)` `w, H = signal.freqs(b, a, 2*pi*f)` `plot(f, abs(H)) or angle(H) or` `semilogx(f, 20*log10(abs(H))) or` `semilogx(f, angle(H)*180/pi)`	Find linear or log frequency plots as magnitude, magnitude in dB, and phase in degrees or radians. May choose f or $\omega = 2\pi f$ as frequency axis.

Figure 15-2: Working with LCC differential equations across domains using PyLab.

Here's what you can find in Figure 15-2:

✔ The time-domain rows show a recipe to solve the differential equation numerically by using `signal.lsim((b,a),x,t)` for a step function input. The arrays `b` and `a` correspond to the coefficient sets $\{b_k\}$ and $\{a_k\}$. Input signals of your own choosing are possible, too. The time-domain simulation allows you to characterize a system's behavior at the actual waveform level.

✔ In the _s_-domain rows, find the pole-zero plot of the system function $H(s)$ by using `ssd.splane(b,a)`. Also find out how to solve the partial fraction expansion (PFE) of $H(s)$ and $H(s)/s$ to get a mathematical representation of the impulse response or the step response.

✔ The frequency-domain section offers a recipe for plotting the frequency response of the system by using `signal.freqs(b,a,2*pi*f)`. Options include a linear or log frequency axis, the frequency response magnitude, and the phase response in degrees.

Discrete time

Just like for differential equation systems described in the previous section, the LCC difference equation system has three representations: time, frequency, and _z_-domains, and the same coefficient sets, $\{b_k\}$ and $\{a_k\}$, exist in all three representations. Here are the corresponding input and output relationships in these domains:

✔ Time domain (from difference equation): $\displaystyle\sum_{k=0}^{N}a_k y[n-k]=\sum_{k=0}^{M}b_k x[n-k]$

✔ Time domain (from impulse response): $\displaystyle y[n]=\sum_{k=-\infty}^{\infty}x[k]h[n-k]$

✔ Frequency domain: $Y\!\left(e^{j\hat\omega}\right)=X\!\left(e^{j\hat\omega}\right)H\!\left(e^{j\hat\omega}\right)$, where $H\!\left(e^{j\hat\omega}\right)=\dfrac{\displaystyle\sum_{k=0}^{M}b_k e^{j\hat\omega k}}{\displaystyle\sum_{k=0}^{N}a_k e^{j\hat\omega k}}$

✔ _z_-domain: $Y(z)=X(z)H(z)$, where $H(z)=\dfrac{\displaystyle\sum_{k=0}^{M}b_k z^{-k}}{\displaystyle\sum_{k=0}^{N}a_k z^{-k}}$

In the second line of the difference equation, the impulse response, $h[n]$, along with the convolution sum produce the output, $y[n]$, form the input, $x[n]$. In the third line, the convolution theorem for Fourier transforms (Chapter 9) produces the output spectrum, $Y\left(e^{j\hat{\omega}}\right)$, as the product of the input spectrum, $X\left(e^{j\hat{\omega}}\right)$, and the frequency response, $H\left(e^{j\hat{\omega}}\right)$, which is the discrete-time Fourier transform of the impulse response. In the fourth line, the convolution theorem for z-transforms (Chapter 14) produces the z-domain output, $Y(z)$, as the product of the input, $X(z)$, and the system function, $H(z)$, which is the z-transform of the impulse response.

Figure 15-3 highlights the key functions in PyLab and the custom `ssd.py` code module you can use to work across discrete-time domains.

	PyLab with `scipy.signal` and the module `ssd` from the book	Purpose / Description
Time domain	`y =signal.lfilter(b, a, x)`	Solve the difference equation by recursion.

	Recipe:
	`n = arange(0, nstop)`
	`x = 1* ones(len(t))`
	`y = signal.lfilter(b, a, x)`
	`stem(n, y)` or use `plot(n, y)`
		Simulate system with input $u[n]$. Other inputs are possible. You may also define the input x of your choice.

z-domain	`ssd.zplane(b, a)`	z-plane pole-zero plot.
	`R, p, K= signal residuez (b, a)`	Partial fraction expand $H(z)$ to get exact $h[n]$.
	`R, p, K= signal.residuez (b, signal.` `convolve(a,[1,-1]))`	Find exact step response with $1-z^{-1}$ included in the denominator.
Frequency domain	`w, H = signal freqz(b,a,2*pi*f)`	Frequency response given array of frequency values.

	Recipe:
	`f = linspace(f1, f2, Npt)` or
	`f = logspace(log f1,log f2,Npt)`
	`w, H = signal freqz(b,a,2*pi*f)`
	`plot(f, abs(H))` or `angle(H)`
	`semilogx(f, 20 * log10(abs(H)))`
	`semilogx(f, angle(H)*180 / pi)`
		Find linear or log frequency plots as magnitude magnitude dB, and phase in degrees or radians. May plot using $\hat{\omega}$ or $f = \hat{\omega} / (2\pi)$. Nominally $f_1 = 0$ and $f_2 = 0.5$.

Figure 15-3: Working with LCC difference equations across domains by using PyLab.

Figure 15-3 parallels the content of Figure 15-2, so I simply point out differences in the content here:

- ✔ In the time domain rows, you solve the difference equation exactly, using `signal.lfilter(b,a,x)`.

- ✔ In the *z*-domain rows, you can find the pole-zero plot of the system function $H(z)$, using `ssd.zplane(b,a)`, and the partial fraction expansion, using `signal.residuez` instead of `signal.residue`.

- ✔ The frequency domain rows show you how to find the frequency response of a discrete-time system with `signal.freqz(b,a,2*pi*f)`, where *f* is the frequency variable $\hat{\omega}$ normalized by 2π.

Mashing Domains in Real-World Cases

This section contains two examples that show how analysis and modeling across domains works. The examples include the time, frequency, and *s*- and *z*-domains. The first problem sticks to continuous-time, but the second one works across continuous- and discrete-time systems. A third example problem is available at www.dummies.com/extras/signalsandsystems.

Problem 1: Analog filter design with a twist

You're given the task of designing an analog (continuous-time) filter to meet the amplitude response specifications shown in Figure 15-4. You also need to find the filter step response, determine the value of the peak overshoot, and time where the peak overshoot occurs.

The objective of the filter design is for the frequency response magnitude in dB $(20\log_{10}|H(f)|)$ to pass through the unshaded region of the figure as frequency increases. The design requirements reduce to the passband and stopband critical frequencies f_p and f_s Hz and the passband and stopband attenuation levels A_p and A_s dB.

Additionally, the response characteristic is to be *Butterworth*, which means that the filter magnitude response and system function take this form:

$$\left|H_{\text{BU}}(f)\right| = \frac{1}{\sqrt{1+\left(f/f_c\right)^{2N}}} \text{ and } H_{\text{BU}}(s) = \frac{1}{\prod_{k=1}^{N}(s-p_k)}$$

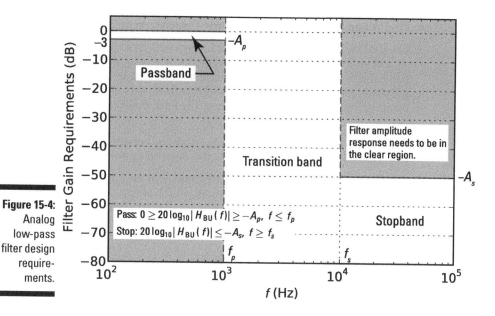

Figure 15-4: Analog low-pass filter design requirements.

The figure shows a filter gain requirements plot with:
- y-axis: Filter Gain Requirements (dB) with values 0, −3, −10, −20, −30, −40, −50, −60, −70, −80
- x-axis: f (Hz) with values 10^2, 10^3, 10^4, 10^5
- Labels: $-A_p$, Passband, Transition band, Stopband
- "Filter amplitude response needs to be in the clear region."
- $-A_s$
- Pass: $0 \geq 20\log_{10}|H_{BU}(f)| \geq -A_p,\ f \leq f_p$
- Stop: $20\log_{10}|H_{BU}(f)| \leq -A_s,\ f \geq f_s$
- f_p, f_s

Here, N is the filter order, f_c is the passband 3 dB cutoff frequency of the filter, and the poles, located on a semicircle is the left-half s-plane, are given by $p_k = 2\pi f_c \exp\left[j\pi(0.5) - (2k-1)/(2N) \right],\ k = 1,\dots,N$.

This problem requires you to work in the frequency domain, the time domain, and perhaps the s-domain, depending on the solution approach you choose.

From the requirements, the filter frequency response has unity gain (0 dB) in the passband. The step response (a time-domain characterization) of the Butterworth filter is known to overshoot unity before finally settling to unity as $t \to \infty$.

To design the filter, you can use one of two approaches:

1. **Work a solution by hand, using the Butterworth magnitude frequency response $|H_{BU}(f)|$ and the system function, $H_{BU}(s)$.**
2. **Use the filter design capabilities of the SciPy signal package.**

I walk you through option two in the next section.

Finding the filter order and 3 dB cutoff frequency

Follow these steps to design the filter by using Python and SciPy to do the actual number crunching:

1. **Find N and f_c to meet the magnitude response requirements.**

   Use the SciPy function N,wc=signal.buttord(wp,ws,Ap,As,analog= 1) and enter the filter design requirements, where wp and ws are the pass-band and stopband critical frequencies in rad/s and Ap and As are the passband and stopband attenuation levels (both sets of numbers come from Figure 15-4). The function returns the filter order N and the cutoff frequency wc in rad/s.

2. **Synthesize the filter — find the $\{b_k\}$ and $\{a_k\}$ coefficients of the LCC differential equation that realizes the desired system.**

   If finding circuit elements is the end game, you may go there immediately, using circuit synthesis formulas, which aren't described in this book. Call the SciPy function b,a=signal.butter(N,wc,analog=1) with the filter order and the cutoff frequency, and it returns the filter coefficients in arrays b and a.

3. **Find the step response in exact mathematical form or via simulation.**

Here's how to use the Python tools with the given design requirements and then check the work by plotting the frequency response as an overlay to Figure 15-4. *Note:* You can do the same thing in MATLAB with almost the same syntax.

```
In [379]: N,wc =signal.buttord(2*pi*1e3,2*pi*10e3,3.0,
                     50,analog=1) # find filter order N
In [380]: N # filter order
Out[380]: 3
In [381]: wc   # cutoff freq in rad/s
Out[381]: 9222.4701630595955
In [382]: b,a = signal.butter(N,wc,analog=1) # get coeffs.
In [383]: b
Out[383]: array([7.84407571e+11+0.j])
In [384]: real(a)
Out[384]: array([1.00000000e+00, 1.84449403e+04,
          1.70107912e+08,7.84407571e+11])
```

The results of Line [379] tell you that the required filter order is $N = 3$ and the required filter cutoff frequency is 9,222.5 rad/s ($2\pi f_c$). The filter coefficient sets are also included in the results.

I use the real() function to safely display the real part of the coefficients array a because I know the coefficients are real. How? The poles, denominator roots of $H_{BU}(s)$, are real or occur in complex conjugate pairs, ensuring that the denominator polynomial has real coefficients when multiplied out. The very small imaginary parts, which I want to ignore, are due to numerical precision errors.

Checking the final design frequency response

To check the design, use the frequency response recipe from Figure 15-2.

```
In [386]: f = logspace(2,5,500) # log frequency axis
In [387]: w,H = signal.freqs(b,a,2*pi*f)
In [388]: semilogx(f,20*log10(abs(H)),'g')
```

Figure 15-5 shows the plot of the final design magnitude response along with the original design requirements. I use the frequency-domain recipe from Figure 15-2 to create this design.

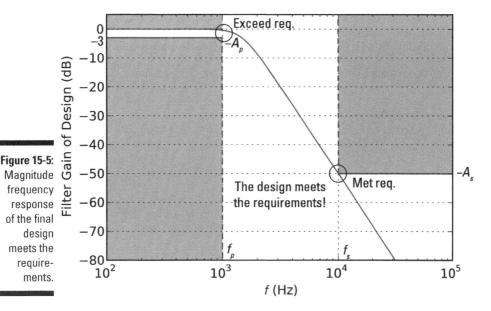

Figure 15-5: Magnitude frequency response of the final design meets the requirements.

Finding the step response from the filter coefficients

The most elegant approach to finding the step response from the filter coefficients is to find $\mathcal{L}^{-1}\{H(s)/s\}$. The s-domain section of Figure 15-2 tells you how to complete the partial fraction expansion (PFE) numerically. You have the coefficient arrays for $H(s)$, so all you need to do is multiply the denominator polynomial by s. You can do this by hand or you can use a relationship between polynomial coefficients and sequence convolution.

When you multiply two polynomials, the coefficient arrays for each polynomial are convolved, as in sequence convolution (see Chapter 6).

Here, I work through the problem, using `signal.convolve` to perform polynomial multiplication in the denominator. To convince you that this really works, consider multiplication of the following two polynomials:

$$\left(x^2 + x + 1\right)\cdot\left(x+1\right) = x^3 + 2x^2 + 2x + 1$$

If you convolve the coefficients sets [1, 1, 1] and [1, 1] as arrays in Python, you get this output:

```
In [418]: signal.convolve([1,1,1],[1,1])
Out[418]: array([1, 2, 2, 1])
```

This agrees with the hand calculation. To find the PFE, plug the coefficients arrays b and `convolve(a,[1,0])` into R,p,K = `residue(b,a)`. The coefficients [1, 0] correspond to the s-domain polynomial $s + 0$.

```
In [420]: R,p,K = signal.residue(b,signal.
              convolve([1,0],a))
In [421]: R #(residues) scratch tiny numerical errors
Out[421]:
array([ 1.0000e+00 +2.3343e-16j,   # residue 0, imag part 0
       -1.0000e+00 +1.0695e-15j,   # residue 1, imag part 0
        1.08935e-15 -5.7735e-01j,  # residue 2, real part 0
        1.6081e-15 +5.7735e-01j])  # residue 3, real part 0
In [422]: p #(poles)
Out[422]:
array([ 0.0000 +0.0000e+00j,       # pole 0
       -9222.4702 -1.5454e-12j,    # pole 1, imag part 0
       -4611.2351 -7.9869e+03j,    # pole 2
       -4611.2351 +7.9869e+03j])   # pole 3
In [423]: K #(from long division)
Out[423]: array([ 0.+0.j]) # proper rational, so no terms
```

I crossed out parts of Lines Out[421] and Out[422] to indicate that in absence of small numerical errors, the true values should be zero.

You have four poles: two real and one complex conjugate pair — a bit of a mess to work through, but it's doable. Refer to the transform pair $e^{-at}u(t) \xleftarrow{\ \mathcal{L}\ } 1/(s+a)$ (see Chapter 13) to calculate the inverse transform for all four terms.

For the conjugate poles, the residues are also conjugates. This property always holds.

You can write the inverse transform of the conjugate pole terms as sines and cosines, using Euler's formula and the cancellation of the imaginary parts in front of the cosine and real parts in front of the sine: $R_2 e^{p_2 t} + R_2^{*} e^{p_2 t} = 2\left[\operatorname{Re}\{R_2\}e^{\sigma_2 t}\cos(\omega_2 t) + \operatorname{Im}\{R_2\}e^{\sigma_2 t}\sin(\omega_2 t)\right]$, where $p_2 = \sigma_2 + j\omega_2$. Putting it all

together, you get $y_{step}(t) = u(t) - e^{-9{,}222.47t}u(t) - 2 \cdot 0.5774e^{-4{,}611.74t}\sin(7{,}986.89t)u(t)$.
Having this form is nice, but you still need to find the function maximum for $t > 0$ and the maximum location. To do this, plot the function and observe the maximum.

A more direct approach is to use simulation via `signal.lsim` and the time-domain recipe from Figure 15-2. The system input is a step, so the simulation output will be the step response of Figure 15-6. From the simulated step response, you can calculate the peak overshoot numerically and see it in a plot. The IPython command line code is

```
In [425]: t = arange(0,0.002,1e-6) # step less than
                smallest time constant
In [426]: t,ys,x_state = signal.lsim((b,a),ones(len(t)),t)
In [428]: plot(t*1e3,ys)
```

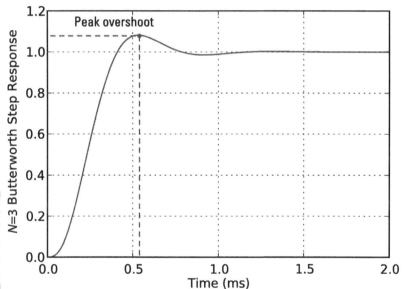

Figure 15-6: Third-order Butterworth filter step response.

Using the time array `t` and the step response array `ys`, you can use the `max()` and `find()` functions to complete the task:

```
In [436]: max(real(ys)) # real to clear num. errors
Out[436]: 1.0814651457627822 # peak overshoot is8.14%
In [437]: find(real(ys)== max(real(ys)))
Out[437]: array([534]) # find peak to be at index 534
In [439]: t[534]*1e3 # time at index 534 in ms
Out[439]: 0.5339
```

Problem 2: Solving the DAC ZOH droop problem in the z-domain

The zero-order-hold (ZOH), that's inherent in many digital-to-analog convert-ers (DACs), holds the analog output constant between samples. The action of the ZOH introduces *frequency droop,* a roll off of the effective DAC frequency response on the frequency interval zero to one-half the sampling rate f_s, in reconstructing $y(t)$ from $y[n]$. Two possible responses are to

✔ Apply an inverse sinc function shaping filter in the continuous-time domain.

✔ Correct for the droop before the signal emerges from the DAC.

The system block diagram is shown in Figure 15-7.

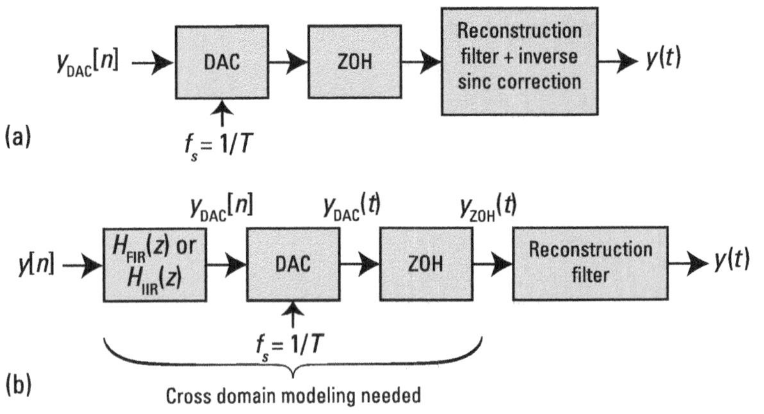

Figure 15-7:
DAC ZOH
droop com-
pensation
system block
diagram:
pure analog
filter solu-
tion (a)
and hybrid
discrete-time
and analog
solution (b).

Imagine that a senior engineer asks you to investigate the effectiveness of the simple infinite impulse response (IIR) and finite impulse response (FIR) digi-tal filters as a way to mitigate ZOH frequency droop. You need to verify just how well these filters really work. The filter system functions are

$$H_{FIR}(z) = \frac{1}{16}\left(-1 + 18z^{-1} - z^{-2}\right)$$

$$H_{IIR}(z) = \frac{9}{8 + z^{-1}}$$

To solve this problem, you need to use the frequency-domain relationship from the discrete- to continuous-time domains. As revealed in Figure 15-1, the relationship, relative to the notation of Figure 15-7, is $Y_{DAC}(f) = Y_{DAC}\left(e^{j2\pi f/f_s}\right)$ for $0 \le f \le f_s/2$. You can assume that the analog reconstruction filter removes signal spectra beyond $f_s/2$.

The frequency response of interest turns out to be the cascade of $H_{\text{FIR/IIR}}\left(e^{j\hat{\omega}}\right)$ and $H_{\text{ZOH}}(\omega)$. Follow these steps to justify this outcome:

1. **Let $Y\left(e^{j\hat{\omega}}\right)=1$. From the convolution theorem for frequency spectra in the discrete-time domain, get $Y_{\text{DAC}}\left(e^{j\hat{\omega}}\right)=Y\left(e^{j\hat{\omega}}\right)\cdot H_{\text{FIR/IIR}}\left(e^{j\hat{\omega}}\right)=1\cdot H_{\text{FIR/IIR}}\left(e^{j\hat{\omega}}\right).$**

2. **Use the discrete to continuous spectra relationship to discover that the output side of the DAC is $Y_{\text{DAC}}\left(e^{j2\pi f/f_s}\right)=H_{\text{FIR/IIR}}\left(e^{j2\pi f/f_s}\right).$**

3. **Use the convolution theorem for frequency spectra in the continuous-time domain to push the DAC output spectra through the ZOH filter:**

$$Y_{\text{ZOH}}(f)=Y_{\text{DAC}}(f)\cdot H_{\text{ZOH}}(f)=H_{\text{FIR/IIR}}\left(e^{j2\pi f/f_s}\right)\cdot H_{\text{ZOH}}(f)=H_{\text{cascade}}(f)$$

The cascade result is now established.

To view the equivalent frequency response for this problem in the discrete-time domain, you just need to change variables according to the sampling theory: $\hat{\omega}=\omega T$ or $\omega=\hat{\omega}/T$, where $\omega=2\pi f$ and $T=1/f_s$. Rearranging the variables in the cascade result viewed from the discrete-time domain perspective is $H_{\text{cascade}}\left(e^{j\hat{\omega}}\right)=H_{\text{FIR/IIR}}\left(e^{j\hat{\omega}}\right)\cdot H_{\text{ZOH}}(\hat{\omega}/T)$. The ZOH frequency response is

$$H_{\text{ZOH}}(\omega)=\frac{2\sin\left(\omega T/2\right)}{\omega}e^{-\omega T/2}$$

Putting the pieces together and considering only the magnitude response reveals this equation:

$$\left|H_{\text{cascade}}(\hat{\omega})\right|=\left|H_{\text{FIR/IIR}}\left(e^{j\hat{\omega}}\right)\right|\cdot T\underbrace{\left|\frac{\sin\left(\hat{\omega}/2\right)}{\hat{\omega}/2}\right|}_{\text{sinc}[\hat{\omega}/(2\pi)]}$$

To verify the performance, evaluate the sinc function and the FIR responses by using the SciPy `signal.freqz()` function approach of the frequency domain recipe in Figure 15-3. Check out the results in Figure 15-8.

```
In [393]: w = linspace(0,pi,400)
In [394]: H_ZOH_T = sinc(w/(2*pi))
In [395]: w,H_FIR = signal.freqz(array([-1, 18,-
          1])/16.,1,w)
In [396]: w,H_IIR = signal.freqz([-9/8.],[1, 1/8.],w)
In [402]: plot(w/(2*pi),20*log10(abs(H_ZOH_T)))
In [403]: # other plot cammand lines similar
In [412]: plot(w/(2*pi),20*log10(abs(H_FIR)*abs(H_ZOH_T)))
```

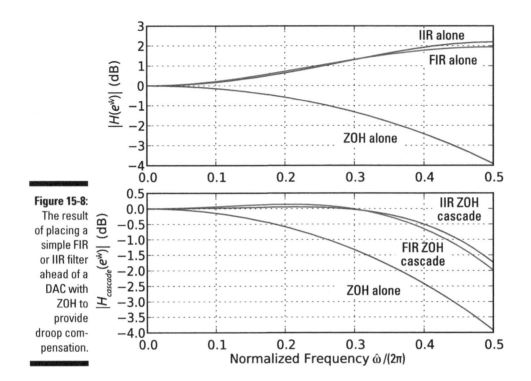

Figure 15-8:
The result
of placing a
simple FIR
or IIR filter
ahead of a
DAC with
ZOH to
provide
droop com-
pensation.

I think these results are quite impressive for such simple correction filters. The goal is to get flatness that's near 0 dB from 0 to π rad/sample (0 to 0.5 normalized). The response is flat to within 0.5 dB out to 0.4 rad/sample for the IIR filter; it's a little worse for the FIR filter.

Find an additional problem that shows you how to take the RC low-pass filter to the *z*-domain at www.dummies.com/extras/signalsandsystems.

Part V

The Part of Tens

Enjoy an additional signals and systems Part of Tens chapter online at www.dummies.com/extras/signalsandsystems.

In this part . . .

✔ As you ride the signals and systems train, avoid derailment by reading about ten common mistakes when solving problems.

✔ Rock your engineering class with confidence when you take in the sights and sounds of ten properties you never want to forget.

Chapter 16

More Than Ten Common Mistakes to Avoid When Solving Problems

*O*ne of my math professors from graduate school made an offhand comment one afternoon that has stuck with me. He said, "The obvious is often difficult, if not impossible, to prove." What does this have to do with making mistakes on signals and systems problems, you ask? Well, here's a not-so-hidden secret: Professors often put a few problems on each exam thinking the answers are obvious. Imagine the surprise when almost every student misses the gimmes.

Don't be one of the students who misses the obvious. Slow down enough to think through solutions, and make sure your fundamental understanding of the core material is at least as good as your ability to work through detailed problems.

In this chapter, I point out more than ten common mistakes students make when trying to solve problems, and I tell you how to avoid them.

Miscalculating the Folding Frequency

In sampling theory, the alias frequencies fold over $f_s/2$ (known as the *folding frequency*), where f_s is the sampling frequency in hertz. An error in the calculation of the principle alias or the alias frequency results when you use the folding frequency improperly.

Consider $f_s = 10$ Hz and the calculation of the principle alias frequency relative to $f = 7$ Hz. You may quickly reason that the principle alias frequency is $7 - f_s/2 = 7 - 5 = 2$ Hz, because 7 folds about 5 to produce 7. This is wrong! This isn't the folding frequency interpretation. Because 7 Hz is 2 Hz above 5 Hz, the corresponding folded frequency is 2 Hz below 5 Hz, or 3 Hz.

This same concept can be misunderstood when you're given a principle alias frequency and need to find the alias frequency on the interval $[f_s/2, f_s]$. With 10 Hz, suppose the principle alias is $f_0 = 4$ Hz. The nearest alias frequency isn't $f_s/2 + 4 = 5 + 4 = 9$ Hz; the principle alias sits 1 Hz below the folding frequency so the corresponding alias frequency is 1 Hz above the folding frequency or 6 Hz.

Finding alias frequencies is best done with respect to integer multiples of the sampling frequency. The principle alias frequency band is $[0, f_s/2]$. If $f_0 \in [0, f_s/2]$ (principle alias frequency), then you find other alias frequencies as $f = |k \cdot f_s \pm f_0|$, where k is a nonnegative integer. When you're given f and need to find the principle alias f_0, first find k so f is within $\pm f_s/2$ of kf_s.

Getting Confused about Causality

In a causal system, only the *present* and *past* values of the input can form the *present* output. When given a system input/output relationship, such as $y(t) = 5x(t-2) + u(t+5)$, don't be thrown off by the $u(t+5)$. The system is causal because the input two seconds in the past forms the present value of the output.

The system also contains a time-varying bias that turns on at $t = -5$. This bias is part of the system and isn't related to the input $x(t)$.

Extra credit: Is this system time-invariant? No, the system contains the time-varying bias term.

Plotting Errors in Sinusoid Amplitude Spectra

Plotting the two-sided amplitude spectra of sinusoidal signals seems so easy, but students too frequently ignore or forget about the 1/2-amplitude scaling factor from Euler's formula.

Consider a signal composed of a single sinusoid (in a real problem, you may have more sinsuoids) and a direct current component (DC): $x(t) = B + A\cos(2\pi f_0 t + \phi)$. Create the two-sided line spectra by expanding the cosine and using Euler's formula: $\cos(2\pi f_0 t) = \left(e^{j2\pi f_0 t} + e^{-j2\pi f_0 t}\right)/2$. Applying the expansion to $x(t)$, you get $x(t) = B + A/2 \cdot e^{j\phi} \cdot e^{j2\pi f_0 t} + A/2 \cdot e^{-j\phi} \cdot e^{-j2\pi f_0 t}$.

There's a spectral line of amplitude $\left|A/2 \cdot e^{j\phi}\right| = A/2$ at f_0 due to the positive frequency complex sinusoid, a spectral line of amplitude $\left|A/2 \cdot e^{-j\phi}\right| = A/2$ at $-f_0$ due to the negative frequency complex sinusoid, and a spectral line of amplitude $|B|$ (absolute value in case the DC component is negative) at 0 Hz (DC). Did you notice the 2 in $A/2$ for the spectral lines at $\pm f_0$? Don't forget it in your plot!

Missing Your Arctan Angle

Stumbling with angle calculations on basic scientific calculators is an easy mistake to make. Although you may be thinking that this is a first-grade error, carelessness can swoop in unexpected when you're under pressure.

For instance, to find the angle of complex number $z = x + jy$, maybe you start by finding $\arctan(y/x)$, but you need to make note of which quadrant of the complex plane the number is actually in. For Quadrants I and IV, arctan faithfully returns the correct angle (preferably in radians). For a Quadrant II complex number, arctan thinks you're in Quadrant IV, so you need to add $\pm\pi$ to the arctan result. For a Quadrant III complex number, arctan thinks you're in Quadrant I, so you must add $\pm\pi$ to the arctan result. The $\pm$ is your choice depending on how you like your angle.

Being Unfamiliar with Calculator Functions

When manipulating complex numbers on your calculator, I have two recommendations to help you avoid making careless mistakes:

- ✔ Be aware of the angle mode you've set for your device. Use radians mode for all your angle calculations, and be consistent. If you need a final answer in degrees, do that at the end by multiplying by $180/\pi$.

- ✔ Know how to use your calculator. You may be tempted to borrow a friend's super calculator but fail to spend any time using it until you're under the pressure of a quiz or exam. Bad idea.

Foregoing the Return to LCCDE

When you want to find the linear constant coefficient (LCC) difference or differential equation starting from the system function, you may end up swapping the numerator and denominator polynomials by being careless.

The case in point here is for the z-domain. Say you're given the following equation and asked to find the difference equation from $H(z)$:

$$H(z) = \frac{1-2z^{-1}}{1-3/4z^{-1}} \overset{\text{you know}}{=} \frac{Y(z)}{X(z)}$$

You notice $Y(z)$ across from $1-2z^{-1}$ and $X(z)$ across from $1-3/4z^{-1}$ and may think $y[n] - 2y[n-1] = x[n] - 3/4x[n-1]$. But that approach is wrong. To get back to the difference equation, you need to cross-multiply: $Y(z) \cdot \left(1-3/4z^{-1}\right) = X(z) \cdot \left(1-2z^{-1}\right)$. And then you can correctly write $y[n] - 3/4\,y[n-1] = x[n] - 2x[n-1]$.

Ignoring the Convolution Output Interval

When convolving two functions or two sequences, you need to consider a lot of details. On the heels of forgetting to slow down and take a deep breath, many people forget to first find the convolution output interval from the input signals/sequences x_1 and x_2.

This simple calculation tells you where you're going with your final answer. Without it, you can still get a nice answer, but the support interval may be wrong due to other errors.

Given that $x_1(t)$ has support interval $[t_1, t_2]$ and $x_2(t)$ has support interval $[t_3, t_4]$, the convolution $y(t) = x_1(t) * x_2(t)$ has support interval no greater than $[t_1 + t_3, t_2 + t_4]$. Similar results hold for sequences with t replaced by n.

Forgetting to Reduce the Numerator Order before Partial Fractions

When working with inverse Laplace transforms (ILTs) and inverse z-transforms, you typically deal with a rational function, such as $N(s)/D(s)$ or $N(z)/D(z)$. Before you can begin your partial fraction expansion, make sure the function is proper rational.

In other words, make sure you check the order of the numerator polynomial. Is it one or more less than the order of the denominator? If not, you need to use long division to reduce the numerator order.

The surprise with making this careless error is that you'll get an answer, and you may leave the exam feeling good — until your buddy comments on the need for long division on a problem. Uh oh.

Forgetting about Poles and Zeros from H (z)

When finding the poles and zeros of a finite impulse response (FIR) filter for a problem such as $H(z) = 1 - \frac{3}{4}z^{-1} + \frac{1}{8}z^{-2}$, forgetting about the two poles at $z = 0$ is easy. If you just factor the polynomial as $(1 - 0.25z^{-1})(1 - 0.5z^{-1})$ and plot zeros at $z = 0.25$ and $z = 0.5$, then your solution is wrong.

Find the poles $z = 0$ by switching to positive powers of z:

$$H(z) = (1 - 0.25z^{-1})(1 - 0.5z^{-1}) = \frac{(z - 0.25)(z - 0.5)}{z^2}$$

There, the poles are now visible. The number of poles and zeros is always equal, but some may be at infinity.

Missing Time Delay Theorems

When applying the time delay theorem in the Fourier domain, the time shift theorems apply everywhere that the independent variable occurs. Too often, students apply the theorem partially, so some t or n values are left unmodified. For $z^{-3}/(1 - 0.5z^{-1})$, the inverse z-transform is $(0.5)^n u[n]\big|_{n \to n-3} = (0.5)^{n-3} u[n - 3]$. Notice the time shift in two places!

Disregarding the Action of the Unit Step in Convolution

In both the continuous- and discrete-time convolution, you may need to flip and slide a signal containing a unit step function. The error occurs when you don't carefully consider the action of the unit step function with respect to the integration or sum variable. You may ignore the fact that the flipped and shifted unit step function turns *off* at some point, rather than *on*, as the integration or sum index variable increases. Your integration or sum limits likely depend on the turning off behavior, so the problem solution drives off course with one or more errors.

For example, consider $x(t-\lambda)$ or $x[n-k]$ along the λ- or k-axis. In the continuous-time case, suppose $x(t)=e^{-2t}u(t-4)$ in the convolution integral $y(t)=\int_{-\infty}^{\infty}h(\lambda)x(t-\lambda)d\lambda$. You need to sketch $x(t-\lambda)=e^{-2(t-\lambda)}u[(t-\lambda)-4]=e^{-2(t-\lambda)}u(t-4-\lambda)$ to determine the integration limits. You have to deal with $u(t-4-\lambda)$ in the integrand with respect to the λ-axis.

Remember, t is just a parameter. The step function runs backward due to the $-\lambda$; that is, it turns off at some point. In this case, with t fixed, the step function turns off when $t-4-\lambda<0$ or $\lambda>t-4$. You now know that the upper limit of integration is in part set by $\lambda=t-4$. The same concept applies when evaluating the convolution sum.

Chapter 17

Ten Properties You Never Want to Forget

In This Chapter

▷ Looking at significant properties you can't live without as a signals and systems engineer

▷ Seeing properties side by side between continuous and discrete signals and systems

A big wide world of properties is associated with signals and systems — plenty in the math alone! In this chapter, I present ten of my all-time favorite properties related to signals and systems work.

LTI System Stability

Linear time-invariant (LTI) systems are bounded-input bounded-output (BIBO) stable if the region of convergence (ROC) in the *s*- and *z*-planes includes the $j\omega$-axis and unit circle $e^{j\tilde{\omega}}$, respectively. The *s*-plane applies to continuous-time systems, and the *z*-plane applies to discrete-time systems. But here's the easy part: For causal systems, the property is poles in the left-half *s*-plane and poles inside the unit circle of the *z*-plane.

Convolving Rectangles

The convolution of two identical rectangular-shaped pulses or sequences results in a triangle. The triangle peak is at the integral of the signal or sum of the sequence squared. Figure 17-1 depicts the property graphically.

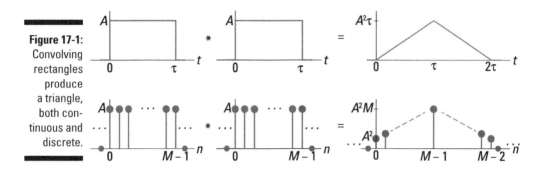

Figure 17-1:
Convolving
rectangles
produce
a triangle,
both con-
tinuous and
discrete.

The Convolution Theorem

The four (linear) convolution theorems are Fourier transform (FT), discrete-time Fourier transform (DTFT), Laplace transform (LT), and z-transform (ZT). *Note:* The discrete-time Fourier transform (DFT) doesn't count here because circular convolution is a bit different from the others in this set.

These four theorems have the same powerful result: Convolution in the time domain can be reduced to multiplication in the respective domains. For x_1 and x_2 signal or impulse response, $y = x_1 * x_2$ becomes $Y( ) = X_1( ) \cdot X_2( )$, where the function arguments may be f, ω, $e^{j\hat{\omega}}$, s, or z.

Frequency Response Magnitude

For the continuous- and discrete-time domains, the frequency response magnitude of an LTI system is related to pole-zero geometry. For continuous-time signals, you work in the s-domain; if the system is stable, you get the frequency response magnitude by evaluating $|H(s)|$ along the $j\omega$-axis. For discrete-time signals, you work in the z-domain; if the system is stable, you get the frequency response magnitude by evaluating $|H(z)|$ around the unit circle as $e^{j\hat{\omega}}$. In both cases, frequency response magnitude nulling occurs if $j\omega$ or $e^{j\hat{\omega}}$ passes near or over a zero, and magnitude response peaking occurs if $j\omega$ or $e^{j\hat{\omega}}$ passes near a pole. The system can't be stable if a pole is on $j\omega$ or $e^{j\hat{\omega}}$.

Convolution with Impulse Functions

When you convolve *anything* with $\delta(t - t_0)$ or $\delta[n - n_0]$, you get that same anything back, but it's shifted by t_0 or n_0. Case in point:

$$e^{-6t}u(t) * \delta(t-2) = e^{-6t}u(t)\Big|_{t \to t-2} = e^{-6(t-2)}u(t-2)$$

$$\left(u[n] - u[n-5]\right) * \delta[n+2] = u[n+2] - u[n+2-5]$$

Spectrum at DC

The direct current (DC), or average value, of the signal $x(t)$ impacts the corresponding frequency spectrum $X(f)$ at $f = 0$. In the discrete-time domain, the same result holds for sequence $x[n]$, except the periodicity of $X(e^{j\hat{\omega}})$ in the discrete-time domain makes the DC component at $\hat{\omega} = 0$ also appear at all multiples of 2π.

Frequency Samples of N-point DFT

If you sample a continuous-time signal $x(t)$ at rate f_s samples per second to produce $x[n] = x(n/f_s)$, then you can load N samples of $x[n]$ into a discrete-time Fourier transform (DFT) — or a fast Fourier transform (FFT), for which N is a power of 2. The DFT points k correspond to these continuous-time frequency values:

$$f_k = \frac{k}{N} \cdot f_s \text{ (Hz)}, k = 0, \dots N-1$$

Assuming that $x(t)$ is a real signal, the useful DFT points run from 0 to $N/2$.

Integrator and Accumulator Unstable

The integrator system $H_i(s) = 1/s$ and accumulator system $H_{acc}(z) = 1/(1-z^{-1})$ are unstable by themselves. Why? A pole at $s = 0$ or a pole at $z = 1$ isn't good. But you can use both systems to create a stable system by placing them in a feedback configuration. Figure 17-2 shows stable systems built with the integrator and accumulator building blocks.

Figure 17-2:
Making stable systems by using integrator (a) and accumulator (b) subsystems.

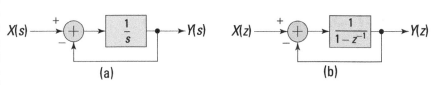

You can find the stable closed-loop system functions by doing the algebra:

$$Y(s) = X(s)\frac{1}{s} - Y(s)\frac{1}{s} \Rightarrow H(s) = \frac{Y(s)}{X(s)} = \frac{1/s}{1+1/s} = \underbrace{\frac{1}{s+1}}_{\text{poles at } s=-1}$$

$$Y(z) = X(s)\frac{1}{1-z^{-1}} - Y(z)\frac{1}{1-z^{-1}} \Rightarrow H(z) = \frac{1/(1-z^{-1})}{(2-z^{-1})/(1-z^{-1})} = \underbrace{\frac{1/2}{1-1/2\,z^{-1}}}_{\text{poles at } z=1/2}$$

The Spectrum of a Rectangular Pulse

The spectrum of a rectangular pulse signal or sequence (which is the frequency response if you view the signal as the impulse response of a LTI system) has periodic spectral nulls. The relationship for continuous and discrete signals is shown in Figure 17-3.

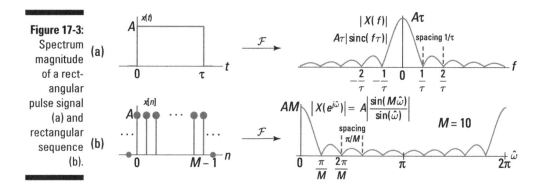

Figure 17-3: Spectrum magnitude of a rectangular pulse signal (a) and rectangular sequence (b).

Odd Half-Wave Symmetry and Fourier Series Harmonics

A periodic signal with odd half-wave symmetry, $x(t \pm T_0/2) = -x(t)$ where T_0 is the period, has Fourier series representation consisting of only odd harmonics. If, for some constant A, $y(t) = A + x(t)$, then the same property holds with the addition of a spectra line at $f = 0$ (DC). The square wave and triangle waveforms are both odd half-wave symmetric to within a constant offset.

Index

Apple & Mac

iPad For Dummies,
5th Edition
978-1-118-49823-1

iPhone 5 For Dummies,
6th Edition
978-1-118-35201-4

MacBook For Dummies,
4th Edition
978-1-118-20920-2

OS X Mountain Lion
For Dummies
978-1-118-39418-2

Blogging & Social Media

Facebook For Dummies,
4th Edition
978-1-118-09562-1

Mom Blogging
For Dummies
978-1-118-03843-7

Pinterest For Dummies
978-1-118-32800-2

WordPress For Dummies,
5th Edition
978-1-118-38318-6

Business

Commodities For Dummies,
2nd Edition
978-1-118-01687-9

Investing For Dummies,
6th Edition
978-0-470-90545-6

Personal Finance
For Dummies,
7th Edition
978-1-118-11785-9

QuickBooks 2013
For Dummies
978-1-118-35641-8

Small Business Marketing Kit
For Dummies,
3rd Edition
978-1-118-31183-7

Careers

Job Interviews
For Dummies,
4th Edition
978-1-118-11290-8

Job Searching with
Social Media
For Dummies
978-0-470-93072-4

Personal Branding
For Dummies
978-1-118-11792-7

Resumes For Dummies,
6th Edition
978-0-470-87361-8

Success as a Mediator
For Dummies
978-1-118-07862-4

Diet & Nutrition

Belly Fat Diet For Dummies
978-1-118-34585-6

Eating Clean For Dummies
978-1-118-00013-7

Nutrition For Dummies,
5th Edition
978-0-470-93231-5

Digital Photography

Digital Photography
For Dummies,
7th Edition
978-1-118-09203-3

Digital SLR Cameras &
Photography For Dummies,
4th Edition
978-1-118-14489-3

Photoshop Elements 11
For Dummies
978-1-118-40821-6

Gardening

Herb Gardening
For Dummies,
2nd Edition
978-0-470-61778-6

Vegetable Gardening
For Dummies,
2nd Edition
978-0-470-49870-5

Health

Anti-Inflammation Diet
For Dummies
978-1-118-02381-5

Diabetes For Dummies,
3rd Edition
978-0-470-27086-8

Living Paleo For Dummies
978-1-118-29405-5

Hobbies

Beekeeping
For Dummies
978-0-470-43065-1

eBay For Dummies,
7th Edition
978-1-118-09806-6

Raising Chickens
For Dummies
978-0-470-46544-8

Wine For Dummies,
5th Edition
978-1-118-28872-6

Writing Young Adult Fiction
For Dummies
978-0-470-94954-2

Language &
Foreign Language

500 Spanish Verbs
For Dummies
978-1-118-02382-2

English Grammar
For Dummies,
2nd Edition
978-0-470-54664-2

French All-in One
For Dummies
978-1-118-22815-9

German Essentials
For Dummies
978-1-118-18422-6

Italian For Dummies
2nd Edition
978-1-118-00465-4

Available in print and e-book formats.

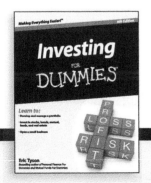

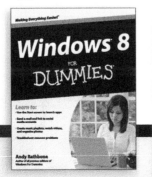

Math & Science

Algebra I For Dummies,
2nd Edition
978-0-470-55964-2

Anatomy and Physiology
For Dummies,
2nd Edition
978-0-470-92326-9

Astronomy For Dummies,
3rd Edition
978-1-118-37697-3

Biology For Dummies,
2nd Edition
978-0-470-59875-7

Chemistry For Dummies,
2nd Edition
978-1-1180-0730-3

Pre-Algebra Essentials
For Dummies
978-0-470-61838-7

Microsoft Office

Excel 2013 For Dummies
978-1-118-51012-4

Office 2013 All-in-One
For Dummies
978-1-118-51636-2

PowerPoint 2013
For Dummies
978-1-118-50253-2

Word 2013 For Dummies
978-1-118-49123-2

Music

Blues Harmonica
For Dummies
978-1-118-25269-7

Guitar For Dummies,
3rd Edition
978-1-118-11554-1

iPod & iTunes
For Dummies,
10th Edition
978-1-118-50864-0

Programming

Android Application
Development For
Dummies, 2nd Edition
978-1-118-38710-8

iOS 6 Application
Development For Dummies
978-1-118-50880-0

Java For Dummies,
5th Edition
978-0-470-37173-2

Religion & Inspiration

The Bible For Dummies
978-0-7645-5296-0

Buddhism For Dummies,
2nd Edition
978-1-118-02379-2

Catholicism For Dummies,
2nd Edition
978-1-118-07778-8

Self-Help & Relationships

Bipolar Disorder
For Dummies,
2nd Edition
978-1-118-33882-7

Meditation For Dummies,
3rd Edition
978-1-118-29144-3

Seniors

Computers For Seniors
For Dummies,
3rd Edition
978-1-118-11553-4

iPad For Seniors
For Dummies,
5th Edition
978-1-118-49708-1

Social Security
For Dummies
978-1-118-20573-0

Smartphones & Tablets

Android Phones
For Dummies
978-1-118-16952-0

Kindle Fire HD
For Dummies
978-1-118-42223-6

NOOK HD For Dummies,
Portable Edition
978-1-118-39498-4

Surface For Dummies
978-1-118-49634-3

Test Prep

ACT For Dummies,
5th Edition
978-1-118-01259-8

ASVAB For Dummies,
3rd Edition
978-0-470-63760-9

GRE For Dummies,
7th Edition
978-0-470-88921-3

Officer Candidate Tests,
For Dummies
978-0-470-59876-4

Physician's Assistant Exam
For Dummies
978-1-118-11556-5

Series 7 Exam
For Dummies
978-0-470-09932-2

Windows 8

Windows 8 For Dummies
978-1-118-13461-0

Windows 8 For Dummies,
Book + DVD Bundle
978-1-118-27167-4

Windows 8 All-in-One
For Dummies
978-1-118-11920-4

Available in print and e-book formats.

Printed and bound by CPI Group (UK) Ltd, Croydon, CR0 4YY

27/10/2024

14580322-0002